Serial Textbooks of Ethiopian ATVET Supported by the Aid Program of Chinese Government

DAIRY CATTLE PRODUCTION

China Agriculture Press

EDITORIAL BOARD

Zhu Wenqiang 朱文强	Embassy of the People's Republic of China in Ethiopia 中华人民共和国驻埃塞俄比亚大使馆	Secretary 秘书

Leaders in the Ministry of Agriculture and Natural Resources of Ethiopia

Tesfaye Mengiste	Ethiopian Ministry of Agriculture and Natural Resources	State Minister
Wondale Habtamu	Extension Directorate of Ethiopian Ministry of Agriculture and Natural Resources	Director-General
Kebede Atiseb	Agriculture Advisory and Training Directorate of Ethiopian Ministry of Agriculture and Natural Resources	Director
Kebede Beyecha	Federal Alage ATVET College	Dean
Chala Feyera	Federal Alage ATVET College	Vice Academic Dean
Tamirat Tesema	Federal Agarfa ATVET College	Dean
Akele Molla	Federal Agarfa ATVET College	Vice Academic Dean
Debela Bersisa	Holeta TVET College	Dean
Getachew Cibsa	Holeta TVET College	Outcome-based Training Vice Dean

2. Author Committee

Chief Author:

Tong Yu'e 童玉娥	Center of International Cooperation Service, MARA, P. R. China 中华人民共和国农业农村部国际交流服务中心	Director-General 主任

Associate Authors:

Chinese Associate Authors:

Luo Ming	Center of International Cooperation Service, MARA, P. R. China	Deputy Director-General
罗鸣	中华人民共和国农业农村部国际交流服务中心	副主任
Lin Huifang	Center of International Cooperation Service, MARA, P. R. China	Deputy Director-General
蔺惠芳	中华人民共和国农业农村部国际交流服务中心	副主任
Wang Jing	Center of International Cooperation Service, MARA, P. R. China	Division Director
王静	中华人民共和国农业农村部国际交流服务中心	处长
Guo Su	Center of International Cooperation Service, MARA, P. R. China	Deputy Division Director
郭粟	中华人民共和国农业农村部国际交流服务中心	副处长
Li Ronggang	Jiangsu Internet Agriculture Instructing Center, China	Professor
李荣刚	江苏省互联网农业指导中心	推广研究员
	Term 15, 16, 17 Chinese ATVET Project to Ethiopia	Coordinator
	援埃塞俄比亚农业职教组第15、16、17期	协调员

Ethiopian Associate Authors:

Ayele Gizachew	Agriculture TVET Office of Ethiopian Ministry of Agriculture and Natural Resources	Former Office Head
Getachew Demisie	Agriculture TVET Office of Ethiopian Ministry of Agriculture and Natural Resources	Office Head

Coauthors:

Chinese Coauthors:

Zhang Junyou 张君友	Center for Animal Disease Prevention and Control of Nanyang City，Henan Province，P. R. China 河南省南阳市动物疫病预防控制中心	Senior Livestock Engineer 高级畜牧师
	Term 15，16，17 Chinese ATVET Project to Ethiopia 援埃塞俄比亚农业职教组第 15、16、17 期	Livestock Expert 畜牧专家
He Wang 何望	Hunan Fisheries Science Institute，P. R. China 湖南省水产科学研究所	Vice Researcher 副研究员
	Term 3，4，14，15，16，17 Chinese ATVET Project to Ethiopia 援埃塞俄比亚农业职教组第 3、4、14、15、16、17 期	Aquaculture Expert 水产专家
Chen Xiongzhen 陈雄珍	Agriculture Bureau of Huanjiang County，Guangxi Zhuang Autonomous Region，P. R. China 广西壮族自治区环江县农业局	Senior Agronomist 高级农艺师
	Term 16，17 Chinese ATVET Project to Ethiopia 援埃塞俄比亚农业职教组第 16、17 期	Silkworm Expert 蚕桑专家
Lei Yingping 雷英平	Agriculture Bureau of Zixing City，Hunan Province，P. R. China 湖南省资兴市农业局	Aquaculture Engineer 水产工程师
	Term 13，14，15，16 Chinese ATVET Project to Ethiopia 援埃塞俄比亚农业职教组第 13、14、15、16 期	Aquaculture Expert 水产专家

Ethiopian Coauthors:

Gizachew Delilo	Animal Science Department，Alage ATVET College，Zeway，Ethiopia	Animal Science (MSC)

Shimelis Tsegaye	Animal Science Department, Alage ATVET College, Zeway, Ethiopia	Animal Science (MSC)
Habtom Negussie	Animal Science Department, Alage ATVET College, Zeway, Ethiopia	Animal Science (MSC)
Ibrahim Edao	Animal Science Department, Agarfa ATVET College, Robe, Ethiopia	Animal Science (BSC)
Demelash Kifle	Animal Science Department, Agarfa ATVET College, Robe, Ethiopia	Animal Science (BSC)
Milion Bulo	Agriculture TVET office of Ethiopian Ministry of Agriculture and Natural Resources	Expert

3. Reviser Committee

Chinese Revisers:

Yang Yang	Center of International Cooperation Service, MARA, P. R. China	Programme Officer
杨飏	中华人民共和国农业农村部国际交流服务中心	项目官员
Fu Yan	Center of International Cooperation Service, MARA, P. R. China	Programme Officer
付严	中华人民共和国农业农村部国际交流服务中心	项目官员
Zhou Min	Center of International Cooperation Service, MARA, P. R. China	Programme Officer
周敏	中华人民共和国农业农村部国际交流服务中心	项目官员
Li Jun	Center of International Cooperation Service, MARA, P. R. China	Programme Officer
李俊	中华人民共和国农业农村部国际交流服务中心	项目官员

Ethiopian Revisers:

Gebre Michael Meles	Agriculture TVET office of Ethiopian Ministry of Agriculture and Natural Resources	Expert
Yassin Jamal	Animal Science Department, Alage ATVET College, Zeway, Ethiopia	Senior Instructor

FOREWORD

Ethiopia is the second-most populous nation on the African continent and houses the headquarters of the African Union. With the agriculture and livestock sector playing a pivotal role in its national economy, Ethiopia boasts many advantages in developing agriculture. China and Ethiopia have established diplomatic relations for almost half a century. Building on the profound traditional friendship, the two countries have generated fruitful outcomes in cooperation in various fields. In particular, China-Ethiopia agricultural cooperation is featured by many highlights as follows:

The China-Ethiopia Agricultural Technical Vocational Education and Training Program (the ATVET Program), as a key program in China-Ethiopia cooperation, was jointly launched by the Ministries of Agriculture of China and Ethiopia in 2001 and transformed into a foreign aid program of the Chinese government in 2012. Up to now, the Chinese government has dispatched 405 instructors in 16 batches to Ethiopia for the program. These Chinese instructors have devoted themselves to Ethiopia for 15 years. Stationed in 13 ATVET colleges, they have given lectures on 56 subjects in 5 disciplines, namely plant science, animal science, natural resources, veterinary science and agricultural cooperatives. Thanks to their dedication, 2,100 local teachers, 13,000 agricultural technicians and 39,000 students have been trained and imparted with over 70 practical and advanced technologies. These Chinese teachers also helped Ethiopia establish a tailor-made agricultural vocational education system. This program has won recognition from the Ministry of Agriculture and Natural Resources of Ethiopia and the beneficiaries, the warm welcome by local students, teachers and farmers, and extensive coverage by Chinese and foreign core media. It is undoubtedly a role model of China's pro-Africa agricultural cooperation.

After years' of joint study and exploration, Chinese instructors and Ethiopian partners have realized that the absence of professional textbooks is a serious constraint to the sustainable development of Ethiopian agricultural vocational education system. Therefore, the Chinese and Ethiopian governments attach great importance to the development of textbooks. The two countries conducted a range of studies and drafted a textbook compilation and publishing plan. According to the plan, the textbook series consist of 13 volumes, involving poultry production, animal health, horticulture, small-scale irrigation, dairy production, apiculture, water and soil conservation, beef cattle, field crop, coffee, tea and spice crops, cotton, forestry, farm machinery, etc. Building upon the experience of the ATVET program and taking into account the practical demand of agricultural development in Ethiopia, the textbooks have summarized the advanced agricultural theories and technologies of China and will serve as practical guidance to more agricultural practitioners in Ethiopia. Authors of the textbooks are long-term practitioners in the ATVET program and have gone through rigorous review by the Leader Committee. In order to make sure that the textbooks are localized and operable, authors have had thorough communication with the Ethiopian Ministry of Agriculture and Natural Resources, received overall guidance from the Leader Committee and worked wholeheartedly to develop and revise each and every volume.

The publication of the textbooks have obtained strong support from the Chinese Ministry of Agriculture and Rural Affairs, Chinese Ministry of Commerce, Embassy of the People's Republic of China in Ethiopia, Ethiopian Ministry of Agriculture and Natural Resources and relevant Ethiopian vocational education colleges. The textbooks will certainly provide effective means for the training of practical agricultural professionals in Ethiopia, promote the sustainable development of Ethiopian agricultural vocational education, and enhance the capacity and level of agricultural development in Ethiopia.

PREFACE

Dairy production is biologically an efficient system of converting large quantities of roughage, the most abundant feed resources into milk in the tropics. Milk is one of the most nutritious and highly important food items for human consumption all over the world.

Ethiopia is known for its huge and genetically diverse dairy cattle resources and favorable production environment. In Ethiopia, milk production plays a significant role as a source of food, income, and creating employments. Dairy product consumption is expected to grow in Ethiopia at 3% to 4% annually until 2020 (Holloway et al., 2000). However, milk production in Ethiopia is still traditional and underdeveloped as the result of several technical and non-technical constraints, among which the lack of trained and skilled manpower is the most underlined.

Despite the scenarios, booming rapid growth of population and urbanization and growth in income, corroborated with a stable political system and due government priority in Ethiopia, are expected to create more than double demand and greater market opportunities for milk production (Azage et al., 2013; Yoseph et al., 2003).

The aim of this textbook is, therefore, to equip students and trainees, instructors/trainers, agricultural development agents, and other professionals and producers engaged in commercial and/or small-scale dairy farming with appropriate knowledge and skills in dairy cattle production and could serve as a reference for higher education and other related institutes. The contents of this book cover relevant information necessary for dairy cattle production in an output-based system in line with the Ethiopian Occupation Standard (EOS). The book consists of ten modules entitled: selecting dairy herd breeds and foundation stocks, identifying dairy cattle housing facilities and requirements, developing and implementing feeding plan, performing feeding and management of dairy cattle, managing reproduction and breeding in dairy cattle, implementing calf and heifer rearing, carrying out milking operations, performing milk handling and processing operations, implementing dairy cattle health and disease prevention activities, and performing dairy farm record keeping.

Since 2001, Ethiopia has been practicing an output-based Agricultural Technical Vocational Educational and Training (ATVET) Program to boost agricultural productivity and to achieve the food security program of the country. The ATVET colleges in Ethiopia have been producing middle level skilled agricultural development workers every year. However, ATVET Colleges are limited in standard teaching, training and learning materials (TTLM) to deliver the program. Accordingly, this textbook is prepared in partnership with the government of the People's Republic of China to be used as a textbook and a reference material for the ATVET students.

The book is written by instructors of Animal Science Department from Alage and Agarfa ATVET Colleges. Final draft of the book is given to other professionals in related fields for further review and comment. Through all these efforts, we believe that we have produced a book at our best level.

ACKNOWLEDGEMENT

We would like to express our heartfelt gratitude and acknowledgement to the People's Republic of China for the multi-faceted support providing to the people of Ethiopia and its overall development. People in Ethiopia nowadays recognize China as a true friend due to its provision of support in various economic sectors based on mutual benefit of the two sisterly countries.

The publication of this textbook wouldn't have been possible without that genuine support we are getting from China. Therefore, we would like to thank the Chinese government—the embassy of China in Addis Ababa and the Ministry of Agriculture and Rural Affairs; Dr. Li Ronggang, the Chinese instructors coordinator in ATVET Office; and our Chinese colleagues in Alage and Agarfa ATVET Colleges, for their strong support with the guiding principle of mutual benefit.

The production of this book couldn't have been achieved without the approval and support of the State Minister of Agriculture development sector, Agriculture Extension Directorate and the Coordination, support and guidance of the ATVET Coordination Office of the Ministry of Agriculture and Natural Resources. The support of the top-managements in Alage and Agarfa ATVET Colleges was vital. Therefore, our most sincere and especial thanks go to Dr. Kebede Beyecha & Mr. Tamirat Tesema, and Dr. Chala Feyera and Mr. Akele Molla, dean and academic and training vice dean of the aforementioned colleges, respectively.

Our special gratitude goes to Mr. Gebre Michael Meles, one of the initiators and supporter for the production of the textbook and who contributed a lot in editing the textbook.

We hope that this book will remain a living witness for the current all rounded economic support of China to Ethiopia. Similarly, this acknowledgement will remain witness that how much we respect their genuine intention of cooperation for our mutual benefit.

CONTENTS

MODULE 1:

SELECTING DAIRY HERD BREEDS AND FOUNDATION STOCKS

>>> INTRODUCTION

Proficiency in the selection of dairy cattle cannot be attained through the casual inspection of a herd. Long experience or pains-taking detailed study of the animals is necessary. While records of production and reproduction of a particular animal or its offspring are a necessary condition for selection of animals, such records are only rarely available, especially in our country. Therefore, the selection is usually based on physical appearances.

If the productive and reproductive quality of the existing herds is to be improved, the ability to judge the performance of animals based on their appearance will be necessary and essential. Long experience and the careful observation of dairymen, together with the results of several scientific studies, are possible to determine the productivity of animals. This indicates that there is a fairly well-defined relationship between the form or the proportionate size of certain parts the animals and their production and reproduction abilities.

Buying dairy cows on the basis of their physical appearance alone is, of course, not the ideal way, because it has got some disadvantages. Records show that many cows displaying good dairy cattle body conformations are sometimes low producers, owing chiefly to the lack of persistency in the assumed relationship between body conformation and productivity.

In the selection of bulls for breeding purposes, the most serious mistakes are likely to be made. Too often, the appearance of the animal is the only deciding factor in the choice of the animal. Bulls of good reproductive merits tend to transmit their good merits to their male offspring, but there is no certainty when they are used to produce female offspring. In fact, the records show definitely that the female offspring of certain bull which has good dairy characteristics yield less milk and butterfat than their respective dams. The most reliable method of choosing a sire that is likely to produce calves of high milk and butterfat capacity is to select the one that has already sired and produced such offspring. In the case of using a "proved sire", that produced known productive cows, the reliability of producing the best producing offspring cannot be secured. The next best procedure is to use a bull from high producing dams and that has female siblings of high production records. If there are no female siblings with high milk and butterfat production records, the records of the dams and grand dams are the third best basis for estimating the value of the bull from the production standpoint. Thus, in the selection of dairy bulls, while it is possible to use the physical observations for selection of the dam, this cannot be relied upon as a complete index of the ability of the bull to transmit the productive capacity to its entire offspring.

The current level of genetic improvement in dairy herds is the result of advances in methodologies able to identify genetically superior animals with greater efficiency. However, the traditional selection depends strongly on the number of phenotypic records used during selection to increase the accuracy of estimates. Thus, traits difficult to measure, such as those expressed later in the life of the animal, limited to sex or of low heritability are more difficult to be improved. From the bovine genome sequencing in 2009, a new breakthrough was possible in selection methodologies, mainly for production traits. The advent of molecular biology, that accesses genetic information allowing the distinction between individuals by differences in their nucleotide sequences, is called molecular markers. Molecular markers associated with production traits enabled the selection of individuals, excluding the influence of the environment by marker-assisted selection (MAS).

A new technology called genomic selection is revolutionizing dairy cattle breeding, allowing estimation of breeding value (GEBV) based on the genotype of hundreds of thousands of SNPs (single nucleotide polymorphisms) densely distributed throughout the genome. The GEBV is predicted as the sum of the effects of all markers. Traditional selection methodologies depend on the phenotypic record of the animal and its offspring to estimate the breeding value of animals. In contrast, the genomic selection allows the use of the genomic value of each individual without using phenotypic data, based only on the previously estimated SNP effect in a reference population. The reference population is a herd with phenotype data, pedigree and genotypes known (Mello et al., 2014).

Furthermore, operating a profitable dairy farm business requires that the factors of production such as land, labor and capital must be combined and managed to achieve a value of production that is greater than the cost of production. Therefore, the goal of every dairy manager should be to maximize the efficiency of high producing dairy cows so that profitability will increase.

This module identifies and describes the important high producing breeds of dairy cattle, along with the best selection and judging methods of dairy cattle. However, the advanced selection methods such as marker-assisted selection and genomic selection are not included in this book as they are beyond the scope of this book.

1 BREEDS OF DAIRY CATTLE

The breed is defined as a group of similar animals within a species with a common origin. Animals within a breed have physical characteristics that distinguish them from other breeds or groups of animals within that same species (Solomon, 2010). Cattle breeds generally are classified into two major groups:

- *Bos taurus*: A subspecies of cattle often referred to as "European or Continental" breeds or exotic breeds. Most dairy and beef breeds belong to this group. The term dairy breed is used to differentiate those cattle that are bred primarily to produce milk against those that are used for meat production and other purpose. The Dairy breeds include: Holstein, Jersey, Ayrshire, Guernsey, Brown-Swiss, and Milking Shorthorn. Dual purpose dairy breeds include Simmental, Red Poll, and Milking Devon.
- *Bos indicus*: A subspecies of cattle of south Asian origin also known as Zebu cattle or indigenous breed. These breeds are widely found in the tropics, including Ethiopia. They are considered as dual-purpose breeds producing milk, beef and draft power. Tropical dairy breeds (*Bos indicus*) with high milk production ability include Sahiwal and Red Sindhi. Each breed is described in the subsequent sections.

Breeds from *Bos indicus* and *Bos taurus* origin do not have the same potential in production ability. These variations are due to environmental and genetic variation. According to Payne and Wilson (1999), the characteristic features in which exotic (*B. taurus*) and indigenous (*B. indicus*) cattle differ include the following:

Conformation

- Indigenous cattle differ from the exotic in that they have a much narrower body, longer legs and a well-developed dewlap, which in some breeds is very prominent. All of them have a large hump over the top of the shoulder and neck.
- Spinous processes below the hump are extended, and there is considerable muscular tissue covering the processes.
- The other characteristics of these cattle are their horns, which usually curve upward and are sometimes tilted to the rear, their ears, generally large and pendulous, and the throatlatch and dewlap, which have a large amount of excess skin. They also have more highly developed sweat glands than European cattle (*Bos taurus*) and so can perspire more freely.
- *B. indicus* cattle produce an oily secretion from the sebaceous-glands, which has a distinctive odor and is reported to assist in repelling insects.

Fitness characters

- The metabolic rate of indigenous cattle, at high temperature, is by far less than temperate cattle, i. e. the indigenous cattle generate less heat. The Zebu cattle sweat more readily than temperate cattle, and their sweat glands are larger.
- Indigenous breeds can exist on poor quality diets. High temperature reduces intake more in temperate cattle than local ones. Indigenous cattle are better able to retain

water and food in the large intestine.

- Disease resistance and mortality: indigenous cattle are more resistant to disease than the temperate ones.

Production characteristics

- Slow to reach sexual maturity
- Long calving interval
- Slow growth rate
- Short lactation

1.1 Selecting a Breed

It is good practice for a dairy farmer to select a breed that suits best and will thrive in that geographic location. It is more important to select individual cow within a selected breed that are high producers than to put too much emphasis on which breed to select. There are some general breed differences that are discussed in more detail as each breed is described. Some general guidelines (factors) for the selection of a breed include (Milk South Africa, 2014; Gillespie and Flanders, 2010):

- Selecting a breed that is common in the area: Selecting a breed common in the area increases availability of breeding stock. There will also be a better market for surplus animals. If you have to transport your cattle over a long distance, it can cost you a lot of money. Transporting animals can also cause injury.
- Adaptable to prevailing environmental conditions: The breed to be selected should be adaptable to the prevailing environmental conditions such as disease resistance, ambient temperature, and grazing/pastureland conditions. Some breeds perform better in hot areas, while others perform better in cooler areas.
- Economy: Choose a breed that will give you the income and profit that you need. Base your choice on three very important things: the highest quality milk, the highest quantity of milk (high production) and the lowest possible cost.
- Feeding: Before buying cattle, make sure of the feeding system that you can provide for your cattle. Some farms are suitable for zero grazing systems, while others are more suitable for grazing systems. Some breeds are more suited to grazing than others.
- Volume: Decide whether you want to produce a large amount of milk (high volume), or whether you would prefer milk with a higher butterfat or protein content. These factors are very important when you start selling milk to a milk buyer. Some buyers prefer milk with a high butterfat content, as this milk is used to make cheese and other dairy products.
- Learn from your neighbors: If other farmers are farming successfully with a certain breed of cattle in your area, then you should probably follow their example. Do not

try to farm with a breed that has not been proven successful in your vicinity.

1. 1. 1 Pure Dairy Cattle Breeds

Dairy cattle may be defined as a particular group of animals developed in a certain area for a definite purpose and having the same general characteristics such as color, conformation and quality of milk (Mosielele, 2006). A purebred dairy cow is one whose ancestry traces back to the same breed. Thus, cattle that are considered dairy breeds are selected for breeding on the basis of their ability to produce large quantities of milk for a long period of time. Generally, six major dairy cattle breeds are found worldwide, such as Holstein Friesian, Jersey, Guernsey, Ayrshire, Brown Swiss, and Dairy/Milking Shorthorn.

Holstein Friesian

The Holstein Friesian breed was originally developed in the northern part of the Netherlands in the province of Friesland and northern Germany. The color pattern is varying proportions of black and white. There are occasionally red and white, born from black and white parents that carry the red factor as a recessive gene. The switch (tail) has white on it. Solid black or solid white animals are not registered. Off colors include black on the switch, solid black belly, one or more legs encircled with black that touches the hoof at any point. The horns are medium in length, incline forward and curve inward (Gillespie and Flanders, 2010; Holstein Association Inc. , 2012) (Fig. 1).

Holstein has a large body frame and they are the largest of the dairy breeds. Mature cows weigh about 680 kg and stand 147 cm at the withers, and bulls weigh about 998 kg. Holstein cows rank the first among the dairy breeds in average milk production per cow at 6,577 kg. They average about 3. 5% milk fat and rank the sixth among the dairy breeds in average milk fat produced per cow (Mosielele, 2006).

The cows are generally docile, but the bulls can be mean and dangerous. Cows have large udders. Holstein cows have excellent grazing ability and a large feed capacity (vigorous appetites). Holstein cows are adaptable to a wide range of conditions. A healthy, newborn Holstein calf will weigh 40 kg or more at birth. Holstein cows have gained increased popularity around the world. Holstein cows can be found on every continent and in almost every country. The reason for the Holstein cow's popularity is that Holstein cows excel in total milk production and income per cow, and adaptability to a wide range of conditions (Paulson et al. , 2012).

Jersey

The Jersey breed was developed on the island of Jersey in the Channel Islands off the coast of France. Jerseys are cream to light fawn to almost black in color, some have white markings. The muzzle is black encircled by a light color ring, and the tongue and switch may be either white or black. The horns curve inward and incline forward. They are of medium length and taper toward the tips. Heifers of this breed develop more rapidly than any other breed (early maturing breed). Jersey cattle have excellent udders that are well attached and are excellent dairy type. They are adaptable and efficient users of feed and have excellent grazing ability even on poor pastures. Jerseys are very nervous and react quickly to both good and bad treatment.

Jerseys rank the fifth among the dairy breeds in average milk production per cow at 4,536 kg with a butterfat content of 5 - 5.4%, which is rich in color, and rank the first among the dairy breeds in average milk fat. The Jersey is the smallest of the dairy breeds. Mature cows weigh about 453 kg and bulls weigh about 725 kg (Fig. 2).

Generally, strengths of the breed include excellent heat tolerance, high fertility, young age at first calving (early maturing), calving ease, and high kg of milk solids produced per kilogram of feed consumed (Paulson et al., 2012).

Guernsey

The breed originated in Channel Islands near the north coast of France. They have a shade of fawn, either solid or with white markings, with golden yellow pigmentation of coat color. A clear or buff muzzle is preferred over smoky or black. The horns incline forward, are refined and medium in length and taper towards the tips. The Guernsey is an early-maturing moderate sized breed. Mature cows weigh 500 kg and bulls about 816 kg.

Guernsey rank the fourth among the dairy breeds in average milk production per cow at 4,808 kg, with 4.5% milk fat and rank the second among the dairy breeds in average milk fat. The Guernsey is noted for the superior flavor of its golden color milk (due to higher level of beta-carotene content). Guernsey is widely known for its adaptability, calm and docile dispositions and relatively good calving ease (Gillespie and Flanders, 2010).

Ayrshire

The breed was developed in County Ayr in southwestern Scotland. Ayrshire is characterized by its red and white color. The red color ranges from cherry red to mahogany red, which is different from the reds found in other breeds. The proportion of the two colors varies greatly. Each color should be clearly defined. The horns spread long and curve up at the ends.

Ayrshire is characterized by strongly attached, evenly balanced, well-shaped udders and the teats are medium in size. They are vigorous and have strong constitutions, sometimes are quite nervous and hard to manage, keep good body conditions when kept under poor breeding conditions. They have excellent grazing ability.

This is a medium size breed, and mature cows weigh about 544 kg and bulls weigh about 816 kg. Ayrshire rank the third among the dairy breeds in average milk produced at 5,307 kg/ (lactation • cow) with average milk fat of 4%. They are known for their fertility, overall health, and resistance to mastitis (American Ayrshire Dairy Breed cattle association,

2015).

Brown Swiss

The Brown Swiss breed originated in Switzerland and was one of the oldest of the dairy breeds in the world. Brown Swiss is solid brown, ranging from light to dark. The nose and tongue are black. The horns incline forward and slightly upward.

The unusual physical exertion and high altitude under which the breed evolved over centuries has probably played an important role in the strength, hardiness and ruggedness found in the Brown Swiss breed today.

The Brown Swiss cattle are large-framed. Mature cows weigh about 680 kg and bulls about 907 kg. The heifers mature more slowly than other dairy breeds. Brown Swiss cattle have a quiet, docile temperament. They are good grazers. Brown Swiss rank the second among the dairy breeds in average milk production per cow at 5,488 kg with the average of about 4.1% milk fat.

Brown Swiss cattle are usually docile, slower moving cows with more heat tolerance than other breeds. Also, they tend to be heavier muscled than other dairy breeds. For this reason, they have been used in beef cattle crossbreeding programs to produce docile cows with increased milk production. Because of their ruggedness and heat tolerance, they have been in high demand for export to tropical countries. Brown Swiss are known for their outstanding feet and legs, udders that last, and longevity (Mosielele, 2006; Gillespie and Flanders, 2010).

Milking Shorthorn

The Milking/Dairy Shorthorn originated in northeastern England in the valley of the River Tees (Fig. 3). Their common colors are red or deep roan, although red and white are also found. Roan is a very close mixture of red and white and is found in no other breed of cattle. The breed has a small head while the neck is thin towards the head, rapidly thickening as it approaches the shoulder. Horns are short, blunt and creamy. This breed is considered as a multipurpose breed offer meat, milk and power.

The breed is intermediate to large with cows being approximately 138 cm at the withers and weighing 546 – 636 kg in average condition. Milking Shorthorns are known for their excellent reproductive efficiency and longevity. They also are fairly heat tolerant. Bulls not kept for breeding are successfully fed for beef and hung beef, high-quality carcasses. Milking Shorthorns rank the sixth among the dairy breeds in average milk production per cow at 4,340 kg with average about 3.9% milk fat.

1.1.2 Dual Purpose Cattle Breeds

Dual purpose breeds are breeds which can be used for both beef and milk production. They have characteristics intermediate between those of dairy and beef types in conformation and in the production of both meat and milk. They are medium sized or large.

Dual purpose dairy breeds could be grouped into *B. taurus* (temperate developed) and *B. indicus* (tropical developed/originated). Dual purpose dairy breeds of *B. taurus* origin include: Simmental, Milking Devon and Red Poll breeds. Dual purpose dairy breeds of *Bos indicus* (zebu) origin include: Sahiwal, Red Sindhi. Common breeds of dual purpose cattle are the following:

Simmental

The Simmental originated in central western Switzerland in the Simme River Valley, from which it took its name (Fig. 4). They have deep reddish-brown color with white spots and they are horned. They are known for their ability to do work and to produce milk and meat. In its native country, Switzerland, Simmental cows average about 805 kg in weight and bulls about 1,304 kg.

The cows could produce 4,500 kg of milk per lactation. The Simmental was a popular breed worldwide because it was so adaptable, and it crossed well with existing native cattle. They are good foragers and are capable of withstanding harsh environmental conditions (Berry and Buckley, 2014).

Milking Devon

Milking Devon originated in north Devonshire, England (Fig. 5). The breed is red in color, varying in shade from deep red to light red or chestnut color. They may show white on the tail switch, udder or scrotum. They have medium sized curving horns that are light colored with dark tips.

The Milking Devon is a medium sized triple-purpose breed (milk and beef production, and draft power). Mature bulls weigh up to 900 kg and cows up to 730 kg. They adapt to survive on a low-quality, high forage diet under severe climatic conditions. Devon is known for its speed, intelligence, strength, willingness to work.

The milk was good for cream and cheese making; and the carcass developed fine beef on poor forage. They are healthy, long-lived, and possess a docile temperament (American Milking Devon Association, 2014).

Red Poll

The Red Poll cattle were developed as a dual-purpose breed in their native counties in England. Its body is a brownish-red and the tip of the tail is white.

This naturally hornless breed produces good meat and excellent milk. Bulls weigh up to 900 kg and cows up to 680 kg (Briggs and Briggs, 1980). They are the preferred cattle with a combination of milk production and high-quality carcasses.

Sahiwal

They originated in the dry Punjab region which lay along the Indian-Pakistani border. Females have reddish dun color; males may have a darker color around the neck and hindquarters. Males have stumpy horns; females are often dehorned. Ears are medium-sized and drooping. The hump in male is massive, but in female, it is nominal. The udder is large and strong and occasionally has white patches.

Milk yield is 1,500 - 2,200 kg per lactation with a fat content of 4.5%. The breed is medium-sized, adult males weigh 400 - 500 kg, while females weigh 300 to 350 kg. Average lactation length is 235 days. Age at first calving is 45 months. Service period is 155 days. Dry period is 205 days. Calving interval is 440 days.

The Sahiwal is the highest milk producer of all zebu breeds and noted for their hardiness under unfavorable climatic conditions. This would appear to be an ideal low maintenance, dual purpose meat/milk breed for harsh tropical and sub-tropical environments. Due to their heat tolerance and high milk production, they have been exported to Africa and the Caribbean (Jassar, 2015).

Red Sindhi

The Red Sindhi originated from a mountainous region in Sindhi, Pakistan. This is a medium-sized breed, with a compact build and red body color. It has a large head with an occasional bulge in the forehead. Horns are thick and stumpy in males, but thin in females. Ears are fine and small. The hump is well-developed in males. The dewlap is moderate in both sexes. Hindquarters are round and drooping tail switch is black. Udder is medium to large and strong. Red Sindhi cattle are hardy and adapt very well to stress-full environment.

Under reasonable management conditions, Red Sindhi cows average about 1,700 kg of milk after suckling their calves, but under optimum conditions there have been milk yields of over 3,400 kg per lactation.

Adult males weigh 400 - 500 kg, while females weigh 300 - 350 kg. Average lactation length is 265 days. Age at first calving is 45 months. Service period is 210 days. Calving interval is 495 days (Purdy and Dawes, 1987; Ritchie, 2002; Jassar, 2015).

1.1.3 Ethiopian Cattle Breeds

In Ethiopia there are 27 recognized indigenous cattle breeds/types which vary in size and color, all grouped into *B. indicus* (Addisu et al., 2010). Among the indigenous cattle, 8 breeds are widely distributed in the country. It is a well-known fact that *B. indicus* is adapted to tropical environment through natural selection. No interference has been made by man except for few attempts tried in few research sites. They are not well characterized. Thus, they are called with references to the place where they dominate, such as *Horro* (dominating western part around Horro in Wollega), *Boran* (dominating southern part around Borena), Fogera (northern part of Ethiopia around Fogera).

In general, indigenous cattle are multipurpose in their function, i.e. meat, milk and draft. Multipurpose type exhibits not only the properties typical of dual purpose cattle, but also some of the features characteristics of working cattle (strong, bones, sound constitution, and quiet temperament). Some of the distinctive characteristics of indigenous cattle breed include:

Boran

Boran cattle are the famous Ethiopian breed, originated from Borena, southeastern part of the country; today it is also found in Somalia and Kenya (Haile Mariam et al., 1998). They are larger than the short-horned zebu with good body conformation; color is normally white or gray, but also red or pied animals occur; horns are usually small, and the hump is thoracic (Fig. 6). These have been subjected to selection both in Ethiopia and Kenya. They are considered an important dam breed for crossbreeding with temperate cattle for dairy production. The mature cows weigh 300 - 450 kg and bulls weigh 550 - 675 kg under ranch conditions. Under pastoral conditions, the corresponding weights are 225 and 400 kg, respectively. Their milk yield is 440 - 680 kg per lactation in pastoral condition. In ranch, cow can produce 1,400 kg of milk with butterfat percentage of 6% over the lactation period of 280 days. Average calf birth weight is 23.5 kg with average weaning weight of 170 kg (8 months basis), and average daily body weight gain is 540 g and have dressing percentage of 54 - 57%.

Fogera

Fogeras are found in the northwestern part of the country around Fogera (Fig. 7). They have pied coat of black and white or black and gray color; short, stumpy and pointed horns; hump ranging from thoracic to cervicothoracic; dewlap folded and moderate to large; docile temperament; are used for draft, milk and meat. They produce about 281 kg of milk per lactation with milk fat of 5.8%. Average mature body weights 232 kg.

The reproductive performance of Fogera cattle indicated that the age at sexual maturity, age at first calving, lactation length and calving interval were 47.3, 59.9, 10.5 and 25.5 months, respectively. Furthermore, the early, mid and late stages of daily milk yield were 4.2, 3.6 and 2.9 kg, respectively (Damitie et al., 2015).

Horro

Horro originates from the western parts of the country, Wollega (Fig. 8). These are medium sized, mainly raised for meat and draft power, and have poor temperament (highly aggressive). Selected Horro produce up to 543 kg per lactation, Mekonnen et al. (2012) reported that age at puberty, age at first calving and calving interval of Horro cows were 46.6, 58.1 and 21.1 months, respectively.

Arsi

Arsis dominate the highlands of Arsi, Bale, Harar, Shoa and Sidamo, Ethiopia (Zewdu, 2004). The breed is characterized by small, compact and well-proportioned body size; coat color is variable, but includes red, black, roan, white, gray and various color combinations; horns are small and short, and usually crescent shaped; hump is of medium size; dewlap is long and thin (Fig. 9). The average body weight ranges 232 - 245 kg for females. Selected Arsis produce up to 500 kg/lactation with fat percent of 5.4 - 5.8%.

Begait/Barka

They were found in medium to high altitude, originated in the west part of Eritrea, but abundant in Tigray and Gondar. They are classified as Abyssinian short horns (Fig. 10). Coat color is variable, but black pied is common; head is small and short; horns are variable in shape and short to medium in size; hump is very large in the male, chest/thoracic in position and may fall to one side but is small in the females; dewlap is large. They are good milk cattle; selected cattle produce about 647 liters of milk per lactation. Weight at maturity is 290 - 310 kg for male, and 230 - 250 kg for females.

Sheko

Found in the humid southwestern part of the country around Keffa, Bench zone. Sheko cattle do have a specific morphological appearance, which can be utilized in identifying this breed (Takele, 2005) (Fig. 11). Sheko cattle are commonly polled and are generally short and have a compact body; they are alert and strong. The coat color is dominated by red with a glossy appearance. Their eyes are prominent and have folded eyelid. They have broad and short horizontally oriented ears and their muzzle is broad and their facial profile is predominantly straight. They have a small cervicothoracic hump. They are believed to have some level of trypanotolerance. Age at sexual maturity, age at first calving and calving interval were 42, 54.1, and 17 months, respectively.

Ogden

These breeds originated in the eastern part of the country around Ogden (Fig. 12). The breed is characterized as compact body conformation, short horns, small head and long facial profile, well developed hump and a large dewlap with white to gray coat color. They are similar to the Ethiopian Boran in conformation, mainly used for milk, but are also good beef animals.

Senga

Sengas dominate the central part of the country (Fig. 13). They have a cervicothoracic hump (neck humped); they are mixed type not identified yet. Age at maturity, age at first calving and calving interval were 40, 54.1 and 18 months, respectively.

1.2 Crossbred Dairy Cattle

A cross-breed usually refers to an animal with purebred parents of two different breeds. The basic objective of crossbreeding of high performing cattle breeds with tropical breeds is to combine the milk producing ability of the temperate breeds with the climatic adaptability, heat tolerance and disease resistance traits of the tropical breeds (Peters, 1991). Crossbred animals have heterosis or hybrid effect. It is the amount by which merit in crossbreds deviated from the additive component and is fully exploited only when non-related breeds are crossed.

Under small holder production systems, farmers prefer crossbred cow. Because of their smaller size, they have lower feed requirements and relatively less management requirement. Crossbred animals with exotic inheritance of about 50% are preferable. This preference is based on comparison of performance of the animals with different percentage of exotic inheritance. However, the delicate balance between genetic performance ability and adaptability is determined by the degree of exotic inheritance.

Crossbreeding indigenous Ethiopian Boran with Holstein has resulted in improvement of milk production performance traits. For example, 50% Holstein crosses have a fourfold increase over the Ethiopian Boran breed in terms of daily milk yield and lactation yield; they also milk for 97 more days than Ethiopian Boran. However, higher exotic inheritance levels (for example, 87.5% Holstein) could be justified under intensive production systems. As the level of management achievable under most smallholders' conditions in Ethiopia is rather unfavorable to higher exotic inheritance levels (Haile et al., 2008).

The lactation milk yield and days in milk of indigenous, crossbred and exotic cattle in Ethiopia are reported by researchers (Table 1.1). The milk production potential of indigenous breeds of cattle is very low. In addition, milk production potential of temperate breeds in the tropical environments is higher than the indigenous breeds, but this yield is still far below the genetic potential. Therefore, crossbreeding of the indigenous breeds with temperate exotic breeds has been practiced in the country, since the establishment of the IAR, with the aim of making use of the high genetic potential for milk production of exotic breeds and the adaptability to the local environment of indigenous breeds (Kelay, 2002) (Table 1.2).

Table 1.1 Milk production performances of indigenous, crossbred and exotic cattle in Ethiopia

Breed		Lactation		Study location
		Milk yield (kg)	Length (days)	
Indigenous	Boran	494	155	On station
	Horro	559	285	On station
	Arsi	809	272	On station
	Barka	552	128	On station
	Fogera	613	353	On station

(Continued)

Breed		Lactation		Study location
		Milk yield (kg)	Length (days)	
Crossbred	Holstein × Boran (50%)	1,554	350	On farm
	Holstein × Arsi	1,040	350	On farm
	Holstein × Arsi	1,977	356	On station
	Holstein × Arsi (25 - 62.5%)	1,547	366	On farm
	Holstein × Arsi (> 75%)	2,924	361	On farm
	Holstein × Barka	1,488	301	On farm
	Jersey × Barka	970	257	On farm
	Jersey × Arsi	1,741	334	On station
	Friesian × Boran	1,554	350	On farm
Exotic	Friesian	3,796	323	On station
	Jersey	1,619	276	On farm

Source: Kelay, 2002.

Table 1.2 Unadjusted means of different breed groups

Breed groups	Total milk (kg)		Annual milk yield (kg)		Average daily milk yield (kg/day)		Calving interval (days)		Breeding efficiency (%)	
	*N**	*Mean*	*N**	*Mean*	*N**	*Mean*	*N**	*Mean*	*N**	*Mean*
Barka	35	869	21	1,099	32	4.46	43	397	—	—
1/2 Barka×1/2 Holstein	109	2,055	93	1,903	103	6.66	91	415	91	92.45
1/2 Boran×1/2 Holstein	87	1,740	54	1,752	87	5.93	59	440	59	98.99
1/4 Barka×3/4 Holstein	87	2,214	74	1,797	79	6.17	77	474	77	86.55
1/4 Boran×3/4 Holstein	129	2,044	102	1,689	126	5.77	107	471	107	87.08
1/8 Barka×7/8 Holstein	36	2,381	15	1,511	24	5.84	16	512	16	85.54
1/8 Boran×7/8 Holstein	35	1,902	28	1,420	23	5.55	31	493	31	81.21
Holstein Friesian	90	3,028	75	2,611	81	9.99	82	460	82	82.61

* Number of animals.

Source: Tadesse et al., 2003.

2 SELECTING FOUNDATION STOCK

Now that you know what to look for and which breed will suit your farm best, it is time to choose the cows or heifers as foundation stock. This process is called selection and is done by looking at the physical appearance of the animals and different records. Generally, proper selection is the first and the most important step to be adopted in dairying.

2.1 Factors in Selecting Dairy Cattle

Variation between and among/within breeds is the most important factor for selection. The selection of desirable dairy animals for breeding and production is based on the animal's physical appearance, milk production records, production records, health records, and pedigree records. Records are the basis of selection and hence proper identification of animals and record keeping is essential (Stamschorr, et al., 2000; Paulson et al., 2012).

Basically, selection should focus on the important production and reproduction traits relevant to the breed of animal and to the breeding objectives.

2.1.1 Physical Appearance of Dairy Cattle

The physical appearance of an animal determined by an evaluation of important characteristics is referred to as type. Type is an ideal or standard of perfection, combining all the characteristics that contribute to the animal's usefulness for a specific purpose.

Dairy type

Cattle of this type are usually not large and are of somewhat lean build or are characterized by a lean, angular form and a well-developed mammary system.

Type and milk production are closely related. Cows with good dairy type or conformation usually produce more milk for a longer period. Since conformation traits are heritable and have been shown to be linked with functionality, selection for conformation traits is an effective tool to increase milk yield and facilitate genetic improvement of dairy cattle in functionality (Atkins et al., 2008). It takes a great deal of practice and study to learn the ideal dairy animal type. Animals are judged by comparing with an ideal dairy type as described on the Dairy Cow Unified Score Card (Gillespie and Flanders, 2010).

In Ethiopia, some conformation traits are traditionally in use by smallholder farmers to select best dairy cows. Pure size traits, such as stature and heart girth, are closely related to body weight (Zewdu et al., 2006). Related conformation traits include wide hindquarter, long and thin tail, longer naval flap, thin and long neck, concave face, reduced hump, attractive appearance, drooping vulva (for easiness of calving), bushy tail end, thick skin (to withstand the infliction of biting flies) and big body size. Other relevant traits include temperament, non-black hair coat, better growth rate; good mothering ability and being in good health condition are also taken in mind (Takele, 2005). Kelay (2002) also reported that the important conformation traits such as straight back and large udder of the cows were the main selection criteria to select the dairy cows.

2.1.2 Milk Production Records

This is the important trait among the selection criteria, which have a tight connection to the economic effectiveness of dairy cattle production (Meszaros et al., 2008). Milk production records show the kilograms of milk produced and the percent of milk fat. Milk production records that show past performance may or may not be available for the individual animal. If the cow has been in production and records were kept, the record should be evaluated. Such records give some indication of the possible production of the offspring of the cow.

Since milk and butterfat records are a much more reliable guide than physical appearance in determining a cow's productive ability, the question naturally arises, why not buy cows upon the basis of milk records? There are two answers (Nevens and Kuhlman, 1997):

- First, the characters that determine milk production are inherited independently of those that determine body form, and therefore production records alone cannot be relied upon as a satisfactory guide for the selection of animals in building up a good dairy herd.
- Second, even if production records could be used as the sole guide in cattle selection, the fact remains that production records of few cows are being kept.

Despite the importance of milk production record, dairy cow survival is influenced by many genetic and non-genetic factors. Non-genetic factors include house and feed availability, milk quota restriction and the availability of replacement heifers. Genetic factors include the capability for high production and desirable milk components, the functional conformation necessary for a cow to express her potential to give high milk yield and the ability to maintain adequate body condition to resist metabolic disorders, and the ability to move with sound locomotion (Atkinset al. , 2008).

Production records should show the kilograms of milk produced, the kilograms of milk fat produced, and the percent of milk fat. To properly evaluate production records, more information is needed. This includes the number of times milked per day, growth rate, age, feed and care received, and the lactation length.

Production records for young cows may be used as the basis for predicting future performance. Younger animals that have not been in production will not have production records. Bulls are evaluated on the basis of the production records of their daughters. Young bulls may not have such records available.

The production record can only give an estimate of the ability of the cow to transfer high production ability to her daughters. The best indicator of a cow's transmitting ability is in the records of her offspring. The best foundation stock is a cow with records of the daughters with high production indicated below:

- Health records: Health records include a history of vaccinations and the general health of the herd from which the animal comes. The apparent health of the animal being considered is also important. All health records should be evaluated in selecting dairy cattle.
- Pedigree records: The pedigree is the record of the animal's ancestors. Pedigrees that are most valuable as a basis for selection give the name, registration number, type rating, production record, and show-ring winnings of each ancestor for three or four generations. Such a record gives a more complete picture of the possible inheritance of

type and production than information on only the sire and dam. Pedigrees must be studied carefully. Sometimes they contain misleading information (Gillespie and Flanders, 2010).

Generally, the selection criteria for dairy cattle should be based on: milk production, fat and protein yield, feet and legs, udders, body capacity, and dairyness (Goitom et al., 2015).

2.1.3 Dairy Cow Judging and Selection

Judging and selecting dairy cattle is a comparative evaluation of cattle in which animals are ranked based on their closeness to the ideal dairy type/conformation. Desirable dairy conformation involves functional traits associated with high milk production over a long productive life. Before beginning to judge and select dairy cattle, you should become familiar with the parts of the dairy cow. The next step is to recognize characteristics of ideal dairy type and comparing and selecting cows using dairy cow unified score card.

Recognizing body parts of dairy cows

In learning to select dairy cattle, first familiarize yourself with the names of the different parts of a cow's body. A list of these names, together with their location on the cow's body, is given in Fig. 14. Be able to locate all these parts of a living cow and be able to make a diagram and indicate the parts from memory. After this is done, you are ready to proceed with a study of selection of dairy cattle (Paulson et al., 2012).

Characteristics of an ideal dairy type

The descriptions of an ideal dairy cow with the ability to produce milk for a long period of time, in relation to physical appearance are:

- Breed characteristics: True to particular dairy breed
- Head: Moderate in length, clean-cut and alert
- Trunk: Wedge-shaped in side view, the rear portion being deeper than the front
- Shoulder blades: Blend tightly into the body
- Back: Straight and strong
- Rump: Long and wide; level from hooks to pins with refined, level tail-head; thurls high and wide
- Legs and feet: Strong; forelegs straight and squarely set; hind legs straight with slight set when viewed from the side; strong pasterns
- Neck: Long and clean-cut, blends smoothly with shoulders
- Withers: Sharp
- Ribs: Wide apart, highly sprung
- Thighs: Thin and flat
- Skin: Loose and pliable
- Barrel: Long and deep, increasing in spring of rib towards the rear
- Heart girth: Deep, full crops; wide chest floor

- Udder: Symmetrical with evenly balanced quarters; moderate crease between halves when viewed from rear; strongly attached
- Teats: Uniform and squarely placed; moderate size
- Mammary veins: Large, twisting and branching

2.1.4 Dairy Cow Unified Score Card

Judging dairy animals is a process of comparing the individuals being judged with an ideal dairy type. The ideal dairy type is described in the Dairy Cow Unified Score Card (DCUSC), which is developed by the Purebred Dairy Cattle Association. This score card can be used with any of the dairy breeds, as all dairy cows of good production possess the characteristics of dairy type to a greater or less degree regardless of breed. This score card is called a utility score card because it attempts to place a value upon the cow both as a useful milk producer and as an animal which has desirable type for breeding purposes. The DCUSC describes the general traits of a good dairy cow that focuses and evaluates five major categories; these are: frame/general appearance, dairy character, body capacity, feet and legs, and udder (Holstein Foundation, 2012).

General description of the score card:

Frame (15%)

The frame is defined as the skeletal structure of the cow, except the feet and legs. In priority order, the areas considered when evaluating the cow's frame are rump, stature, front end, back, and breed characteristics.

The rump is the highest priority because it is closely related to reproductive efficiency, and the support and placement of the udder.

The width of the pelvic region affects the ease of calving. The animal should be properly proportioned throughout with a strong & straight top-line.

Dairy Character (15%)

Dairy character is an indication of milking ability. The priority order for evaluating dairy character characteristics is ribs, thighs, withers, neck, and skin.

- Excellent dairy character indicates an animal that is converting feed to milk with maximum efficiency.
- Animals with poor dairy character are usually coarse and too fat (over-conditioned).
- The ribs should be wide, and the thighs should be lean and flat.
- When viewed from the rear, the thighs should be wide enough to provide plenty of space for the udder attachment.

Body Capacity (10%)

Good body capacity is needed so the animal can consume the amounts of feed needed for high milk production. The priority order for evaluating body capacity characteristics is barrel and chest.

- Animals with good body capacity can use more roughage in their ration.
- Adequate body capacity allows proper development of the heart and lungs.
- Animals with poor body capacity will not be able to maintain high milk production over a long period of time. The age of the animal is taken into consideration when evaluating the length, depth, and width of the body.

Feet and Legs (20%)

Feet have a little higher priority than rear legs when evaluating feet and legs.

- The ability of an animal to reproduce efficiently over a long period of time is closely related to the structure and strength of its feet and legs.
- Proper placement of the legs improves the ability of the animal to move about with ease. The width between the rear legs provides room for a large udder.
- Correct set to the hocks affects the ability of the animal to stand and walk on concrete surfaces over a long period of time. Too much set at the hocks (sickle hock) will cause the legs to weaken as the animal becomes older. Legs that are too straight place too much stress on the hocks (Fig. 1. 1).

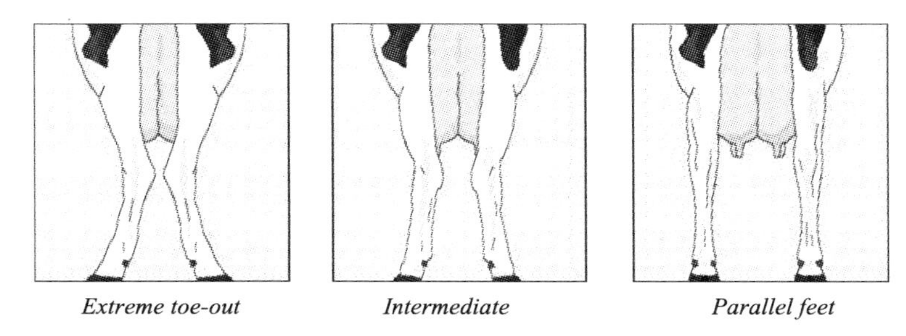

Fig. 1. 1 Rear leg rear view

Udder (40%)

The udder is the most important part of her body, since the main purpose of the dairy cow is to produce milk. The priority order for evaluating the udder is udder depth, teat placement, rear udder, udder cleft, fore udder, teats, and balance and texture.

- Udders that have poor conformation, are weakly attached, and are poorly balanced do not stand up well under the stress of high production.
- The size of the udder is generally related to milk-producing capacity. Cows with small udders are usually not high-producing cows.
- The udder should be soft, pliable, and elastic. If it is still quite firm and large after milking, it is probably full of fibrous or scar tissue. An udder that is still firm after milking is referred to as "*meaty*" .
- The size and placement of the teats are important for ease of machine milking. Teats

that are uneven in size or poorly placed make it more difficult to use the milking machine. Teats should be 3. 8 to 6. 4 cm long. When the udder is full they should hang straight down.

➢ The mammary veins are blood vessels that circulate blood to the udder. The size of these mammary veins indicates the amount of blood circulation to the udder. Large mammary veins are desirable.

➢ The capacity of the mammary system is reduced by a small udder, deep cuts between the quarters or halves, meaty texture, and small mammary veins. A cow with a poor mammary system is not a good foundation cow for the dairy farmer who wants a high-producing herd.

The official scores for Holsteins are as follows: Excellent, 90 - 100 points; very good, 85 - 89 points; good plus, 80 - 84 points; good, 75 - 79 points; fair, 65 - 74 points; poor, 50 - 64 points (Table 1. 3).

Table 1. 3 Dairy cow unified score card

Major Trait Description	Score
1. Frame 15% The skeletal parts of the cow, except for the feet and legs, are evaluated.	
Rump: Long and wide throughout with pin bones slightly lower than hip bones. Thurls need to be wide apart and centrally placed between hip bones and pin bones. The tail head is set slightly above and neatly between pin bones, and the tail is free from coarseness. The vulva is nearly vertical	5
Stature: Height, including length in the leg bones. A long bone pattern throughout the body structure is desirable. Height at the withers and hips should be relatively proportionate	2
Front end: Adequate constitution with front legs straight, wide apart and squarely placed. Shoulder blades and elbows need to be firmly set against the chest wall. The crops should have adequate fullness	5
Back: Straight and strong; Loin: Broad, strong, and nearly level	2
Breed characteristics: overall style and balance. The head should be feminine, clean-cut, slightly dished with a broad muzzle, large open nostrils, and a strong jaw is desirable	1
Rump, stature, and front end receive primary consideration, when evaluating frame	
2. Dairy Character 15% The physical evidence of milking ability is evaluated. Major consideration is given to general openness and angularity while maintaining strength, flatness of bone and freedom from coarseness. Consideration is given to stage of lactation	
Ribs: Wide apart. Rib bones are wide, flat, deep, and slanted toward the rear	8
Thighs: Lean, incurving to flat, and wide apart from the rear	2
Withers: Sharp with the chine prominent	2
Neck: Long, lean, and blending smoothly into the shoulders. Clean-cut throat, dewlap, and brisket are desirable	2
Skin: Thin, loose, and pliable	1

(Continued)

Major Trait Description	Score
3. Body Capacity 10% The volumetric measurement of the capacity of the cow is evaluated with age taken into consideration	
Barrel: Long, deep, and wide. Depth and spring of rib increase toward the rear with a deep flank	4
Chest: deep and wide floor with well-sprung fore ribs blending into the shoulders	6
The barrel receives primary consideration, when evaluating body capacity	
4. Feet and Legs 20% Feet and rear legs are evaluated. Evidence of mobility is given major consideration	
Movement: The use of feet and rear legs, including length and direction of step. When walking naturally, the stride should be long and fluid with the rear feet nearly replacing the front feet	5
Feet: Steep angle and deep heal with short, well-rounded closed toes	3
Rear legs rear view: Straight, wide apart with feet squarely placed	3
Rear legs side view: A moderate set (angle) to the hock	3
Thurl position: Thurls need to be centrally placed between hip bones and pin bones	2
Hocks: Cleanly molded, free from coarseness and puffiness with adequate flexibility	2
Pasterns: Short and strong with some flexibility	1
Bone: Flat and clean with adequate substance	1
Slightly more emphasis placed on feet than on rear legs when evaluating this breakdown	
5. Udder 40% The udder traits are the most heavily weighted. Major consideration is given to the traits that contribute to high milk yield and a long productive life	
Udder depth: Moderate depth relative to the hock with adequate capacity and clearance. Consideration is given to lactation number and age	10
Teat placement: Squarely placed under each quarter, plumb and properly spaced from side and rear views	5
Rear udder: Wide and high, firmly attached with uniform width from top to bottom and slightly rounded to udder floor	9
Udder cleft: Evidence of a strong suspensory ligament indicated by adequately defined halving	5
Fore udder: Firmly attached with moderate length and ample capacity	5
Teats: Cylindrical shape and uniform size with medium length and diameter	3
Milk veins: Large, long, crooked and much branched; milk wells large and numerous (age of a cow to be considered)	3
The quarters should be evenly balanced; soft, pliable and well collapsed after milking	
Total	100

Source: Holstein Foundation, 2012.

Judging and selecting procedure

A systematic and logical method will make judging cows much easier. A technique that works well is to always keep in mind your priorities. Your priorities are udder, dairy character, feet and legs, body capacity, then the rest of the general appearance / frame. The procedures of judging and selection of dairy cows:

Step 1: Observe moving animal

View the animal or group of animals from6 to 9 m. Observe walking animals for ease of movement and set to the rear legs and strength of pasterns.

Step 2: Stand back and get an overall view

The animals should be standing on the level or with the front feet slightly higher than the rear. Look at the side, front, and rear of the animals.

A side view: it permits to look at and evaluate the top-line, rump, barrel, heart girth, shoulders, udder, teats, flank, legs, and neck, spring of rib and firmness of udder attachments.

A front view: it shows width of chest, head, thickness of the shoulders and front legs.

A rear view: it aids in evaluating height, width, and strength of rear udder attachment, depth of rear udder, amount of udder cleft, shape and length of treats, thickness of thighs, straightness of rear legs, width of rump, width of pins and hips.

Step 3: Examine individual parts of the animal

After observing the animals from a distance, move in for a closer inspection.

Observe the shape of the withers, quality of hair, mellowness of hide, texture of udder, and development of the mammary veins in this close inspection.

Step 4: Compare animal to the "ideal type"

Compare the animal with the ideal dairy type as described in Table 1.3 of DCUSC.

Example on judging and selection of dairy cows

First, look at the four Holstein cows in Fig. 15 and make your judgment and selection based on DCUSC (Table 1.3) and judging procedures described in the previous sections. Second, compare your result with the result provided by Hoard's Dairyman (2012). Third, follow the reasoning and justification provided.

The final ranking provided by the judge is: first cow D, second cow B, third cow A and fourth cow C. D easily wins this class with her combination of modern dairy strength and the fact she has the best udder in this class. When you view the cows from the side, D holds her udder floor higher above the hock than does B; meanwhile, D has a snugger and longer fore udder attachment and has more width at the top of the rear udder compared to B. D also is wider in her chest, deeper in her fore rib, and has more depth and spring of rear rib. Additionally, D is wider about her hips, pins, and thurls. Lastly, D stands with her legs more squarely beneath her rump structure with less set to the hock. B has her teats placed more centrally on the quarters.

In the middle placing, B follows the dairy pattern established by the winning cow. B is longer and leaner in her neck, cleaner in her brisket, sharper over the shoulder, flatter and more incurving in the thigh. B has much more bloom and capacity to the mammary system, especially when the cows are viewed from behind where B has a dominating advantage in width of rear udder. B also exhibits a more open dairy rib and spells more milk from end to end than A. A does hold her udder floor higher above the hock than B.

In the final placing, A goes over the similarly-made cow C due to her advantage in overall strength and correctness to the udder. A is a stronger-made cow from end to end by being wider in her muzzle, chest, and rump.

Furthermore, A has more spring of rib with a more desirable tail head setting. As you look at the mammary systems, A has a longer fore udder attachment with a more nearly level udder floor. C is sharper over the shoulders, but she lacks the height of rear udder and correctness of rear feet and legs to place higher.

2.2 Dairy Heifer Judging and Selection

A high-producing herd is built up by careful selection of young stock. Selection of young calves should not be limited to production records, but such records should be supplemented by a careful study of the type and individuality of the animals. Most of the desirable characteristics sought in the mature cow can be visualized in the young heifer. Interest and enthusiasm of father and son in the development of the herd can be maintained by mutual discussion and selection of the calf herd.

The objective in selecting dairy heifers is to choose those which will develop into good dairy cows. Dairy heifers are judged on many the same points as dairy cows. Heifers do not have as much development as mature cows. Therefore, the judge must visualize how the heifer will develop as she matures. This is especially true for the mammary system. When examining the udder, place emphasis on uniformity of quarters, placement and size of the teats, length and width, and rear and fore attachments.

A heifer does not have the depth of barrel and dairy temperament of a mature cow. Pregnant heifers often carry some surplus fat. This may make them appear to be coarse over the withers. This accumulated fat disappears when the heifer comes into production. The selection characteristics are:

Healthy and growing properly

The first essential is that the heifer is a healthy, growing properly with promise of developing into a cow of at least average size for the breed.

- A heifer with a rough, uneven coat of hair that is harsh and lacks luster and with a "pot belly" shows evidence of improper feeding and a stunted condition or of ill health.

➢ A bright, alert appearance, and a healthy coat of hair and pliable hide, together with a trim body having good length and depth and a full, rounded development at the heart girth, indicate health and vigorous growth.

Good dairy form

➢ She should be without blackness, such as thick, heavy brisket; thick withers and thighs; and meaty loin.

➢ The heifer shows a greater dairy tendency, as indicated by angularity and freedom from fleshiness, after calving than before.

Udder

The udder increases in size with age, but even in small heifers having a good dairy tendency.

➢ The udder shows loose folds of skin, indicating the possibility of expansion.

➢ The teats should be uniform in size and symmetrically placed.

Feeding capacity

It is shown largely by length of body, with ribs which are well arched in the upper part with a well-rounded barrel and deep flank.

General appearance and style should be considered:

➢ The bones should not be unusually large or coarse.

➢ Straight, strong backs and top lines, with heads well held up contribute to the pleasing appearance of an animal.

➢ Broad, level rumps are desirable for the same reason and also because the breadth of rear quarters is thought to be related to milk capacity.

2.3 Dairy Bull Judging and Selection

Selection of a young bull is more important than the selection of a heifer. Young dairy bulls are selected in much the same way as heifers. Basically, herd reproductive performance is determined by the fertility of cows and bulls, and the capacity of bulls in identifying and servicing cows in heat. If one of these factors is inadequate, reproductive rates will be significantly impaired. Consequently, reproductive ability of bulls should be evaluated and prior to breeding. Therefore, bull should be evaluated and selected for their physical appearance (special attention should be given to the masculine features), breeding soundness (fertility) and serving capacity (*libido*) (Cooke, 2011).

2.3.1 Physical Appearance

the general physical appearance or conformation and masculine features to be evaluated are discussed below:

Masculine features

After 6 to 8 months of age, the head of the bull becomes larger than that of the heifer and the neck becomes thicker and more rounded.

- As the bull becomes older, his neck assumes the characteristic arch, or crest, which is lacking in the cow. Feminine-like heads in bulls are undesirable.
- The shoulders also become heavier, but the body as a whole should exhibit considerable angularity and freedom from blackness. Heavy fleshing of brisket, loin, and thighs is undesirable.
- Viewed from the rear, the thighs should be only medium heavy with a tendency toward an incurving outer surface.

Good feeding capacity

In the young bull with increased age, the barrel should become deeper and the flank should be carried down well, but, as a rule, the mature bull should not be as "paunchy" in appearance as the mature cow. This is partly a matter of so adjusting the roughage and concentrates in the ration that the bull will not be forced or allowed to consume a large proportion of roughage.

Good body size

Mature bulls, as a rule, should weigh not less than 50% more than mature cows of the same breed. This means that the young bull should show evidence of rapid growth and large body size because the herd sire contributes his characteristics to all the calves born in the herd, care should be taken to choose young bulls with rugged constitution, good top lines, general refinement, and good breed type.

Good style is shown by a clean-cut appearance, that is, freedom from folds of loose skin about neck and brisket, medium-sized bones, neat head and shoulders, and a straight top line. Shoulders which join smoothly to the body without deep depressions back of them, and a full development back of the elbows, showing a broad, capacious chest, are desired.

Strong, straight legs and good feet are especially desirable in the mature bull. Legs and pasterns should be so strong that the animal stands squarely on his feet.

Common faults

In the bull are rough shoulders, narrow heart girth, sloping rump and small rear quarters. As in the cow, a sloping rump detracts greatly from the appearance.

2.3.2 Reproductive Organs

The scrotum, testicles, epididymis, and penis should be physically evaluated for evidence of disorders. The internal accessory glands should be evaluated via rectal palpation to check for inflammation and other concerns. The bull should be free from any reproductive disease and history of any genetic defects.

2.3.3 Semen Evaluation

Semen samples are evaluated for sperm concentration, motility (ability of sperm cells to move), and morphology (incidence of misshapen sperm cells). The circumference of the testicle should also be measured to ensure that only bulls with adequate capacity for sperm production are being used.

***Libido* and serving capacity exam**

It should be evaluated to determine the potential of the bull in servicing cows. Restrain 2 females 8 m apart in a small pen, introduce the bull, and record their sexual activity for 10 min. A bull that has an active sexual interest makes two mounting, and one service should be selected.

Bulls of good type tend to transmit good type to their offspring, but there is no certainty when good type is used as the only guide that the daughters will be good milk producers. The most reliable method of choosing a herd sire (bull) is based on the performance of his daughters. The next best procedure is to secure one that has a dam and sisters with records of high production. If there are no sisters with milk and butterfat records, the records of the dam and grand-dams are the third best basis for estimating the value of the bull from the production standpoint (Nevens and Kuhlman, 1997).

Progeny testing program

It is an important tool for selection of bulls for the best milk production. Be sure to ask any competitor how their bulls are tested. The most effective manner to identify bias is to divide the number of effective daughters (Eff. Daugs.) by the number of herds. This figure should be between 1.0 and 3.0, meaning that there is effectively between one and three daughters of the bull in many dairy herds around the country. This rules out any preferential treatment the daughters may get, which if this were the case would bias the results. In our example, the number of effective daughters per herd is 1.25, which means that only a few herds have 2 daughters of this bull while most of the herds have 1 daughter each.

2.3.4 Calving Ease

A rule of thumb in this regard is that one calving out of ten is difficult. If this figure falls to one difficult calving out of every three, then there is a large problem. Therefore, the ideal easy calving figure is between 5% and 12% whilst bulls with between 15% and 25% should be avoided. While calving difficulty is often closely related to feeding levels, some bulls are problematic. In South Africa, the "A" rating is easy calving, "B" somewhat more difficult, and "C" are bulls to use with caution. The latter two ratings should be exclusively used on cows, whilst the former can be used on both heifers and cows (Fig. 1.2).

RBV (ADSRI 6/91)

BUTTERFAT	MILK	PROTEIN
RVI: 118		
124	116	119

Milk				Butterfat		Protein	
R (%)	Herds	Eff. Daugs.	EPD (kg)	Perc. BF	Diff. (%)	Pero. Prot.	Diff. (%)
67	24	30	+537	3.52	−0.10	3.20	−0.06

CALVING EASE B RED FACTOR

Fig. 1. 2 An example of a South African sire analysis
Source: Ewing, 1989.

2. 3. 5 Linear Score

This summarizes the conformation points listed earlier in the article. The above example shows that this bull improves all these traits, especially heel depth and teat placement.

- Strive for all these characteristics to be positive (to right of 0, viz. +1 or +2). This means that the bull improves these characteristics.
- Deviations from normal occur, e. g. if a bull breeds very shallow heels, the chances are that the A. I. marketing organization will only show his best daughter, with exceptional depth of heel.
- Beware of negative bulls. Form follows function. If a cow cannot walk, she will be unable to graze and therefore unable to produce milk (Fig. 1. 3).

		−2	−1	0	1	2
Height	0.5					
Strength	0.5					
Legs	1.5					
Heel Depth	2					
Fore Udder	0.5					
Rear Udder	0.5					
Teat Placement	1.5					

Fig. 1. 3 A South African sire linear score

>>> SELF-CHECK QUESTIONS

Part 1. Multiple choices.

1. The major variable cost that affects dairy farm profitability is/are

   A. Feed cost B. Labor cost

C. Treatment cost D. Miscellaneous cost

2. Which one of the following is considered as disadvantage of dairy production?
   A. Efficient roughage utilization
   B. Efficient labor utilization
   C. High capital investment
   D. Provide steady income
3. A factor that affects profitability of dairy production, but it is less influenced by the producer decision is/are
   A. Milk production per cow
   B. Milk price
   C. Variable cost
   D. Replacement cost
4. Which of the following is not feed cost reduction strategies?
   A. On-farm forage production
   B. Forage conservation
   C. Use of low quality forage & high amount of concentrate
   D. Use of high quality forage & low amount of concentrate
5. Which of the following increases labor use efficiency in dairy farm?
   A. The use of loose housing system instead of tie-stall barns
   B. Planning of chores
   C. The use of milking machine
   D. All
6. The factor that is given less attention in selecting cattle breed for dairy production is
   A. Availability of the breed in the area
   B. Body color of the breed
   C. Adapting to the environment
   D. High milk production capacity
7. A breed characterized by black & white coat color, largest body frame, highest milk producer and highly distributed worldwide is
   A. Holstein B. Jersey
   C. Guernsey D. Ayrshire
8. A breed characterized by cream to fawn coat color, the earliest maturing, and highest milk fat content is
   A. Holstein B. Jersey
   C. Guernsey D. Ayrshire
9. Which of the following is considered as dual-purpose dairy breed?
   A. Brown Swiss B. Simmental
   C. Ayrshire D. Milking Shorthorn
10. A dual-purpose zebu dairy breed developed in the tropics is

A. Sahiwal B. Simmental
C. Red Poll D. Milking Devon

11. The Ethiopian cattle breed originated in southeastern part of the country, characterized by white or gray coat color is?
A. Sheko B. Barka
C. Fogera D. Boran

12. The Ethiopian cattle breed found in the northwestern part of the country, characterized by having a pied coat of black and white or black and gray color is
A. Sheko B. Ogden
C. Fogera D. Boran

13. Which of the following traits has the highest possible score in the dairy unified score card?
A. Dairy character B. Body capacity
C. Body frame D. Udder

14. Which of the following is a distinguishing feature of dairy type?
A. Angularity of form
B. Development of milking organ
C. Development of feeding capacity
D. All

15. A trait that indicates a cow's ability to consume large quantities of feed and thus greater milk production potential.
A. Dairy character B. Body capacity
C. Body frame D. Udder

Part 2. Match Column 'A' with the appropriate words/phrases from column 'B'.

A	B
1. Breed	A. Zebu/tropical cattle breeds
2. *Bos taurus*	B. European/continental breeds
3. *Bos indicus*	C. Body conformation and dairy character
4. Dairy type	D. Animals in a species with common origin
5. Angularity of form	E. Performance of animal's ancestor/parents
6. Crossbred cattle	F. Hybrid vigor or heterosis
7. Pedigree record	G. Physical appearance of an animal

Part 3. Match the parts of a dairy cow on the left (Column A) to the description of how they should look in an ideal animal on the right (Column B).

A	B
1. Breed characteristics	A. Long and deep, increasing in spring of rib towards the rear
2. Head	B. Straight and strong
3. Shoulder blades	C. Long and clean-cut, blends smoothly with the shoulders
4. Back	D. Flat
5. Rump	E. Sharp
6. Legs and feet	F. True to particular dairy breed
7. Neck	G. Symmetrical with evenly balanced quarters; moderate crease between halves when viewed from rear; strongly attached
8. Withers	H. Deep, full crops; wide chest floor
9. Ribs	I. Long and wide; level from hooks to pins with refined, level tail-head; thurls high and wide.
10. Thighs	J. Wide apart, highly sprung
11. Skin	K. Strong; forelegs straight and squarely set; hind legs straight with slight set when viewed from the side
12. Barrel	L. Large, twisting and branching
13. Heart girth	M. Blend tightly into body
14. Udder	N. Moderate in length; clean-cut and alert
15. Teats	O. Uniform and squarely placed; moderate size
16. Mammary veins	P. Loose and pliable

Part 4. True or false.

1. For ideal cow pin bones should be slightly lower than hip bones.

2. A slight to moderate slope from hips to pins is associated with less calving difficulty, fewer reproductive problems, and greater longevity.

3. Withers, shoulders, backbone, hips and pin bones should be projected and sharp in good dairy cow.

4. Desirable features of milk veins are: small sized veins with small branches extending forward from the udder.

5. Judging dairy animals is a process of comparing the individuals being judged with an ideal dairy type.

6. Dairy character is an indication of milking ability.

7. The size of the udder is generally unrelated to milk producing capacity.

8. Dairy heifers are not judged on the same points as dairy cattle.

Part 5. Give short and precise answer to the following questions.

1. Discuss the advantages and disadvantages of dairy cattle production.
2. What does it mean by dairy farming is an ecofriendly business?
3. Briefly describe feed cost reduction strategies?
4. Briefly discuss dairy breed selection criteria.
5. Compare *Bos indicus* and *Bos taurus* cattle.
6. Name and briefly describe the common defects of a dairy cow.
7. Briefly describe how each of the following is related to the milk producing ability of the dairy cow: general appearance/frame, dairy character, body capacity, and mammary system.
8. Describe the steps to follow when judging a dairy cow.
9. How does a judging dairy heifer differ from judging dairy cows?

>>> REFERENCES

Addisu B, Mengistie T, Adebabay K, 2010. Milk yield and calf growth performance of cattle under partial suckling system at Andassa Livestock Research Centre, North West Ethiopia [J]. Livestock Research for Rural Development, 22.

Atkins G, Shannon J, Muir B, 2008. Using Conformational Anatomy to Identify Functionality and Economics of Dairy Cows [J]. WCDS Adv Dairy Technol, 20: 279-295.

Berry D, Buckley F, 2014. Dairy Cow Breeding [E]. https://www.teagasc.ie/media/website/animals/dairy/Breeding.pdf.

Blauw H, den Hertog G, Koeslag J, 2008. Dairy cattle husbandry: More milk through better management [M]. Wageningen, Netherlands: Agromisa Foundation and CTA.

Briggs H M, Briggs D M, 1980. Modern Breeds of Livestock [M]. 4th ed. Macmillan Publishing Co.

Damitie K, Kefyalew A, Endalkachew G, 2015. Reproductive and Productive Performance of Fogera Cattle in Lake Tana Watershed, North Western Amhara, Ethiopia [J]. Journal of Reproduction and Infertility, 6 (2): 56-62.

Gillespie J R, Flanders F B, 2010. Modern Livestock and Poultry Production [M]. 8th ed. Delmar Cengage Learning.

Goitom Y, Tesfom K, Addis G, 2015. The role of conformational traits on dairy cattle production in Gondar town, Ethiopia [J]. Journal of Agriculture and Biotechnology Research, 1 (2): 64-69.

Haile A, Joshi B K, Ayalew W, et al., 2008. Genetic evaluation of Ethiopian Boran cattle and their crosses with Holstein Friesian in central Ethiopia: milk production traits [J]. Animal Consortium, 3 (4): 486-493.

Mekonnen A, Haile A, Dessie T, et al., 2012. On farm characterization of Horro cattle breed production system in western Oromia, Ethiopia [J]. Livestock Research for Rural Development, 24 (6).

Milk South Africa, 2014. Guide to Dairy Farming [M]. 2nd ed. Cape Town, South Africa: Agri Connect (Pty) Ltd.

Moran J, 2009. Business management of tropical dairy farms in Asia [M]. Thailand: Landlinks press.

Payne W J A, Wilson R T, 1999. An Introduction to Animal Husbandry in the Tropics [M]. 5th ed. United Kingdom: Blackwell Publishing Ltd.

MODULE 2: IDENTIFYING DAIRY CATTLE HOUSING FACILITIES AND REQUIREMENTS

>>> INTRODUCTION

Good dairy housing is important to produce quality milk. A well-designed barn provides a clean, comfortable environment for the herd, and for the operator. The principles of what a cow needs to be comfortable, healthy and achieve optimum production are central to good housing design and good animal welfare. Badly designed, overstocked or poorly managed housing systems will lead to a decline in health, to poor welfare and less comfort of the animals, which quickly affects production.

Irrespective of the production system selected, to maximize performance of the cows the accommodation must provide the cows with the behavioral as well as most basic needs. To achieve this, it is important to understand how an animal behaves when performing routine activities such as drinking, feeding, lying, rising and walking. At a minimum, it must provide a comfortable, well drained lying area, shelter from adverse weather and space to allow the animal to move freely around without undue risk of injury. The cows also need access to wholesome food and water. The design of the system and the level of management applied to the system can affect the health of the cows. The cleanliness of the housing and animals can have an impact on both lameness and mastitis.

Careful planning for the storage and handling of milk, feed, bedding and manure, is very important as this account for most of the farm labor. It should be understood also that a dairy building must satisfy a number of regulations, which needs investigation before construction begins. Thus, this module generates useful information on: housing and facilities, factors to determine site selection, different systems of dairy production, design of dairy farm structure, occupational health and safety (OHS) hazard assess and control risks, material and chemicals used for sanitation and waste treatment activities.

1 UNDERSTANDING DAIRY CATTLE PRODUCTION SYSTEMS IN ETHIOPIA

Livestock production refers to production of farm animals such as dairy and beef cattle, sheep and goat, poultry, swine, and equine species. This can be classified into different production systems based on criteria such as integration with crops, relation to land, agro-ecological zones, intensity of production and type of product (Sere and Steinfield, 1995).

Among all other livestock production systems, the mixed farm rain fed temperate (MRT) system or the tropical highland system is by far the largest. Globally, it represents 41% of the arable land, 21% of the cattle population, and 37% of dairy cattle (Sere and Steinfield, 1995).

According to Dereje et al. (2005), based on climate, land holdings and integration with crop production, in Ethiopia dairy cattle production systems are recognized as: rural smallholder, peri-urban and urban production systems. Rural smallholder system, including pastoralism, agro-pastoralism and highland mixed smallholder dairying, contributes to about 98% of milk production, while the peri-urban and urban dairy farms produce only 2% of the total milk production of the country (Ketema, 2000).

1.1 Rural Smallholder

This system is non-market oriented and most of the milk produced in this system is retained for home consumption. The level of milk surplus is determined by the demand for milk by the household and its neighbors, the potential to produce milk in terms of herd size and production season, and access to a nearby market. The surplus is mainly processed using traditional technologies and the processed milk products such as butter, ghee, *Ayib* and sour milk are usually marketed through the informal market after the households satisfy their needs (Tsehay, 2001).

Pastoralists raise about 30% of the indigenous livestock population, which serve as the major milk production system for an estimated 10% of the country's human population living in the lowland areas. Milk production in this system is characterized by low yield and seasonal availability (Zegeye, 2003). The main source of food for pastoralists is milk. The herd is dominated with unimproved zebu animals and milk production is of subsistent type.

1.1.1 Mixed Production System

This highland mixed smallholder dairying is found in the central part of Ethiopia, where dairying is nearly always part of the subsistence, smallholder mixed crop and livestock farming. These areas occupy the central part of Ethiopia, over about 40% of the country (approximate 490.000 km^2) and are the largest of their kind in Sub-Saharan Africa (Tedla et al, 1989). In the highland areas, agricultural production system is predominantly subsistence smallholder mixed farming, with crop and livestock husbandry typically practiced within the same management unit. In this farming system, all the feed requirements of animals are derived from native pasture and a balance comes from crop residues and stub grazing.

The majority of milking cows are indigenous animals which have a low production performance with the average age at first calving of 53 months and average calving intervals of 25 months. Cows have 3 – 4 calves before leaving the herd at 11 – 13 years of age. The average lactation yield is 524 liters for 239 days, of which 238 liters is off-take for human use while 286 liters is suckled by the calf (Zegeye 2003), but also a very small number of crossbred animals are milked to provide the family with fresh milk butter and cheese. Surpluses are sold, usually by women, who use the regular cash income to buy household necessities or to save for festival occasions.

1. 1. 2 Pastoral and Agro-pastoral System

The pastoralist livestock production system which supports an estimated 10% of the human population covers 50 – 60% of the total area mostly lying at altitudes ranging from below 1500 meters above sea level. Pastoralism is the major system of milk production in the lowland. However, because of the scarcity of rainfall with poor pattern and shortage of feed availability, milk production is low and highly seasonally dependent.

Agro-pastoralists can be described as settled pastoralists who cultivate sufficient areas to feed their families from their own crop production. Agro-pastoralists hold land rights and use their own or hired labor to cultivate land and grow staples. While livestock is still valued property, agro-pastoralists' herds are usually smaller than those found in other pastoral systems, possibly because they no longer rely solely on livestock and depend on a finite grazing area which can be reached from their villages within a day. Agro-pastoralists invest more in housing and other local infrastructure and, if their herds become large, they often send them away with more nomadic pastoralists.

Agro-pastoralism is often also the key to interaction between the sedentary and the mobile communities. Sharing the same ethno-linguistic identity with the pastoralists, agro-pastoralists often act as brokers in establishing cattle tracks, negotiating the "camping" of herds on farms (when crop residues can be exchanged for valuable manure) and arranging for the rearing of work animals, all of which add value to overall agricultural production.

1. 2 Peri-urban Production System

It includes small and medium dairy farms in backyards in and around towns and cities. It is developed in areas where the population density is high and agricultural land is shrinking due to urbanization around big cities. It possesses mainly animal types ranging from 50% crosses to high grade Friesian in small to medium-sized farms. The peri-urban milk system includes smallholder and commercial dairy farmers in proximity to the capital city and other regional towns. This sector owns most of the country's improved dairy stock (Tsehay 2001).

The main sources of feeds are both agro-industrial by-products such as oil seed cakes, bran, and home produced or purchased roughage; the primary objective is to get additional cash income from milk sale. In recent decades, a peri-urban dairy sector has developed very rapidly around the larger cities of many developing countries, in response to expanding market demand. The system comprises small and medium size dairy farms located mainly in the highlands of Ethiopia. They were practiced by state sector and few individuals on commercial basis. These days there are no as such state owned dairy farms except for research institutes, colleges etc.

Farmers located in these areas are part of this system. Farmers use all or part of their land for home grown feeds. Peri-urban dairy producers benefit from their closeness to markets, but their production is based on purchased inputs and may encounter problems of feed supply and poor waste disposal system. Generally, the primary of the production system is for milk sale as a means of additional cash income.

1.3 Urban Farming

This is a purely market-oriented production system involving highly specialized, state (government institutes' farms) or investors' owned farms, located within and close to the boundaries of big cities. Most them have no access to grazing land. Currently, a few smallholder and commercial dairy farms are emerging mainly in the urban and peri-urban areas of the capital (Azage, 2003) and most regional towns and districts (Nigussie, 2006). Smallholder rural dairy farms are also increasing in number in areas where there are market access. In terms of marketing, 71% of the producers sell milk directly to consumers (Tsehay, 2001).

2 UNDERTAKING SITE SELECTION ACTIVITIES

Appropriate site selection is an essential operation for a successful dairying. Site selection requires careful planning to ensure that your investments allow you to build towards the future rather than continuing the past. A few factors are important to ensure that you have a site suitable for the present and for 20 - 30 years in the future as set-out under the sub-headings below:

2.1 Factors Determining Site Selection

Factors to be considered while the selection of an appropriate site for any sorts of dairy production include:

- Water supply: Water serves many functions in a dairy, such as drinking, washing and cleaning, indirect heating or cooling milk, adjusting product composition and for forage production. Water comes into direct contact with the product. It is important, therefore, to locate the dairy near a plentiful supply of clean water. Water can be collected from the roof of a dairy building or nearby barns by putting guttering around the roof and directing the water to a storage tank.
- Land availability: When selecting a site, one should allow for possible future expansion.
- Proximity to the road: For convenience in collecting milk and for product distribution, the dairy should be located near to all-weather road. However, if the building is too near the road, dust contamination will be a problem. Therefore, the doors and windows should not

face the road. Windows for letting in light only can face the road.

- Availability of feed resources and supply: Feed resource includes forage, crop by-product, and concentrate. Cattle yards should be constructed and situated in relation to feed storages, hay stacks, silo and manure pits as to affect the most efficient utilization of labor. So, before we start dairy farm, we must consider the availability of feed resource and consistent supply.
- Access to markets and transportation: Dairy buildings should only be in those areas from where the owner can sell his products profitably and regularly. Since milk is highly perishable product, it has to be sold before it is spoiled. To do so, availability and access to the market is important. Bulk milk cooling system and refrigerator facility are important to transport milk to distant areas. If the farmers are residing in the remote area and producing milk, they should be advised to change milk to butter and other milk product. Another option is that they could form cooperatives and could transport the milk to the town or process in their areas and sell the product to the town. In addition to the market, means of transporting the products to the market should be given due attention.
- Availability of the appropriate dairy breed in the area: Different breed has different adaptation ability and potential of production in different areas. So, selecting an appropriate breed to the specific area is necessary. After deciding the types of breed, selection of individual animal considering body conformation of dairy cattle should be done.
- Weather condition: In the context of livestock rearing, the climate must be considered as it affects forage production, general agricultural production and the animal itself. High temperature associated with high humidity affects production of animal as it decreases feed intake. High humidity also favors bacterial and parasite multiplication, which affect the health of animals. So, an area having good weather that fits the dairy breed should be selected.
- Availability of veterinary service facilities: Cattle health is of utmost importance. As such, every dairy farm has a cattle hospital where ill animals are examined and treated. Cattle are checked daily for any symptoms of sickness. Pen checkers check the pens regularly and assess the health of the cattle. Any animal showing signs of illness is taken from the pen to the cattle hospital. Careful records are kept of each animal's health treatment history. Thus, access to the health service is a very important consideration and necessary facilities should be available.
- Distance from other farm: There should be sufficient distance from other farm such as poultry farm, beef, swine, etc. The distance should be more than 100 meters to avoid transmission of disease from one farm to another farm.
- Topography and drainage: Proper surface and subsurface drainage are required to divert storm water away from animal housing units and to help prevent frost from

heaving of foundations. A dairy building should be at a higher elevation than the surrounding ground to offer a good slope for rainfall and drainage for the wastes of the dairy to avoid stagnation within. A leveled area requires less site preparation and thus lesser cost of building. Low lands and depressions and proximity to places of bad odor should be avoided.

➢ Soil type: Fertile soil should be spared for cultivation. Foundation soils as far as possible should not be too dehydrated or desiccated. Such a soil is susceptible to considerable swelling during rainy season and exhibit numerous cracks and fissures.

➢ Electric power: Electricity is the most important sanitary method of lighting and heating a dairy. Since a modern dairy has always handled electrical equipment which is also economical, it is desirable to have an adequate supply of electricity.

➢ Labor: Honest, cheap and regular supply of labor available should be considered.

2.2 Dairy Farmstead Structures

During building house for dairy cattle, the following points should be considered.

➢ Material: The foundation should be constructed from stone and concrete. If wood is used, it should be treated to prevent damage by termites and water. The material for the superstructure is best chosen according to availability and cost. The dairy farm can be made from basic materials and does not need extravagant construction.

➢ Proximity of other buildings and activities: It is important to locate the dairy correctly in relation to other buildings. It should not be located near a hay barn or an animal feed store where mould spores and dust may contaminate the raw material and products. It should also be located away from other sources of contamination such as dung heaps or cattle assembly areas to avoid bad odors and flies. This doesn't refer to site selection. It is rather to be considered when you design the farm buildings after site selection. Put it in its appropriate place.

➢ Exposure to the sun and protection from wind: The farm structure should be located to obtain the maximum sun exposure since sunshine kills microorganisms. The house facing east is preferable to get morning sunshine while the house facing north preferable to a maximum exposure to the sun in north and minimum exposure to the sun in south and protection from prevailing strong wind currents whether hot or cold. Buildings should be placed so that direct sunlight can reach the platforms, gutters and mangers in the cattle shed. As far as possible, the long axis of the dairy barns should be set in the north-south direction to have the maximum benefit of the sun. Protection from the strong prevailing wind should be done, so that the house construction against prevailing wind. A site with many trees around is ideal for trees act as wind breaks and provides natural shed (shelter).

➢ Distance from residential area and other social service: The farm managers dwelling as well as other office, often called homestead structures, should be in high area which is well drained and commands a view of other buildings and is accessible near to road.

- Walls: The inside of the walls should have a smooth hard finish of cement, which will not allow any lodgment of dust and moisture. Corners should be rounded. For plains, dwarf walls of 4 to 5 ft in height and roofs supported by masonry work or iron pillars will be best or more suitable. The open space between the supporting pillars will serve for light and air circulation.
- Floor: Where possible, all floors should be constructed of concrete with cement surfacing. The floor should slope (1 – 1.5%) to one end to facilitate drainage and cleaning. The cement should continue up the internal walls (curved at the junction of the floor and wall) for at least one meter if the superstructure of the building is not constructed from concrete.
- Effluent piping: The sloped floor drains to an outlet. The effluent should be piped from the outlet to a soak pit through concrete pipes 10 cm in diameter.
- Light: One or two screened windows (total area 3.5 m^2) should be installed to permit the operation of the dairy without artificial light. The windows can also be used for ventilation but should be screened with mesh to reduce the number of insects entering the building.
- Roof: Roof of the barn may be of asbestos sheet or tiles. Corrugated iron sheets have the disadvantage of making extreme fluctuations in the inside temperature of the barn in different seasons. However, iron sheets with aluminum painted tops to reflect sunrays and bottoms provided with wooden insulated ceilings can also achieve the objective. A height of 8 ft at the sides and 15 ft at the ridge will be sufficient to give the necessary air space to cows. An adult cow requires at least about 800 ft^3 of air space under tropical conditions. To make ventilation more effective, continuous ridge ventilation is considered most desirable.
- Door: The main door should be wide enough to allow for equipment installation and easy access of personnel with milk cans, etc.
- Manger: Cement concrete continuous manger with removable partitions is the best from the point of view of durability and cleanliness. A height of 4 – 12 in for a high front manger and 6 to 9 in for a low front manger is considered sufficient. Low front mangers are more comfortable for cattle, but high front mangers prevent feed wastage. The height at the back of the manger should be kept at 2′6″to 3′. An overall width of 2 to 2.5 ft is sufficient for a good manger.
- Alleys: The central walk should have a width of 5 – 6 ft exclusive of gutters when cows face out, and 4 – 5 ft when they face in. The feed alley, in case of a face out system should be 4 ft wide, and the central walk should show a slope of 1 in from the center towards the two gutters running parallel to each other, thus forming a crown at the center.

Apart from the animal sheds, other ancillary buildings/structures are also required to be constructed, including chaff cutter shed, feed store, implements store, straw store, milking parlor, milk-processing room apart from the office, lawns. The large sized dairy farms may also need an overhead water storage tank, a small sized workshop and parking space.

The farm-building may be arranged in such a manner that results in higher animal productivity and labor efficiency with minimum movement of people and the animals. For example, the milking parlor may be situated close to the milking cows shed; the milk room may be adjacent to the milking parlor. The feed storage room also should be close to the milking parlor as the compound feed in mostly fed is the milking parlor during milking.

3 TYPES OF HOUSING, FACILITY REQUIREMENTS FOR DAIRY UNIT

There are several types of housing design and facilities depending upon the different farming system and agro-ecological condition. An efficient management of cattle will be incomplete without a well-planned and adequate housing and facilities.

3.1 Dairy Cattle Housing

Improper planning in the arrangement of animal housing may result in additional labor charges that curtail the profit of the owner. During erection of a house for dairy cattle, care should be taken to provide comfortable accommodation for individual cattle. Except some organized dairy farms belonging to government, private farm, co-operatives or military where proper housing facilities exist; the most widely prevalent practice is to tie the cows with rope on a katcha floor. It is quite easy to understand that unless cattle are provided with good housing facilities, the animals will be exposed to extreme weather conditions. Dairy cattle may be successfully housed under a wide variety of conditions, ranging from close confinement to little restriction except at milking time. Generally, two types of dairy housing system are described below.

3.1.1 Loose House System

Loose housing with tie-stall sheds

Loose housing may be defined as a system where animals are kept loose except milking and at the time of treatment. The system is most economical. In loose housing, animals are usually kept loose in an open paddock in a group of 40 – 50 throughout the day and night except during milking and some other specific purposes like treatment, breeding, etc. when the animals are required to be tied.

This housing system generally provides continuous manger along with covered standing space, open paddock which is enclosed by brick wall or railing and common water trough. Separate structures of calf pens, milking byres, calving pens, bull pens, etc. are required for this system. Taking into consideration the cost of buildings and the investment required, a loose housing system may be desirable. Although it is applicable to most ecological zones, such type of housing is more ideal for areas of low rainfall and warm temperature. A simple shed is important to protect the animals from heat stress and rainfall. Such houses are cheaper to construct, easier to expand at short notice, more congenial to efficient management, less prone to fire hazards to animals and helps cleaner milk production as a special milking barn/parlor is attached. Some features of the loose housing system are as follows:

- Cost of construction is significantly lower than conventional type.
- It is possible to make further expansion without change.
- It facilitates easy detection of animal in heat.
- Animals feel free and therefore, proves more profitable with even minimum grazing.
- Animals get optimum excise which is extremely important for better health production.
- Overall better management can be rendered.

Loose housing with free stalls (cubicles)

Although simple yard and a shade or yard and bedded shed systems are entirely satisfactory in warm climates, particularly in semi-arid areas, some farmers may prefer a system with somewhat more protection. A loose housing yard and shed with free stalls will satisfy this need. Less bedding will be required, and less manure will have to be removed. Free stalls must be of the right size to keep the animals clean and to reduce injuries to a minimum. When stalls are too small, injuries to teats will increase and the cows may also tend to lie in other areas that are less clean than the stalls. If the stalls are too large, cows will get dirty from manure dropped in the stall and more labor will be expended in cleaning the shed area.

A bar placed across the top of the free stalls will prevent the cow from moving too far forward in the stall for comfortable lying down movements, and it will encourage her to take a step backwards when standing so that manure is dropped outside the stall. The bar must, however, not interfere with her normal lying and rising movements. Table 2. 1 and Table 2. 2 lists recommended dimensions for stalls. The floor of the stall must be of a non-slippery material, such as soil. A good foothold is essential during rising and lying down movements to avoid injury. A 100 mm ledge at the back edge of the free stall will prevent any bedding from being pulled out to the alley. The number of stalls should ordinarily correspond with the number of animals housed, except that in large herds (80 or more), only about 90% of the animals need to be accommodated at one time. Young stock may be held in yards with shade or in sheds with either free stalls or deep bedding. The alley behind the free stalls (cubicles) must be wide enough to allow the cows' smooth passage and the following minimum widths apply:

Table 2. 1　Alley Widths in Conjunction with Free Stalls (Cubicles)

Alley between a row of free stalls and a through (increase to 4 m if there are more than 60 cows in the group)	2. 7 – 3. 5 m
Alley between a row of free stalls and a wall	2. 0 – 2. 4 m
Alley between two rows of free stalls	2. 4 – 3. 0 m
Alley between a feed trough and a wall	2. 7 – 3. 5 m

Source: FAO, 1988.

Table 2. 2　Area for bedded sheds and dimensions of free stalls

Animal	Age (months)	Weight (kg)	Bedded shed area per animal (m^2)		Free stalls dimensions (m)	
			A	B	Length	Width
Young stock	1. 5 – 3	70 – 100	1. 5	1. 4	1. 2	0. 6
Young stock	3 – 6	100 – 175	2. 0	1. 8	1. 5	0. 7
Young stock	6 – 12	175 – 250	2. 5	2. 1	1. 8	0. 8
Young stock	12 – 18	250 – 350	3. 0	2. 3	1. 9	0. 9
Bred heifers and small milking cows	—	400 – 500	3. 5	2. 5	2. 1	1. 1
Milking cows	—	500 – 600	4. 0	3. 0	2. 2	1. 2
Large milking cows	—	> 600	5. 0	3. 5	2. 3	1. 2

Source: FAO, 1988.

3. 1. 2　Conventional Dairy Housing System

In temperate area, partially loose housing along with the closed conventional system of housing is desirable. In this system, due attention is given to protect animal from heavy snow fall, rain and strong wind. Tail to tail system of conventional barn, completely roofed and enclosed with side wall is suggested with adequate provision of lying, feeding, watering and milking inside of the barn. Open paddock area with continuous manger in one side along with covered standing space is provided attached to the barn for housing during warm/comfortable weather.

The conventional dairy barns are comparatively costly and are now becoming less popular day by day. However, by this system cattle are more protected from adverse climatic condition.

Based on number of cow, conventional dairy houses can be arranged in a single row if the numbers of cows are 12 – 16 or in a double row system if the herd is greater than 16 in numbers. In a double row system, up to 50 animals can be maintained in a single shed. The distance between two sheds should be greater than 30 ft or it should be twice the height of the building. Ordinarily, not more than 80 – 100 cows should be placed in one building. In double row housing, the stable should be so arranged that the cows face out (tail to tail system) or face in (head to head system) as preferred.

Advantage of tail to tail (face out) system

- In cleaning and milking the cows, the wide middle alley is of great advantage.
- Less danger of spread of diseases from animal to animal.
- Cows can always get more fresh air from outside.
- The head gowala (cow attendant) can inspect a greater number of milkmen while milking. This is possible because milkmen will be milking on both sides of the gowala.
- Any sort of minor disease or any change in the hind quarters of the animals can be detected quickly and even automatically faced in one building.

Advantages of face to face (head to head) arrangement system

- Cows make a better showing for visitors when heads are together.
- The cows feel easier to get into their stalls.
- Sun rays shine in the gutter where they are needed most.
- Feeding of cows is easier; both rows can be fed without back tracking.
- It is better for narrow barns.

Disadvantages of face to face (head to head) arrangement system

- Milking supervision is difficult.
- Possibilities of transmission of disease are more.
- Not labor friendly (Fig. 16 and Fig. 17).

3.2 Housing Facility Requirements

Dairy cattle will be more efficient in the production of milk and in reproduction if they are protected from extreme heat, and particularly from direct sunlight, cold stress and other hazardous climatic/weather condition. Besides, if they are crowded or confined in the limited housing spaces of the area, their genetic potential of milk production and reproduction become depressed, and the animals are going to be suffering with disease and associated with the health problem. In line with this, the following housing requirements and facilities should be considered.

Calving boxes or calving pen

Pregnant animals are transferred to calving pen 2 to 3 weeks before the expected date of calving. Calving pen of 3 m× 4 m (12 m^2) is essential to keep the animals in advanced stage of pregnancy. It should be located nearer to the farmer's quarters for better supervision. The number of calving pens required is 10% of the number of total breed able female stock in the farm. Allowing cows to calve in the milking cowshed is highly undesirable and objectionable. It leads to insanitary in milk production and spread of disease like contagious abortion in the herd.

Special accommodation in the form of loose-boxes enclosed from all sides with a door should be furnished to all parturient cows. It should have an area of 100 to 150 ft^2. With ample soft bedding, it should be provided with sufficient ventilation through windows and ridge vent.

Isolation pen or boxes

It is the separation of sick animals from apparently healthy animals to avoid transmission of diseases to healthy stock. Animals suffering from infectious disease must be segregated soon from the rest of the herd. Loose boxes of about 150 ft^2 are very suitable for this purpose. They should be situated at some distance from the other barns which should be inaccessible to other animals. Every isolation box should be self-contained and should have separate connection to the drainage disposal system.

Bull or bullock shed

Safety and ease in handling a comfortable shed protection from the weather and a provision for exercise are the key points while planning accommodation for bulls or bullocks separately on a farm. The number of bulls required to be one for every 50 breed able females on the farm, if natural breeding is practiced. When artificial insemination (AI) service facilities are available, it is not necessary to keep the bulls in the farm. A bull should never be kept in confinement, particularly on hard floors. Such a confinement without adequate exercise leads to overgrowth of the hoofs creating difficulty in mounting and loss in the breeding power of the bull. The bull sheds shall have covered 3 m× 4 m dimensions leading into a paddock of 120 m^2 with rough cement concrete floor. An adequate arrangement of light and ventilation and an entrance 4 ft in width and 7 ft in height will make a comfortable housing for a bull. The shed should have a manger and a water trough.

Young stock/heifer pen

For an efficient management and housing, the young stock should be divided into three groups, viz., young calves aged up to one year, bull calves, female calves. Each group should be sheltered in a separate calf house or calf shed. As far as possible the shed for the young calves should be quite close to the cow shed. Each calf shed should have an open paddock or exercise yard. It is useful to classify the calves below one year into three age groups, and calves under the age of 3 months, 3－6 months old calves and those over 6 months for a better allocation of the resting area.

Calf pen

Calves should never be accommodated with adults in the cow sheds. The calf house must have provision for daylight, ventilation and proper drainage. Damp and ill-drained floors cause respiratory trouble in calves to which they are susceptible. It can be located either at the end or on the side of the milking barn. This facilitates taking calves to their dams quickly. If there are large numbers of calves, the separate unit of calf shed should be arranged and located near to the milking barn.

Dry animal shed

In large farms, milk and dry cows are housed separately. The floor in the covered area should preferably be made of cement concrete. In small farms, milk and dry animals can be housed together, if one third of the animals in a farm will be in dry or in dry cum pregnant stage.

Quarantine shed

It should be located at the entrance of the farm. The newly purchased animals entering the farm should be kept in quarantine shed for a minimum period of 30 to 40 days to watch out for any disease occurrence.

Store room

The relative position of the feed stores should be quite adjacent to the cattle barn. Noteworthy features of feed stores are given: Feed storages should be located at hand, near the center of the cow barn. There should be one concrete store room with the feed mixing unit at a distant place and a smaller feed store room behind the milking parlor. The floor and walls of store room should be impervious and damp proof.

Milk center/room

It is essential to keep the milk and to chill the milk in larger dairies having 400 – 700 liters production capacity that requires 3.7 m × 5 m size of room and an additional 0.37 m^2 for every 40 liters of milk production. For a smaller dairy unit below 100 litres a small room with a dimension of 3.75 m × 3 m can be sufficient for storing milk and concentrate feed. Generally, this includes milking shed or parlor, collecting yard (part of exercise yard), milk storage and processing.

Hay or straw shed

An adult animal consumes 5 – 10 kg of hay or straw per day, while a young animal consumes 2 – 5 kg of hay or straw per day. The annual requirement can be calculated, and the space requirement can be arrived.

Security & safety

The site should provide security against theft, vandalism, and fire. It should be located away from the residence to reduce the risk of exposing children to injury or death from equipment and animals. Areas infested with wild animals and dacoits should be avoided. Narrow gates, high manger curbs, loose hinges, protruding nails, and smooth finished floor in the areas where the cows move, and other such hazards should be eliminated. Visitor access should be limited to control disease and to reduce interference with farm work.

Effluent or waste disposal area

The satisfactory disposal of effluent from the dairy is important. Since most effluent comes from washing and from spillage, it can be minimized by careful product recovery, proper processing practice and care to avoid spillage. Rinsing and wash water should be piped away from the building for a distance of at least 15 m and directed into a soak pit. The raw effluent should not be piped directly into a river or stream. If the effluent is not piped away from the building, it will become a source of contamination and foul smells.

Resting area for cows

Paved shade, deep bedding in an open sided barn, or free-stalls in an open sided barn.

Exercise yard (paved or unpaved)

Paved feed area: fence line feed trough (shaded or unshaded), or self-feeding from a silage clamp.

Foot Bath

These are important in protecting animals from contagious diseases and pests. A foot bath is a tank constructed near the entrance; this tank is filled with germicidal solution. Animals and carts entering and leaving the farm have to wade through this solution and in the process, the animals' feet or the wheels of vehicles get disinfected. Thus, no disease producing germs will come into the farm through incoming vehicles and animal.

3.3 Space Requirements

Housing space and facilities requirement is very important for animals being unconfined, un-crowded, and for sanitation and hygienic activities to be carried out easily. Animal housing space is determined by the type of housing and age of animals.

3.4 Material and Chemicals for Sanitation and Treatment Activities

Sanitation is necessary in the dairy farm houses for eliminations of all microorganisms that can cause disease in the animals. The presence of organisms in the animal shed contaminates the milk produced thus reducing its self-life, milk produced in an unclean environment is likely to transmit diseases which affect human health: Dry floorings keeps the houses dry and protects from foot injury. Similarly, the presence of flies and other insects in the dairy farm area does not only disturb the animals, but also spread deadly diseases to the animals, e. g. Babesiosis, Theileriosis.

3.4.1 Sanitizers

Sunlight is the most potent and powerful sanitizer which destroy most of the disease producing organism. Disinfection of animal sheds means making these free from disease producing bacteria and is mainly-carried out by sprinkling chemical agents such as bleaching powder, iodine and iodophor, sodium carbonate, washing soda, slaked lime (calcium hydroxide), quick lime (calcium oxide) and phenol.

- Bleaching powder: This is also called calcium hypo chloride. It contains up to 39 % available chlorine, which has high disinfecting activity.
- Iodine and iodophor: This is commercially available as iodophor and contains between 1 and 2 % available iodine which is an effective germicide.
- Sodium carbonate: A hot 4 % solution of washing soda is a powerful disinfectant against many viruses and certain bacteria.
- Slaked lime and quick lime: White washing with these agents makes the walls of the sheds and the water troughs free from bacteria.
- Phenol: Phenol or carbolic acid is very disinfectants which destroy bacteria as well as fungus.

3.4.2 Insecticides

Insecticides are the substances or preparations used for killing insects. In dairy farms, ticks usually hide in cracks and crevices in the walls and mangers. Smaller quantities of insecticide solutions are required for spraying. Liquid insecticides can be applied with a power sprayer, hand sprayer, a sponge or brush; commonly used insecticides are DDT, Gramoxone wettable powders, Malathion, Sevin 50% emulsifying concentration solutions. These are highly poisonous and need to be handled carefully and should not come in contact with food material, drinking, water, milk etc.

Precautions while using disinfection in insecticide:

- Remove dung and used bedding completely.
- Avoid spilling of dung and used bedding while carrying it out.
- Avoid the use of dirty water in cleaning the sheds.
- Never put the fresh fodder over: the previous day's left-over fodder in the manger.
- Prevent algae to grow in the water troughs.
- Use proper concentration of disinfectant / insecticide solutions to avoid any toxic effects poisoning.
- Avoidance of mat in milking time as milk absorbs these quickly.

Procedure:

- Remove the dung from the floor and the urine channel with the help of a shovel and a basket (iron) and transfer it to the wheelbarrow.

- Remove the used bedding and leftovers from the mangers in a similar way.
- Empty the water trough and scrape its sides and bottom with the help of a floor brush.
- Wash the water trough with clean water and whitewash it with the help of lime mixture once a week.
- Scrub the floor with a brush and broom and wash with water.
- Clean and disinfect the splashes of dung on the side walls, rails and stanchions.
- Remove the cobwebs periodically with the help of a wall brush.
- Sprinkle one of the available disinfecting agents in the following concentration: Bleaching powder should have more than 30% available chlorine; phenol 1 - 2% solution; washing soda (4% solution).
- Allow adequate sunlight to enter the shed.
- Spray insecticides at regular intervals, especially during the rainy season (fly season).
- Whitewash the walls periodically by mixing insecticides in it to eliminate ticks and mites living in cracks and crevices.

4 IDENTIFYING, ASSESSING AND CONTROLLING POTENTIAL HAZARDS

Hazard identification or assessment is an important step in the overall risk assessment and risk management process. It is where individual work hazards are identified, assessed and controlled/eliminated as close to the source (location of the hazard) as reasonably as possible. As technology, resources, social expectation, or regulatory requirements change, hazard analysis focuses controls more closely toward the source of the hazard. Thus, hazard control is a dynamic program of prevention. Hazard-based programs also have the advantage of not assigning or implying.

Modern occupational safety and health legislation usually demands that a risk assessment be carried out prior to making an intervention. It should be kept in mind that risk management requires risk to be managed to a level which is as low as is reasonably practical. This assessment should:

- Identify the hazards
- Identify all affected by the hazard and how
- Evaluate the risk and its level of affection or effect
- Identify, prioritize and apply the appropriate control measure

The assessment should be recorded and reviewed periodically and whenever there is a significant change to work practices. The assessment should include practical recommendations to control the risk. Once recommended controls are implemented, the risk should be re-calculated to determine if it has been lowered to an acceptable level. Newly introduced controls should lower risk by one level, i. e. from high to medium or from medium to low.

The prevention of occupational hazards is much more effective and usually cheaper if it is considered at the planning stage of any work process and workplace, rather than as control solutions of already existing hazardous situations (WHO, 1999). This applies first to the planning of new processes or factories, to ensure that hazardous substances are only used if necessary. If they are necessary, then emissions inside and outside the workplace, as well as waste generation, should be minimized, considering the whole life of the process and the products. The workplace and the job should be planned so that hazardous exposure is either avoided or kept to an acceptable minimum. Incentives should reward work practices which minimize exposure. The same considerations should apply to the introduction of new or modified processes and procedures.

The order of priority should be to "plan out" the exposure, by not using hazardous substances, or using them in such a way that no one is exposed. If it does not completely prevent exposure, then prevent or minimize emission of the substances to the air. If it is not possible to prevent exposure by any other method, then give personal protective equipment, including respiratory protective equipment (RPE), to the workers and other persons, as needed. It is essential to adequately plan for supervision and maintenance, to ensure that controls are used and continue to be effective. Workplace control of exposure must be integrated with other measures, such as control of emissions to the atmosphere and waterways, and waste disposal, so that all these measures work together. (Of course, elimination of the hazardous substances prevents all these problems). Similarly, the control of any hazardous substance in the workplace should be part of an integrated control system encompassing other hazards, such as noise and heat, as well as the ergonomic design of tasks and workplaces.

Control of exposure to dusts, alongside other health and safety measures and environmental protection, should be a key priority of the top-level management, and workers should continually be made aware that this is a management priority. Incentive systems for supervisors and workers should be designed to encourage safe procedures. Prevention and control measures should not be applied in an ad hoc manner, but integrated into a comprehensive, well managed and sustainable program at the workplace level, involving management, workers, production and occupational health professionals.

4.1 Major Cause of Firing Accident and Hazards

- Electrical wiring, electrical outlets, and faulty wiring: Whether it's in an electrical outlet or a short in the wall, many fires are caused by electrical wiring. Older homes are particularly susceptible, as they were not wired for the many, many appliances that we have filled our home with. Many homes that were built in the 1950 – 1979 have aluminum wiring that gets very hot and increases the chance of fire.
- Incendiarism: It pertains to the malicious burning of property, using certain chemicals, bombs, etc. and causing fire start.
- Open flames: Unattended burning candles or kerosene lamps that are placed near

flammable/ combustible material is a fire waiting to happen. Unattended cooking causes most fires in the homes that occur in the kitchen. Oil or fat starting to smoke are near the combustion point and can ignite violently.

- Liquefied petroleum gas (LPG): It is liquid inside the container, but immediately transforms to the gaseous state when released. It is liquefied so it can be stored economically and transported easily. A gallon of liquid PG in a bottle will expand about 270 times as much when suspended in a vapor form. Like any other liquid, LPG expands and contracts with changes in temperature. A gallon of liquid LPG expands at higher temperature and contracts at a lower temperature. Therefore, LPG containers are never filled to their full capacity to give allowance for expansion of the liquid. However, it is heavier than air in vapor form, it pushes out the air inside the room, causing a shortage of oxygen and this might suffocate any person in that room. LPG is colorless, tasteless and odorless, but an odorizing agent called "ethyl mercaptan" is introduced to it so that leaks can easily be detected. LPG in its vapor form is about half as light as water. This has its significance in case of gas leaks, so that the tendency of gas flow is to float at lower levels.
- Fireworks: These are beautiful pyrotechnic displays resulting from the occurrence of certain oxidation, reduction, and reactions. The substances that produce fireworks are hazardous materials. These must be stored, transported, handled as oxidizers, and displayed by experienced and knowledgeable professionals. Implementation of prohibitions in the manufacture, handling and use should be strictly complied with everyone as fire safety precaution.
- Spontaneous ignition: This occurs because of a chemical reaction within the material. It is a reaction independent of any outside source of heat. It begins with spontaneous heating which some of the common materials that may spontaneously heat and ignite are animal oils, mixed fish oils, coal, sawdust, hay, grain, and cotton.
- Static electricity: It involves the movement of electrons between two objects in contact with each other. Electrical charges are produced on the objects when they are separated. If the charge builds up, it will develop enough energy to jump as a spark to a nearby grounded or less highly charged object. This spark can ignite flammable vapors, flammable gases or finely dispersed combustible solid materials.
- Smoking: Careless smoking, especially in bed is a leading cause of fire in the homes. In the Philippines it is the fourth leading cause of fire. Fire protection doesn't have to be difficult. Even the simplest things can help save you and your family from a home fire.
- Accidents and carelessness: Many people believe that if they are careful they are much less likely to have a fire. While it is true that being careful will make you safer, it will not stop the fire from happening. Most fires are not caused by carelessness. They are caused from everyday living that is almost impossible for us to change.

- Appliances: Lamps, toasters and even baby monitors can short out. Be particularly careful with older appliances and extension cords. Even new appliances can be the source of a home fire. To be safe, appliances should be unplugged when not in use. Unfortunately, not all appliances can be unplugged, leaving your home at risk 24 hours a day.
- Heating: Heating is another major cause of residential fire deaths. This is especially true in the southeastern states and among wood stove users in the north.
- Unattended stoves: Another cause of residential fires is cooking, but not due to defective stoves or ovens. Often, it is because of unattended pots or the burner being left on accidentally, and who hasn't done that at least once or twice?
- Children playing with matches: Children and grandchildren playing with matches are a major source of home fires. According to the Burn Awareness Coalition, burns are the number one cause of accidental deaths in children under two, fire and burn injuries are the second leading cause of accidental deaths in children ages 1 - 4, and the third leading cause of injury and death for ages 1 - 18. Matches and lighters in the hands of young children are a significant factor in fire fatalities. Educating parents and grandparents to the seriousness of this issue is paramount.

4.2 First Aid and Other Fire Controlling Procedure

Step 1

Identify the fire and explosion hazards and hazards from similar energetic events. This should be an identification and careful examination:

- The dangerous substances present, including those that may be formed in the workplace
- The potential ignition sources of the dangerous substances
- The work activities involving the dangerous substances
- The possible formation and extent of explosive atmospheres
- The scale of the anticipated effects of the fire, explosion or similar energetic event

The supplier's Material Safety Data Sheet (MSDS) should provide key information on the properties and hazards of the dangerous substance to assist you in this task. It should also provide information on the safe methods for the storage, use and handling of the dangerous substance, or makes reference to where this may be found.

Step 2

Decide who might be harmed and how.

Identify the people at risk from the fire, explosion hazards or similar energetic event involving the dangerous substance. Based on your consideration of the anticipated effects of the incident, determine who might be potentially harmed by it. This includes members of the public who might be put at risk by the work activity.

Step 3

Evaluate the risks and decide on precautions.

You should determine whether the measures taken are adequate to eliminate or reduce the risks from dangerous substances, so far as is reasonably practicable. This should take account of such things as:

- The possible substitution of the dangerous substance by one that is non-hazardous, or one that is less hazardous
- The control measures to prevent a fire, explosion or similar energetic incident from occurring
- The mitigation measures to limit the scale and magnitude of the incident should it occur

Step 4

Record your findings and implement control measures.

If you employ five or more employees, you should record the significant findings of your risk assessment. This should include the location and extent of explosive atmospheres and their classification in terms of zones. The risk assessment should also help you decide on:

- The information, instruction and training you give to your employees.
- This should be sufficient for them to safeguard themselves and others from the risks presented by the dangerous substances.
- The arrangements to deal with accidents, and emergencies, including involvement of the emergency service.

Step 5

Review your risk assessment and update if necessary.

You should carry out a risk assessment regardless of the quantity of dangerous substance present, as it will enable you to decide whether existing measures are sufficient or whether any additional controls or precautions are necessary. As well as assessing the normal activities within the workplace, you will also need to assess non-routine activities, such as maintenance work, where there is often a higher potential for fire and explosion incidents to occur. If there is no risk to safety from fires and explosions, or the risk is trivial, no further action is needed.

5 WASTE MANAGEMENT SYSTEM

In countries where cows are grazed outside year-round, there is little waste disposal to deal with. The most concentrated waste is in the milking shed, where the animal waste may be liquefied (during the water-washing process) or left in a more solid form, either to be returned to be used on farm ground as organic fertilizer.

In the associated milk processing factories, most of the waste is milk, milk products and all dairy processing wastes that do not meet applicable quality standards, have become contaminated, or otherwise have become unusable for human consumption, animal feed, or any other beneficial use.

In dairy-intensive areas, various methods have been proposed for disposing of large quantities of milk. Large application rates of milk onto land, or disposing in a hole, is problematic as the residue from the decomposing milk will block the soil pores and thereby reduce the water infiltration rate through the soil profile. As recovery of this effect can take time, any land-based application needs to be well managed and considered. Other waste milk disposal methods commonly employed include solidification and disposal at a solid waste landfill, disposal at a wastewater treatment plant, or discharge into a sanitary sewer.

Methods of waste management

1) Landfill

Landfill is the most popularly used method of waste disposal used today. This process of waste disposal focuses attention on burying the waste in the land. Landfills are found in all areas. There is a process used that eliminates the odors and dangers of waste before it is placed into the ground. While it is true this is the most popular form of waste disposal, it is certainly far from the only procedure and one that may also bring with it an assortment of space.

This method is becoming less these days, although thanks to the lack of space available and the strong presence of methane and other landfill gases, both of which can cause numerous contamination problems. Many areas are reconsidering the use of landfills.

2) Incineration/Combustion

Incineration or combustion is a type disposal method in which municipal solid wastes are burned at high temperatures to convert them into the residue and gaseous products. The biggest advantage of this type of method is that it can reduce the volume of solid waste to 20% to 30% of the original volume, decreases the space they take up and reduce the stress on landfills. This process is also known as thermal treatment where solid waste materials are converted by incinerators into heat, gas, steam and ash. Incineration is something that is very in countries where landfill space is no longer available, which includes Japan.

>>> SELF-CHECK QUESTIONS

Part 1. Choose the best answer for the following question.

1. Which one of the following dairy production system is mostly practiced in Ethiopia?
   A. Pastoralism B. Intensive
   C. Semi intensive D. All the Above
2. Which one of the following is cause of fire in dairy farm?
   A. Home fire B. Heating
   C. Static electricity D. All
3. All the following are features of the loose housing system except one.
   A. Cost of construction is significantly lower than conventional type.

B. It is possible to make further expansion without change.

C. Difficult in detection of animal in heat

D. Animals feel free and more profitable with even minimum grazing.

4. All are precautions used while disinfecting insecticide except one

A. Remove dung and used bedding completely.

B. Avoid spilling of dung and used bedding while carrying it out.

C. The use of dirty water in cleaning the sheds.

D. Prevent algae to grow in the water troughs.

5. Which one of the following chemicals is used for sanitation and treatment activities?

A. Bleaching powder

B. Iodine and iodophor

C. Sodium carbonate

D. Washing soda

E. All the above

6. Which one of the following is true about site selection?

A. Land availability for forage establishment and future expansion

B. Proximity of other buildings and activities

C. Availability of feed resources and consistent supply

D. All the above

7. Which one of the following dairy production system is practiced around the city?

A. Mixed dairy production system

B. Peri-urban dairy production system

C. Pastoralism

D. None of the above

8. Which one of the following housing system is recommended for temperate region?

A. Loose with free stall

B. Loose with tie stall

C. Conventional type

D. Both loose with free stall and tie stall

9. ________ is a tank constructed near the entrance of dairy farm which protects animals from contagious diseases and pests.

A. Water tank

B. Footbath

C. Cattle crush

D. Milking tank

10. All the following are dairy units that should be considered during designing dairy house except one

A. Calf pen

B. Bull boxes

C. Waste disposal area

D. None of the above

Part 2. Match the appropriate phrase under column A with phrase under column B.

A	B
1. Pre-urban and urban dairy production system	A. Market-oriented production system
2. Sanitation	B. Helps cows get fresh air from outside
3. Intensive dairy production system	C. Main feeds agro-industrial byproducts
4. Tail to tail housing system	D. Eliminations of microorganisms

Part 3. Discuss and describe in detail the following questions.

1. Write types of dairy production system.
2. Compare with each other the two of types housing systems.
3. Write first aid in fire controlling procedure.
4. Write major cause of firing accident and hazards.
5. Write aspects should be considered when selecting a site for the dairy production.

MODULE 3: DEVELOPING AND IMPLEMENTING FEEDING PLAN

>>> INTRODUCTION

The dominant variable on any livestock farm is the supply of feed. Frequently, because of poor planning aggravated by inefficient production practices and adverse weather conditions, basic feed supplies are erratic and inadequate. It is not economic to plug these gaps with concentrates. With the price ratio of milk to concentrate currently near 1 : 1, it is more important than ever to realize that concentrates are supplementary feeds and not staples. A constant supply of roughage of good quality is the solid foundation of profitable dairy farming.

Feeding has the most influence on the amount of milk any cow produces. Proper feeding and care allows the cow to produce closer to her potential ability. Taking a bottom-up approach to develop feed planning, starting by calculating the amount of nutrients the animals need to perform to expected levels, calculating what can be grown on-farm, then filling any gaps with purchased feeds, provides an opportunity for farmers and industry to enhance profitability whilst protecting our surroundings.

Therefore, this module covers the information required to develop dairy cattle feeding plan, in general and setting goals and objectives, planning for forage and pasture establishment and grazing, ration preparation and marketing, feed processing, storage, conservation and utilization including treating of different roughages (particularly crop residues).

1 IDENTIFYING ANIMAL FEEDS

1.1 Chemical Composition of Feeds

Nutrients are substances obtained from feed/food and used in the body to promote growth, maintenance, reproduction and production. Feedstuffs contain the nutrients animals require to perform normal body functions such as breathing, pumping blood, fighting diseases, growing, gaining weight, reproducing and producing milk. The feedstuff must be digestible and the end products of digestion (nutrients) absorbed if the feed is to be useful to the animal.

Some components of a foodstuff have no nutritive value because they are indigestible and not absorbable (e. g. some woody plants) and pass out through feces. In addition, some plants contain compounds that are toxic to the animal.

The nutrients that are useful and utilized by the animals are termed as energy (carbohydrates and lipids), proteins, minerals, vitamins and water (Lukuyu et al, 2012).

1. 1. 1 Energy

The energy portion of the feed fuels all body functions, enabling the animal to undertake various activities including milk synthesis. Energy is not a nutrient but is the basic constituent of all feeds that includes the major nutrients called the carbohydrates and the lipids, in particular. These nutrients that provide energy to the animal must be provided with the feed (in terms of quantity) because dairy cows require it crucially in their feed.

Functions

- Maintenance: simply to maintain itself, an animal requires energy. The body weight does not increase or decrease, the animal does not produce; this energy is only for survival and the amount is affected by body size and the environmental temperature.
- Growth and weight gain: Gain is especially important for young animals, which need to attain the recommended weight for a particular age.
- Reproduction: A cow requires more energy during pregnancy for the fetus to grow and develop normally.
- Milk production: the energy requirement of a lactating cow increases with increase in milk production and butterfat content of the milk.

Sources

Energy can be obtained from several types of feedstuffs that contain either carbohydrate or lipids (fats and oils).

Carbohydrates are the major source of energy in the diet of dairy cows. They are found in the staple foods consumed by humans (e. g. rice, maize, wheat, potatoes). Carbohydrates constitute between 50% and 80% of the dry matter in forages and grains.

The three major types of carbohydrates:

- Sugars: sugars are soluble in water, making them readily available to the animal. The major sources are: molasses, sugar beets and sugar cane.
- Starch: starch is the main form of carbohydrate stored in plants. It is the main component of cereal grains and some roots (potato tubers).
- Fiber: forming the structural part of plants, fiber is present in large quantities in roughages. The fiber is broken down by microorganisms in the rumen (microbial enzymes) into products that the animal can use. It is also important in maintaining high levels of milk fat.

The major sources include grasses, fodder crops and crop residues.

Lipids (fats and oils) contain about 2.25 times more energy than carbohydrates per unit weight. Generally, plants are good sources of oils while animal products contain fats. Most plant seeds contain a small amount of lipids. The exception is oilseed plants, which may contain as much as 20% lipids (cotton, sunflower and soybean seeds) and are better sources of lipids than animal fats.

Energy deficiency

The most obvious sign of energy deficiency is poor body condition due to excessive weight loss. Lactating cows are unable to reach peak milk production in early lactation resulting in low lactation yields.

Excessive amounts of energy

Cows consuming too much energy become too fat, resulting in low conception rates. They are prone to difficult calving, retained placenta, and higher incidence of milk fever and ketosis. In early lactation, feeding too much energy, especially in the form of grain, may lead to too much acid in the rumen (acidosis), increased risk of displaced abomasum, depressed feed intake and low milk fat percentage.

Note: Usually forages are high in fiber and low in energy, and concentrates are low in fiber and high in energy. Therefore, there is the need to balance the two, as too much forage limits the intake of energy while too much concentrate results in milk fat depression, rumen acidosis and other health problems.

1.1.2 Protein

Protein is quantitatively the second most important nutrient in feeding the dairy cow. Proteins are made up of building blocks referred to as amino acids.

Functions

Proteins provide the building material for all body cells and tissues (e.g. blood, skin, organs and muscles). Proteins are also major components of products such as milk and meat. Lack of protein therefore adversely affects milk production.

Sources

It sourced from oilseeds and oilseed cakes—residues after the oil is removed from oilseeds, e.g. cottonseed meal or cake, whole cottonseed, whole soybeans (cracked) meal and sunflower meal or cake. Products of animal origin contain fish meal, blood meal, meat and bone meal, feather meal and by-products from milk processing (e.g. skim milk and whey). Herbaceous legumes contain lucerne, *Desmodium* and fodder trees (e.g. *Calliandra* and *Sesbania*).

Non-protein nitrogen (NPN)

Cows can obtain protein from sources that do not contain true proteins, such as urea and poultry waste (contains uric acid). These sources are referred to as non-protein nitrogen sources. Microorganisms in the rumen use the nitrogen in urea to synthesize protein for their own growth that will in turn be available for the animal.

Protein deficiency

For lactating cows, there is a sudden drop in milk production if the amount of protein in the diet is suddenly reduced. Severe deficiency may cause excessive weight loss in lactating cows, reduced growth rate in calves and heifers, and result in underweight calves being born.

Protein and rumen microbes

Most of the protein in feed is broken down by microorganisms in the rumen (rumen degradable protein) and re-synthesized into microbial protein. Bypass proteins are proteins resistant to microbial breakdown in the rumen (un-degradable protein) and pass intact to the small intestines where they are digested and absorbed directly into the body. The microbial protein, on the other hand, is transported along with the microbes to the small intestine for digestion.

Protein and milk production

Milk contains 3.2 – 3.5% protein. Thus, a cow producing 25 kg milk per day secretes 800 – 875 g protein daily. Cows have little ability to store protein in the body and so it must be supplied in the diet daily to maintain the milk yield. Protein should be 15 – 18% of the total ration of a dairy cow depending on milk yield.

1.1.3 Minerals

Minerals are nutrients required in small amounts in the feed. They are required for the body to function properly, i.e. remain healthy, reproduce and produce milk. Some minerals are required in large quantities in the ration dry matter (macro minerals) while others are required in small quantities (micro minerals). Some minerals are stored in the body (e.g. iron in the liver and calcium in bones) while others are not (e.g. sodium, potassium) and have to be supplied in the diet all the time (Table 3.1).

Table 3.1 Minerals required in ruminant diet

Macro minerals	Micro minerals
Calcium	Cobalt
Copper	Chlorine
Iodine	Magnesium
Iron	Phosphorus
Potassium	Manganese
Sodium	Molybdenum
Sulfur	Selenium and Zinc

Functions

Specific minerals may have different functions in the body. Generally, the minerals are required for:

- Bone formation
- Formation of components of enzymes, vitamins and red blood cells
- Production of hormones that control body functions
- Control of water balance in the body
- Milk synthesis

Requirement for minerals is affected by several factors:

- Age: mineral requirements for young growing animals are higher.
- Physiological status: pregnant animals require more.
- Level of production: High-producing cows require large quantities of calcium; deficiency is more likely to occur in early lactation rather than late.

Sources

Although roughages and concentrates contain minerals, the types and amounts vary widely and hence may not meet the requirements. During ration formulation, macro minerals calcium, phosphorus and magnesium are taken into account. Roughages will supply adequate amounts of potassium and common salt can adequately provide as sodium chloride. Some ingredients (supplements) are added to supply a specific mineral (e.g. limestone, salt, magnesium oxide).

Mineral deficiency

Signs of mineral deficiency may not be obvious, but they include:

- Poor fertility: lack of heat signs and low conception rate
- Low milk production
- Poorly developed bones in young animals (rickets)
- Health disorders, for example, milk fever
- Poor body condition, which may be accompanied by a change in coat color

1.1.4 Vitamins

Vitamins are nutrients in the feed required by the body in tiny amounts for normal functioning of the body, through their involvement in many body processes. Some are synthesized by rumen microbes and/or stored in the body of the animal while others must be supplied in the diet. The vitamins that must be supplied in the diet include vitamin A, vitamin D and vitamin E; those that are produced in the body include vitamin B complex, vitamin C and vitamin K.

Functions

Important functions of vitamins include:

- Maintenance of healthy protective tissues such as skin, stomach, intestinal and cell linings (vitamin A)
- Improvement of appetite, hence feed intake (vitamin B compound)
- Production of red blood cells, hence preventing anemia (e. g. vitamin B_6 and vitamin B_{12})
- Enhanced calcium and phosphorus utilization, hence play a role in bone formation and growth (vitamin D)
- Enhancing immunity (vitamin E)
- Helping in blood clotting (vitamin K)

Vitamin deficiency

Vitamin requirements of dairy cows are normally met through diet, rumen microbial synthesis or tissue synthesis. Deficiencies are rare under normal conditions but may occur under certain condition.

1. 1. 5 Water

Water, though not classified as a nutrient, is essential for life in all animals. Water accounts for 74% of the calf's weight at birth and 59% of that of a mature cow. Every 100 kg of milk contains up to 87 kg of water.

1. 2 Classification of Feeds

In general, animal feeds are classified into two major classes. They are named as roughages and concentrates. This classification is based on the results of the chemical composition of feeds after undertaking the chemical analysis necessary to determine the chemical contents of the feed. The process of chemical analysis, sometimes termed as feed analysis, is further discussed in detail in a separate section within this module a little bit later.

Animal feeds in general and dairy cattle feeds as well are classified into these two major classes on the basis of their chemical composition and their cell-wall fractions. For details, see the discussion below.

1. 2. 1 Roughages

Roughages are feedstuffs that contain relatively large amounts of crude-fiber (>18% CF). They are characterized as bulky feeds with lower amount of total digestible nutrient (TDN <60%), and they are usually of very low digestibility and palatability compared to concentrate feedstuffs. They comprise the largest proportion of the dairy animals' ration. Based on the moisture content, roughages are further subdivided into two groups.

Roughages with very high moisture content (60 – 90%) are known as succulent roughage. Whereas, roughages that have with lower amounts of moisture (10 – 15%) are termed as dry roughages.

Succulent roughages

These are fibrous feed materials with high amount of moisture. Succulent roughages commonly utilized by dairy cattle include:

- Forage trees: These are trees with stems and leaves like *Sesbania sesban*, tree lucerne, *Moringa*, etc.
- Pasture: It is an area of grassland used for grazing.
- Cultivated fodder crops: A cultivated crop grown for use as animal feed
- Tree leaves and fodder: This refers to the herbs and browse of forage plants.
- Root crops: A crop having swollen root for animal feed
- Silage: A feed produced by controlling the fermentation from green fodder

Dry roughages

Dry roughages include hay and crop residues, such as straws (teff, wheat, barley, etc.), and stover (maize, sorghum, and millet stover).

1.2.2 Concentrates

Concentrates are feedstuffs containing the relatively lower amount of crude fiber (CF<18%), and higher nutrient contents required for production. These are characterized by better digestibility and palatability than roughages and generally higher TDN (>60%). Concentrates are also either high in protein or energy (proteins and carbohydrates).

Concentrates are also further subdivided into two categories on the basis of their chemical composition or CP contents. Concentrates containing CP (>18%) are termed as protein rich concentrates (protein supplements) and those containing CP (<18%) are termed as energy rich concentrates.

Energy rich concentrates

The well-known energy rich concentrates are mainly derived from plants such as grains, and agro-industrial by-products. Grains and seeds include rice, sorghum, maize, wheat, barley, etc. Cereal grain by-products are produced when grains are processed for flour production. There are a number of products which are left out as by-products of various feed processing industries. These include:

- Wheat bran: This is the most course by-product of wheat milling.
- Wheat-short: It refers to fine bran (in terms of size of particles). Although coarse bran is also there, but not as much as wheat bran.
- Wheat meddling: It is a product that is a mixture containing both wheat bran and wheat shorts.
- Molasses: Molasses is a by-product that comes out as a by-product during the

processing of sugar cane or sugar beet for sugar production. It is the black syrup sweet solution containing at least 46% sugar.

- Local liquor/local beer residues such as *Tella* residue usually comes from small holder producers or cottage level liquor productions.
- Roots and tubers: Root crops are considered equivalent to carbohydrate rich concentrates when measured in terms of dry matter (DM).

Protein rich concentrates (protein supplements)

These are concentrates derived from either plant or animal origin.

(1) Plant origin protein supplements

Protein supplements of plant origin basically derived from feed processing industries as by-products of oilseeds during the extraction of oils and sugar manufacturing industries. These include oilseed meals and cakes. Oilseed cakes and meals are the residues or by-products of oil industry (remaining after the removal of the oil from oilseeds). These residues are rich in protein (200 – 500 g/kg) and the most are available for farm animals. Some of the oilseed meals/cakes are: rape seed cake/meal, cotton seed cake/meal, sunflower seed cake/meal, soybean seed cake/meal, Noug seed cake/meal, palm seed cake/meal, etc.

(2) Animal Origin protein supplements

These are protein supplements derived from animal tissues. These are generally provided to animals in much smaller amounts and in a controlled basis than plants tissue or oilseed derivative protein sources. Examples include meat meal, blood meal, and fish meal, etc.

1) Fish meal

- The product is highly variable, ranging from 35 – 70% CP, which is due to the material used to prepare fish meal. If it is prepared from the whole fish, it will have a higher CP content. But if the fish meal is prepared from unpalatable portions like the head and fins of the fish, it will have a lower CP content.
- Fish meal is highly palatable but is expensive, so mainly used for monogastric and young animals feeding.
- The problem with fish meal is "fishy taint" or flavor. Therefore, do not feed fish meal to animals immediately before slaughter, milking or laying.
- Fish flavor is not a problem if the fish meal has been properly and promptly processed, avoiding putrefaction.

2) Blood meal

- This is obtained by drying blood of slaughtered animals.
- It contains about 80% CP and is important nutritionally as a source of protein alone.
- Blood meal is unpalatable, and the feed is added in limited amount and beyond a certain limit (about 10%), it tends to cause scouring.
- The meal is best regarded for boosting the dietary nutrient level.

3) Meat and bone meal

- Meat and bone meal is the rendered product from mammalian tissues, including bone, exclusive of blood, hair, hoof, horn, hide trimmings, manure, stomach and rumen contents, except in such amounts as may occur unavoidably in good processing practices.
- It shall contain a minimum of 4% phosphorus and the calcium level shall not be more than 2.2 times the actual phosphorus level.
- It shall not contain more than 14% pepsin indigestible residue and not more than 11% of the CP in the product shall be pepsin indigestible.
- The label shall include guarantees for minimum crude protein, minimum crude fat, maximum crude fiber, and minimum phosphorus.
- Meat meal is defined the same as meat and bone meal except that no minimum phosphorus level is required (Larry and Guthrie, 2001).
- Meat and bone meal should be limited to one to two pounds per cow per day.

2 UNDERTAKING PASTURE ESTABLISHMENT ACTIVITIES

2.1 Planning for Pasture Establishment

Forage refers to grasses, leaves, twigs and/or other parts of a fodder crop or a fodder plant that serves as the basic feed source of animals. Fodder crops or fodder plants, on the other hand, include grazing plants, hay, silage, and other root crops and tubers used as animal feeds. Sometimes these two terms are utilized interchangeably to mean animal feeds, in general. But pasture is the land where these forages or fodder crops come from. Whenever we speak about pasture establishment, we mean nothing but the source of feed for our farm animals. Feed production on the farm land is by far better than running a farm based on purchased feeds.

The objective of fodder/ forage production planning is aimed at matching the production capabilities of the farm with the animals' requirements in order to obtain the greatest margin over feed costs, within safe limits of natural resource utilization. The carrying capacity of the property, not the owner's target income, must determine the size of the herd: specifically, how much suitable fodder can be produced annually for the use of the dairy herd. The annual fodder requirements of every 100 cows and their associated replacement heifers must be known. From this total requirement, and from the assessment of the farm's fodder production capacity, the potential herd size can be calculated. It is neither profitable nor wise to exceed that herd size.

When herd size and farm carrying capacity have been reconciled (and not before), one must consider the costs and returns of the farm. The scheme must show an adequate margin over feed costs to cover overheads, and a return to management. If that test is passed, then one should examine the required feed flow that is the amount of feed required each month of the year (usually fairly constant in a fresh-milk herd). The monthly forage flow is also forecast, with due regard to the kinds and areas of pastures and fodder crops being considered.

Fodder planning requires some "crystal ball" gazing, because we cannot predict future weather and, even if we could, we can hardly claim to know exactly how our pastures would respond to it. The manager must regard the plan as a statement of intent and not as the ultimate truth. Unforeseen circumstances and opportunities will arise, and it is essential for him to respond properly to these, regardless of (but surely the wiser for) the provisions of the plan.

Some points to note:

- Fodder planning maps out a program of development which has a time scale of years. Do not get confused by problems that have time scales of days or even months: ration balancing is a common red herring.
- Do not confuse the future with the present nor the proposed with the existing; especially do not be shackled by present thinking or practices on the farm. Development must strive to identify and then remove, ameliorate, or sidestep the constraints and obstacles.
- Make sure that the plan can be economically implemented with the existing situation as a starting point. Intermediate stages of development may have to be worked out before the plan can be accepted; the first stage of development has to be worked out, in any event.
- The farmer is very much a part of the farm. The plan must take account of his strengths and weaknesses, his interests and dislikes, and especially his financial position.
- Clarify the objectives of planning before starting the exercise.
- Don't split hairs, and don't refine the final answers to several decimal places: the future is not that predictable. If a computer is used, it is for rapid and error-free calculation, not to substitute precise arithmetic for imprecise ignorance.
- Round off the results of calculations in such a way that estimates of fodder requirements have increased rather than decreased; that is the cheaper error to make and will lead to the more profitable decision.
- Although the farmer is concerned primarily with the average year in his forward outlook, he should not forget that non-average, but not impossible, conditions once every few years could ruin everything. The best insurance against such disaster is a bulging fodder bank balance.

Planning and managing the fodder flow is not only one of the most critical of all management functions. It can also be one of the most satisfying—financially, psychologically, and aesthetically. Not only does a good fodder flow provide the soundest basis of a profitable operation, but also, there for anyone to see at any time is highly visible proof, pleasing to look at, of a good job well done. Conversely, a poor fodder flow causes trouble on the grazing, in the cow, at the bank, probably even in the home, and usually it offers ugly evidence of incompetence for all, the world to see and scorn. Fodder production planning can be divided into three major sections, namely, the principles of fodder production planning, planning in practice, and implementing the plan.

2.2 Identifying the Principles of Planning

The soundness of any system, existing or proposed, may be evaluated against the following four criteria in very strict order of precedence:

- Safe use of resources
- Meeting the animals' requirements always, even when feed production falls too low either on a seasonal cycle or due to unpredictable causes
- Marginal overall feed costs
- System's realism and manageability

2.2.1 Safe Use of Resources

The resources available for producing fodder are land (including water), labor, management, and capital. If each part of the farm has been developed to produce as much forage as it can, and there is no weakening of its soil and water resources, then the fodder flow rates well under this criterion. This is not to say that all veld must be replaced by pastures, or that all pastures must be irrigated. Resources could be used too intensively (e.g. monoculture of erodible soil, irrigation with saline water) or they may be undersupplied (e.g. arable land, credit, or managerial time or competence). Far too many farmers undertake too many enterprises, do none of them properly, and end up in a worse position than if they did only two really well. A system which misuses or misjudges the resources available will fail sooner or later and is unacceptable from the outset.

A careful assessment of the land and water resources is needed. Determine the areas of land suitable for annual cultivation or for planting pastures, because that largely limits the quantity of high-quality forage that can be produced. Identify steep slopes, erodible soils, wetlands, shallow soils, and rocky areas, all of which have limited or no value for forage crops although they could be suitable for pastures. Considering soils and water, the area of land that can be irrigated is of vital importance, since it determines how much green grazing will be available to the herd in the dry season.

Resource planning will frequently involve a critical look at the location of waterways, fences, roads and buildings. A sound run-off control plan is fundamental to the safe use of resources. Do not be too much influenced by existing developments, since fences and roads can be relocated, and even buildings do not last forever. Farm owners may need outside help for this task as it is psychologically very difficult to be objective.

2.2.2 Meeting the Animals' Requirements

Adequate feed must be produced, stored, or bought to feed all animals present on the farm at any given time. As a first step, enough feed must be available over an average year to meet the annual total dry matter requirement plus the average input to the fodder bank.

Fodder bank

The fodder bank is a store of conserved fodder (hay or silage) which is deliberately accumulated over and above the normal seasonal requirements, for use in unpredictable, lean times such as an unseasonable dry period, a severe hail storm, or an army worm outbreak. A fodder bank is not a permanent or separate store in the sense that a particular silo or hayshed is the fodder bank. Rather, the total store of conserved fodder is built up year by year, part being for dry-season feed and part for the reserve, the division being merely a book entry. The oldest stored fodder is always fed first, whether for normal use or for emergency, and any fodder actually in store will seldom be more than two or three years old. This is especially important in the case of hay, which deteriorates far more rapidly than silage does.

Stock flow and required feed flow

The herd structure, and hence the feed demand, of the livestock on a given farm is not always static. It normally changes from month to month; giving rise to a stock flow and its corresponding required feed flow. Herds producing fresh milk, however, usually do have a fairly constant herd structure, because of the tendency to calve the year round; the required feed flow is therefore also fairly constant.

Fodder flow

The fodder flow is the sum of fodder available from each source (veld, pasture, stover, etc.) month by month. Ideally, it would exactly match the required feed flow. Rarely does this happen naturally, however, so the match must be forced, by purposely altering the stock flow, e. g. by strategic culling and calving, and/or producing more feed at particular times, and/or transferring excess fodder from one time of the year to another as hay, silage, or forage.

If the match is not achieved by the farmer, it will be forced on him by nature: as a loss in production and reproduction (low fertility), a loss in live-mass (thin animals) and a loss of animals, either by forced selling or, in extreme cases, by death due to starvation. Fortunately, dairy farms rarely retrogress that far downhill, nevertheless, the fodder flow often leaves much to be desired; in fact, it is probably one of the major limiting factors to dairy production. The problem, usually one of "subclinical overstocking", manifests itself in the following syndrome:

- An average milk yield below 5,000 L per Holstein-Friesland cow's lactation (herd average of 17 L per cow in milk per day), even with generous levels of concentrates
- Concentrate usage exceeding 400 g/L of milk, average overall cows over the year,

often associated with low butterfat levels

- Large seasonal fluctuations in milk yield, if these are not caused by the calving pattern
- Thin heifers: Underweight first-calvers (mean mass less than 90% of mature mass) and poor first lactation results (under 4,000 L)
- A disproportionate number of thin cows in the herd (more than 15% of the herd thinner than 2 on Mulvany's scale)
- Low fertility, even among young animals

Individually, of course, these problems often arise from causes other than feeding, but if three or more of them occur together, the first place to look for the trouble is in the fodder flow. Remember that cows may show the effects of previous underfeeding at a time when forage supplies are good; for instance, a high incidence of repeat inseminations may be a result of poor nutrition.

Underfeeding can be blamed, at least partly, on the poor quality of roughage. Quality of forage is more important in dairying than in many other enterprises. A high proportion of cows in the herd need a diet rich in energy (dry matter containing more than, say, 10.5 MJ ME/kg or 70 TDN) and good-quality protein. While it is true that cows are better off with lots of poor roughage than with inadequate amounts of good roughage, the aim must always be to produce enough roughage of the best possible quality. To some extent, the quality of the diet can be improved by feeding concentrates, but that strategy has limits, and it is much more expensive than providing good fodder as a basis.

A common fallacy is that protein is the only consideration in assessing roughage quality. In fact, energy is more critical since it constitutes by far the greater part of the cost of feeding cows. Protein, while more expensive per kg, is needed in smaller amounts (10% to 15% of the energy expressed as kg of digestible organic matter, DOM). One should worry primarily about providing enough cheap energy; then worry about providing enough protein. Good pastures provide both, the latter usually in excess.

Note that the obvious signs of a bad fodder flow (hungry and unhappy animals, chronic shortage of grazing, overgrazed pastures in poor condition) have not been included in the above list of problems. These are the signs of "clinical overstocking", immediately obvious to mere humans, and by the time that they have appeared much damage has already been done to the dairy enterprise. Don't rely only on your own assessment of the feeding regime: ask the cows if they have got enough to eat. Their answer is to be found in the list given above.

The second property of a good fodder flow is, therefore, that the herd is properly fed all year round.

2.3 Undertaking Pasture Establishment

Quality pastures are the powerhouse of any livestock enterprise, whether it is a large commercial operation or a small landholding running limited numbers of livestock. The pasture used by the farm can be a brand-new pasture, an existing pasture land, renovated and used by the farm, or else a farm land/ grazing land now used by the farm as a source of forage.

Successful pasture establishment is the result of proper preparation and planning, starting at least 12 months before any pasture is sown. Therefore, before establishing new pastures or renovating existing pastures, producers must evaluate the forage needs of a farm.

2.3.1 Renovation of an Existing Pasture

Renovating a pasture should be based on existing percentages of the desirable species present in the pasture. The following criteria could be used in such a decision:

If the pasture contains 75% or more of the desirable species, then consider not renovating and concentrate on management. If the pasture contains 40 - 75% of the desirable species, then consider over-seed and concentrate on management. If the pasture contains less than 40% of the desirable species, then consider reestablishment. The following are basic requirements and steps for successful forage and pasture establishment.

2.3.2 Establishing a New Pasture

The quality of the pasture established is desired to be sound, productive and containing the quality forage crops. There are a few factors considered during establishment of a brand-new pasture. These factors that include:

Forage species selection

Decide which forage species or mixture will be seeded. Some species are better suited to certain soil types than others. For example, alfalfa does not tolerate poorly drained or low pH soils, while red clover and reed canary grass perform very well under these conditions. Although it often is difficult and expensive to change soil characteristics, species can be changed easily with little or no expense. Proper matching of forage species to soil characteristics not only makes establishment easier, but also improves production over the life of the stand.

Seed quality

Once a pasture species or combination of species has been chosen, it is important that high-quality seed or planting material be used, so that under good conditions a high proportion of the seed germinates or starts growing. Seed should be bought from a reliable company or agency which sells clean, tested seed. The seller should state the standards of seed purity and viability, which applies to that seed. Poorer-quality seed should be sown at a higher rate per hectare than the above 'certified' seed.

Seed treatment

Pasture legume seeds often do not germinate all at the same time after sowing as some seeds have very hard seed coats which water does not penetrate easily. If a farmer is in an area where drought seldom occurs and is sowing when the rains have already well started, he may wish to have an evenly germinating stand of legumes. Scarifying the seed is very important in treating seed with hard coat and scarifying is undertaken by:

- Place the required amount of seeds in a bag which allows the entry of water.
- Soak the seed in a drum with hot (but not boiling) water at about 70℃, for about 10 minutes, this will soften the seed coat and germination will take place quickly.

Seed Inoculation

Inoculation is required for legume seeds such as lucerne when sown into a field or paddock for the first time. This is because the specific *Rhizobium* bacteria required by the plant for the formation of root nodules are not naturally present in soil. To inoculate the seed, follow the procedures:

- Obtain inoculums and keep it in a cool, dark place, preferably a refrigerator, until required.
- Moisten the dry legume seeds slightly with a 5% sugar solution (1 teaspoon per 100 mL of water).
- Sprinkle with enough inoculums to coat the seeds.
- The seeds should be dried in the shade (because direct sunshine kills bacteria) and sown within 24 hours.

Soil test

Pastures establish better in soils of higher fertility. Legumes respond to higher phosphorus levels by increasing the growth of root nodules which then fix more nitrogen and make plants grow faster and establish better. Sculpture also increases the growth of legume seedlings and it should be applied with phosphorus at the establishment, in the form of single superphosphate (21% P_2O_5, 12% S). Grasses respond well when both phosphorus and nitrogen are present in adequate amounts.

A soil test should be completed, and lime should be added to the soil to correct low pH conditions at least six months prior to forage seeding. Planning a year in advance gives producers several opportunities to apply any nutrients that a soil test recommends.

- Soil pH and fertility adjustments: The six months prior to seeding are the last chance producers have to adjust soil pH before planting. Most agricultural grade limestone requires about six months from the time of application until it effectively changes the soil pH. Consequently, adding lime to raise the soil pH within less than six months of seeding will generally result in forages being seeded into soil with a pH lower than desired.

➢ Moisture conditions: Pasture seed should be sown into moist soil, followed within a day or so by enough rain to wet the surface layers of the soil thoroughly. Good germination, emergence and establishment should result as long as the soil does not subsequently dry out too much due to periods of drought. Good establishment may not be obtained if seed is sown too late in the rainy season. Legumes, being slow-growing plants, should be sown as early as possible. Grasses may be planted later as they grow faster.

Seedbed quality

This is more important for small pasture seeds than for the large seeds of crops such as maize. Usually, for improved pastures grown for dairy cattle, most of the trees are removed by stumping and the whole surface area is plowed and disked ready for sowing early in the wet season. Before sowing, the seedbed should be rolled in one way or another. The sowing should be done after the rains have started.

A fine seedbed tilled well is important because then the small pasture seeds have a better chance of having good contact with soil particles which hold water for germination. Also, depth of sowing can be better controlled—pasture seeds generally should not be sown deeper than 1 cm. Some germinating seeds may not emerge from the soil if there is a hard crust on the soil surface. This depends on the soil type. Often the best germination on a rough seedbed (which is not as it should be) takes place in the furrows made by the tractor wheels, because of better soil/seed contact.

Seeding methods

The ideal seeding method depends on the type of equipment available and whether you plant on a no till or a conventional seedbed. To ensure good soil-to-seed contact, seed germination, and timely emergence, different seeding methods are available. Some of these methods are drilling, cultipacker planter and broadcasting.

1) Drilling

Cuts a thin furrow in the soil, deposits the seed, then covers it and firms the soil with press wheels. A good rule is to plant the seed three to four times as deep as the diameter of the seed.

2) Cultipacker planter

The seed is dropped from a hopper into the soil, where toothed rollers press the seed under the surface. When using a cult packer, be careful not to bury the seed too deeply and decreasing germination.

3) Broadcast seeding

A fertilizer with a spreader can result in an uneven seed distribution if the overlap is too wide. When you apply broadcast seeding:

- Less seed is distributed on the outer third, so adjust spacing to provide double coverage.
- Make sure the spreader is calibrated for the appropriate seeding rate.
- When broadcasting, increase recommended seeding rates by 20%.
- Roll with a cult packer to establish a good soil-to-seed contact.

Seeding time, rate and depth

- Seeding time: Seeding on the correct date is also very important. Seeds planted in season usually have plenty of moisture for germination, but they sustain increased weed pressure. Spring seeding should be made at least 4 weeks after the last killing frost. Late summer seeding is recommended for wet areas because the soil is usually dry enough during the summer and has less weed pressure.
- Seeding rate: Proper seeding rates depend on forage species and seeding method. To obtain a good establishment, use seed that is pure, has a high germination rate, and has not been stored for a long period of time. High quality, certified seed is recommended. Seed cost could be a major portion of the total establishment. If the seed is of poor quality, it must be applied at higher rates to obtain a desirable stand. Making the use of cheap seed with low quality neither ergonomically nor economically sounds. If you seed legumes, make sure the seed is inoculated with the proper bacterial strain (Lemus, 2009).
- Seeding depth: It is vital to have proper seeding depth and seed coverage. When drilling legumes, make sure to plant the seeds no deeper than 0.25 - 2 inches, depending on the seed type and size. Planting depths over 2 inches will decrease seedling emergence as much as 50% in some forage species.

Fertilization

If soil nutrient levels are optimum to high at the time of seeding, then fertilization generally should not be a concern during forage establishment. An exception would be the application of 30 - 50 pounds of nitrogen fertilizer per acre to spring-seeded, pure-grass forage crops in late summer of the seeding year, if production warrants this.

Weed and pest control

- Weed control: Weed control in previous crops can significantly reduce weed infestations during forage seedling establishment. However, herbicide use during the year preceding forage seeding should be monitored closely. Triazine herbicides that carry over in soil used for a previous corn crop will cause yellowing and can kill young legume seedlings. Therefore, producers should avoid using triazine herbicides in the last year of corn. If triazine is used in the year preceding forage seeding, application rates should be less than 1 pound per acre.

➢ Pest control: Insect damage to grass forages during establishment generally is not a concern. However, legume forages, especially alfalfa, can be devastated by insect feeding. The primary insect of concern is the potato leafhopper, which can reduce the vigor and later performance of alfalfa seeding. Proper monitoring and control, when the economic threshold has been reached, is extremely important during alfalfa establishment.

Harvest

The goal of harvest management during forage establishment should be to facilitate the production of a healthy, vigorous crop and to suppress annual weeds that may be in the new seeding. Delaying the initial harvest until the forage plant has flowered will allow adequate root reserves to develop for rapid re-growth and optimum establishment. Harvesting earlier to control weeds will reduce the amount of root reserves and will result in weaker plants. Growers should weigh the consequences of producing slightly weaker plants against the harmful effect of weed competition on forage establishment (Hall, 2012).

2.3.3 General Management of Pasture

➢ Do not allow animals to graze new stands too early or too frequently.
➢ Allow plants to become well established before heavy grazing or set stocking.
➢ Mow or lightly graze pastures when plants are 8 – 12 inches tall.
➢ Most forage crops should not be grazed shorter than 3 – 4 inches.
➢ Maintaining proper grazing height will help to trigger new plants to tiller or producer runners.
➢ Allow plants to grow to 8 – 12 inches before grazing or mowing again.
➢ A rotational grazing approach could be beneficial in ensuring the successful establishment (Lemus, 2009).

3 UTILIZING THE PASTURE FOR DAIRY CATTLE FEEDING

The increased forage production creates a potential for increased dairy output. The way the pastures are managed decides whether the effort and resources invested in improving the feed supply bring returns in the form of better milk output and the raising of more young stock. The principal objectives of forage utilization (Humphreys, 1991) are to:

- Cut or graze the pasture in ways which: 1) maximize the delivery of nutrients to the animal, 2) sustain the vigor and persistence of the pasture, 3) maintain cover as a defense against erosion, and 4) preserve a desirable botanical composition.
- Adjust the animal demand in line with the forage supply.
- Arrange continuity of forage supply in order to maintain animal production and minimize animal stress through 1) providing a sequence of feeds of differing seasonal utility, 2) conserving or purchasing feeds, and 3) modifying the pasture environment through irrigation and fertilizer practice.

Forage utilization is the other indicator of the way the pastures are managed.

3.1 Grazing Plan and Grazing Management

Designing a grazing plan is the first step in your pasture management system. As you follow the planning process, the strengths and weaknesses of your current system will become apparent. The grazing plan should include all the components of the grazing and pasture system and serve as a map for making management improvements.

3.1.1 Components of a Grazing Plan

Components of a typical grazing plan include:

- Goals of the farming operation
- Summary of sensitive areas
- Livestock summary & forage requirements
- Fencing system
- Livestock watering system
- Heavy use area protection
- Forages
- Grazing system management

3.1.2 Principles of Grazing Management

Grazing systems range from continuous grazing of one area over a long period of time to intense rotational grazing on small areas for short periods of time. Livestock systems that use continuous grazing of a pasture experience both overgrazing and under grazing of forages. A rotational system provides a rest opportunity for forage plants so that they may re-grow more quickly. The rotational system provides an opportunity to move livestock based on forage growth, promote better pasture forage utilization, and extend the grazing season (Lemus, 2009).

Success with grass-based dairy farming requires high-quality pasture and livestock adapted to a high-forage diet. Grass-based producers ensure that forages provide the bulk of the energy and protein needed to produce milk by providing high-quality pasture during the grazing season and stored forages in the dormant season. Supplementation is provided to cattle based primarily on mineral and energy, as high-quality pasture tends to be high in protein and energy is required to nourish rumen microorganisms and enable them to metabolize high-protein forages.

This level of nutrient management requires strict attention to pasture management, which in grass-based dairies includes rotational grazing systems to maximize forage intake and pasture health. Grazing management must optimize future pasture production and quality, with milk solids production and reproductive performance. Pasture and herd performance are optimized by having sufficient quality feed on an annual basis to meet cow demand. Therefore, follow the following principles of grazing management practices:

- Control the area grazed each day (or rotation length) to manipulate pasture eaten to meet average pasture cover targets for the farm
- Estimate the area and pre-grazing cover required for the cows based on the target grazing residual and adjust after observing when/ if the cows achieve a "consistent, even, grazing height"
- Make management decisions to maximize per cow production for the season not at any one grazing
- Treat pasture as a crop—remove pasture grown since last grazing and prevent post-grazing height increasing over the season
- Have pasture cover distributed between paddocks in a feed wedge to ensure that high quality pasture is offered on all paddocks

Decisions about grazing frequency are often a result of farmer preference due to specific farm characteristics, i. e. even/uneven paddock sizes, shape of the farm, soil types, labor availability etc. Allowing dairy cows to graze forages versus harvesting and storing the forages can be a very economical means of providing some or all of a dairy cow's forage needs. For a grazing herd, the amount of grazed forage consumed can vary greatly (Amaral, 2006) depending on herd size, pasture acreage, growing conditions or the time of year

3. 2 Grazing Systems and Pasture Management

For profitable milk production, proper management of the pasture is essential. Pastures must be managed so that lactating cows consume extremely high-quality (highly digestible, low to moderate concentrations of fiber, and high concentrations of protein). This goal can be accomplished primarily by maintaining the pasture, so the plants are consumed when they are immature. Proper plant maturity can be obtained by managing the rotation grazing of fields or paddocks using temporary fencing.

If given a choice, livestock will eat the highest quality, most palatable plants in a pasture. In order to ensure that plant biodiversity is maintained in the pasture, it is necessary to set up a grazing management system to control livestock grazing. The elements of grazing that should be controlled are timing and intensity of grazing that includes:

- Controlling animal numbers
- How long animals are in a pasture

- The length of the recovery period the pasture is given before grazing again

Several grazing systems can be employed to ensure sufficient pasture in a stage suitable to graze at all times throughout the grazing season discussed below.

3.2.1 Controlled Grazing

Controlled grazing is when animals stay in an area for a long time, but the size of the area is adjusted by moving fences. Controlled grazing is applied by allowing animals to intensively graze a portion of pasture followed by rotation to a "rested" paddock. This permits plant re-growth on the grazed pasture while letting animals forage on the highly nutritious plants in the rested paddock. Animals typically remain on a given paddock for as little as 12 hours and up to two weeks. The timing of animal rotations is based on forage growth in the paddocks rather than a rigid time schedule. An effective controlled grazing system requires an adequate fencing system that provides the manager control of the grazing herd (Murphy, 1995).

- The grazing area can be increased when forage growth is slow, or it can be decreased when forage growth is fast.
- Forage growth is measured by taking the height of the pasture.
- Controlled grazing requires the manager to check pasture growth daily and have additional land for pasture.

Advantages of controlled grazing

- More produced forage is used.
- Higher number of animals can be supported.
- More meat/milk is produced per unit of land.
- Pasture recovers quickly after being grazed.
- Pasture remains productive for a longer period.

3.2.2 Rotational Grazing

This involves dividing a pasture into several small paddocks using fencing. Livestock graze paddocks in sequence, moving to a new paddock when forage is ready to be grazed. Generally, livestock is put into a paddock when the forage is 25 to 30 cm tall and removed when the pasture has been grazed down to 8 cm and paddocks are rested. Using a relatively high stocking rate forces the sheep to graze the forage more evenly. Rotational grazing does not necessarily mean increased daily live weight gains, but does allow for heavier stocking rates, which increases gains per hectare.

Mob grazing is a form of rotational grazing where large numbers of sheep graze the pasture until forage is grazed down evenly and closely. This is normally used to clean up pastures with coarse, mature forage.

3.2.3 Continuous Grazing

Continuous grazing is putting animals out on a pasture and leaving them there for the majority of the season.

- The number of animals the pasture can support is determined by the forage yield during the period of poorest pasture productivity.
- In most cases, stocking rate needs to be very low or the cattle will lose weight during the summer.
- Individual animals can do well under this type of grazing management if stocking rates are low enough.

Drawbacks of continuous grazing

- Very low milk or meat produced per hectare
- Forage produced during the rainy season can be wasted
- Selective grazing of animals can cause the pasture to become less productive over time.

3.2.4 Strip Grazing

Strip grazing is when animals are given just enough pasture to supply half to one day's requirement. Fences are moved once or twice daily to provide fresh forage. This is the most labor-intensive method of grazing. Strip grazing also results in the highest quality of feed and the least waste.

3.2.5 Forward Grazing

Forward grazing is where the pasture is grazed by two groups of animals. The first group to enter the pasture is those with higher nutritional needs (e. g. milking cows and calves) and grazes the top of the plants. The second group, with lower nutrient requirements (e. g. dry ewes), grazes what is left by the first group. This allows for higher weaning weights when forage is limited or where competition between young stock and dams exist.

3.2.6 Mixed Grazing

Mixed grazing is when different types of livestock graze different plants.

- Two or more types of animals graze the paddock at the same time or follow one another through the pasture.
- Sheep, goats and cattle do not have the same grazing habits-this can be very helpful in pasture management.
- Cattle and sheep will complement each other if grazed on pasture with a high proportion of forbs and browse.

Multi-species grazing can benefit the producer with better economic gains (different markets), predator protection, and improved range health. Rotational grazing systems take full advantage of the benefits of nutrient cycling as well as the ecological balance that comes from the relationships between pastures and grazing animals.

High-density stocking for short periods followed by adequate recovery periods helps to build soil organic matter and develops highly productive, dense, resilient pastures.

In rotational grazing systems, plant recovery time is of crucial importance to pasture health and to the provision of high-quality forage to lactating cattle. Managing grazing according to plant growth and recovery is crucial to successful rotational grazing and this goal can be accomplished primarily by maintaining the pasture so the plants are consumed when they are immature.

There are different components of managing a grazing system for the milking dairy cows. These are:

- Rest periods need to be 28 – 35 days between grazing periods.
- Grasses re-grows from the tillers that are close to the soil surface; thus, it is important for 3 – 4 inches of growth to remain after grazing orchard grass or fescue plants.
- Closer grazing increases the time for re-growth, decreases survivability of the plants, especially during drought conditions, and decreases the nutritional value of the plants.
- In contrast to grasses, alfalfa and red clover plants re-grow from the carbohydrate stores in the plants' roots; thus, a lower residual grazing height is possible.
- In stands in which you want to favor the growth of the legumes, graze the plants lower.
- If you want to favor the growth of the grasses, graze the plants higher.
- Graze young lush plants in their vegetative state because vegetative plants contain less fiber, and the fiber is more digestible than more mature plants. Moreover, vegetative plants also contain more starch, sugars, energy, and protein, which will reduce the amount of supplemental concentrate needed to maintain high milk yields.
- Dairy cows graze about 8 hours daily, with the heaviest grazing periods in the early morning and later in the evening.
- Dairy cows graze forage types selectively (legumes are preferred over grasses).
- In any feeding system, maintaining dry matter intake in dairy cows is critical. Dry matter intake on grazed forages is determined by bite rate, and time spent grazing.
- Forage programs should be designed such that dairy cows have high-quality forage to graze at all times.
- Stored forages should be used when high-quality grazed forages are not available in amounts that match the dairy herd's nutrient needs.
- With grazing legumes, caution needs to be taken to prevent bloat.
- With the Sudan grasses, grazing should be avoided until the plants are 24 inches tall to reduce the risk of prussic acid poisoning.

- Supplemental minerals with Magnesium are important for reducing the risk of grass tetany.
- Remember to provide plenty of cool, clean water in every grazing area.
- Water supply is critical for livestock on a controlled grazing system.
- Water should be available in every paddock and within a walking distance of 400 - 500ft for lactating dairy cows.
- Limiting water intake will decrease milk production quickly.
- Dairy cows producing 50 pounds /day of milk drink about 25 gallons/day of water when the ambient temperature is 60℉. When the temperature increases to 90℉, water intake increases by approximately 5 gallons daily. Make sure water is always close to the cows (Table 3. 2).
- During the daylight hours, provide plenty of shade or allow the dairy cows to return to the barn. Shade areas rotation is important to prevent environmental mastitis (Swisher, 1997).

Table 3. 2 Estimated water needs for grazing livestock during average and hot weather

Livestock	Average consumption (gal/day)	Hot weather consumption (gal/day)
Lactating dairy cow	20 - 25	25 - 40
Dry dairy cow	10 - 15	20 - 25
Dairy calf	4 - 5	9 - 10
Lactating beef cow	12 - 18	20 - 25
Dry beef cow	8 - 12	15 - 20
Feeder calf	10 - 15	20 - 25
Sheep	2 - 3	3 - 4
Horse	8 - 12	20 - 25

4 PERFORMING FEED PROCESSING TECHNIQUES

Feeds are processed to facilitate handling and to improve the feeding value of the different feedstuffs, as well. Feed processing improves the storage stability, palatability, digestibility and intake of the feeds or ration by the animals. The feeds are also processed in preparing the rations for different classes of animals. Feed processing is also important to uniformly mix the different feed stuffs used in ration formulation. Some of these important feed processing techniques are discussed in this section.

4. 1 Physical and Chemical Treatment of Feedstuffs

4. 1. 1 Physical or Mechanical Methods of Feed Treatment

There are different kinds of physical treatment techniques used to improve the nutritional value of different feedstuffs. Some of these techniques are discussed in the following paragraphs in detail.

Heat treatment

This is employed to dry certain plant products and to improve the quality of feeds pelleted. Heat is sometimes employed in the extraction processes to remove oil from some oilseeds. Heat treatment improves the nutritional value of soybean meal by destroying the trypsin inhibitor which is naturally present, and by increasing the utilization of proteins and amino acids, fats, and carbohydrates present in the meal.

Digestibility is enhanced by partial "cooking" and thus the metabolizable energy (ME) value is increased. Heat treatment increases the nutritional value of cereal grains by gelatinizing starches and improving digestibility. This change occurs during steam, when dry steam is added to the mixture of feed ingredients just prior to pelleting to condition the feed so that better quality pellets are produced. Proper conditioning of feeds before pelleting, results in improved pellet durability and a reduced amount of fines in the finished product.

- Cottonseed meal: Cottonseed meal contains gossypol, a compound with undesirable nutritional qualities. Free gossypol is associated with the pigment gland. Puncturing the pigment gland causes free gossypol to be mixed with other compounds in cottonseed meal and becomes bound. Bound gossypol is less harmful than free gossypol. The method used to remove the cottonseed oil affects the gossypol content of the residue which becomes cottonseed meal. Hydraulically pressed and solvent-extracted meals contain higher levels of gossypol than does repressed solvent meal. Glandless meal has lower levels of gossypol. Treatment of the meal with iron salts eliminates the toxicity of gossypol, but also produces a darker meal which is unacceptable for the feed industry. Treatment of cottonseed meal with phytase increases the availability of phosphorus, reduces gossypol toxicity, and increases the availability of some proteins.
- Soybean Meal: Heat destroys the trypsin inhibitor. Wet heat is more effective than dry heat.
- Linseed Meal: The toxic factor for poultry can be removed by soaking the meal in water for 12 – 18 hours.

Grinding

This method may increase the nutritional value of the feeds by reducing the particle size of the ingredients and thereby increasing the surface area of the feeds ingested and facilitating digestion. In addition, the grinding process adds metals to feeds from the grinding machinery and can prevent a micro mineral deficiency. The metals that are added to feed are Fe, Zn, Cu, Mn, and Na.

Pelleting and crumbling

The pelleting and crumbling processes compact the mixed feed ingredients and increase nutrient density and bulk density. For some species, pelleting improves palatability or acceptability of feeds. As mentioned above, the steam conditioning that occurs just before pelleting may improve digestibility. The heat generated during compaction of the pellet may also improve digestibility and destroy thermolabile toxic factors that naturally occur in some plant products. Pelleting enhances the availability of phosphorus in wheat bran and also permits the use of low density, bulky, unpalatable feeds that might not otherwise be practical to use (Walker, 1973) (Table 3. 3).

Table 3. 3 Toxins and inhibitor destruction by processing

Feedstuff	Inhibitor	Deactivation process
Cottonseed meal	Gossypol: Cyclopropane fatty acids, phytate	Add iron salts; rupture pigment gland
Soybean meal	Trypsin inhibitor	Heat, autoclaving
Linseed meal	Crystalline water-soluble substance	Water treatment
Raw fish	Thiaminase	Heat
Alfalfa meal	Saponins, pectin methyl esterase	Limit amount of fed
Rye	5 - N-alkyl resorcinol	Limit amount of fed
Sweet clover	Dicoumarol	Limit amount of fed
Wheat germ	Unidentified	Heat
Rapeseed	Isothiocyanate, thyroactive materials	Heat

4. 1. 2 Chemical Methods of Feed Treatment

Alkali treatment of straws and other forages

Alkaline treatment as a method to increase the value of low quality roughage has been re-evaluated. The technology is not new but is worth revisiting due to increased feed costs. One version of alkaline treatment is the quick lime or hydrated lime treatment. How does the lime treatment work? It works by increasing the digestibility of the fiber. The treatment swells the cellulose, which makes it easier for the enzymes to work. Additionally, the hydrogen bonds between lignin and hemi-cellulose are hydrolyzed (Julie, 2013).

When straw is exposed to an alkali, the ester linkages between lignin and the cell wall polysaccharides, cellulose and hemicelluloses, are hydrolyzed, thereby causing the carbohydrates to become more available to the microorganisms in the rumen. This effect was first used to improve the digestibility of straw in Germany in the early 1900s, by a process that involved soaking straw for 1 - 2 days in a dilute solution (15 - 30 g/L) of sodium hydroxide and then washing to remove excess alkali. In the process currently used, chopped or milled straw is sprayed in a mixer with a small volume of concentrated sodium hydroxide (typically 170 L/t of the straw of a solution of 300 g/L of NaOH, supplying 50 kg NaOH).

The product is not washed and the alkali forms sodium carbonate, which gives the product a pH of 10 - 11. This process gives a product that may be mixed with other foods and may also be pelleted.

An alternative alkali to sodium hydroxide (NaOH) is ammonia, which may be applied to straw in the anhydrous form or as a concentrated solution. As both forms are volatile, the process has to be carried out in a sealed container, which may be formed by wrapping a stack of straw bales in plastic sheeting. As ammonia is a weaker alkali than sodium hydroxide, it reacts slowly with the straw; the time required for treatment ranges from 1 day, if heat is applied to raise the temperature to 85℃, to 1 month at winter temperatures. The ammonia is added at 30 - 35 kg/t of straw, and when the stack is exposed to the air, about two-thirds of this is lost by volatilization. The remainder is bound to the straw and raises its crude protein content by about 50 g/kg. In addition to this advantage over sodium hydroxide, ammonia does not leave a residue of sodium (which increases the water intake of animals). Both sodium hydroxide and ammonia have been used on a wide range of low quality forages, including straws, husks and hays.

It should be noted that, in addition to improving digestibility, the alkali treatments caused increases in intake. A danger arising from ammonia treatment is that it may cause the production of toxic imidazoles, which arise from reactions between ammonia and sugars. Forages containing more sugars than straws, such as hays, are more likely to form imidazoles, and their production is encouraged by high temperatures. The toxins cause a form of dementia, which in cattle is sometimes called "bovine bonkers".

Urea treatment of low quality roughages

A chemical for treatment of forages that is easier to handle, and often cheaper than ammonia is urea. When exposed to the enzyme urease, urea is hydrolyzed to yield ammonia. Straw normally carries bacteria that secrete the necessary urease; it is important that the straw should be wet enough (about 300 g water per kilogram) to allow the hydrolysis to take place. After the application of urea, the straw is sealed in the same way as for treatment with ammonia. Urea ammonization of straw has proved reasonably effective in improving its nutritive value but is not as consistently effective as ammonia or sodium hydroxide. Urea can also be used simply as a supplement to straw (i. e. be added at the time of feeding).

With the abundant supply of rice straw throughout Southeast Asia, considerable effort has been directed towards improving its nutritive value. This had led to a plethora of feeding systems utilizing chemically treated rice straw developed for the wide variety of ruminants, ranging from breeding stock to growing stock to milking cows. The high nutrient demands of lactating cow's limits the proportion of their diet based on modified rice straw, particularly when fed to produce high milk yields. Alkali treatment improved digestibility of the dry matter and of energy and metabolizable energy content but reduced dry matter intake. even when alkali treated and supplemented with 30% *Leucaena*, the metabolizable energy content only improved from 5.4 (untreated rice straw) to 7.2 MJ/kg DM (treated rice straw plus Leucaena), still lower than that of Napier grass (8.9 MJ/kg DM).

The *Leucaena* supplement improved the nitrogen balance in the cattle above that of Napier grass (20 : 14 g/d). The metabolizable energy content of rice straw, both untreated and alkali treated, are relatively low, indicative of the large variability in nutritive value of forage by-products. The poorer metabolizable energy content of the best diet based on alkali-treated rice straw, compared to the Napier grass, which is only moderate quality roughage, indicates its limited potential role as a basal ration for milking cows fed to produce 15 to 20 L/d of milk.

Urea treatment of rice straw provides additional nitrogen, some of which will be converted into microbial protein if sufficient dietary energy is provided, although improvements in metabolizable energy content are likely to be relatively low. However, if fed *ad libitum*, intakes of urea treated rice straw would be greater than the untreated straw. Dietary crude protein levels decreased with rice straw feeding, but there was little difference in dietary metabolizable energy contents. Milk yields were similar on all four diets, although milk fat contents increased with feeding urea-treated rice straw. Milk responses to chemical treatment of rice straw may vary with the type of rice straw (John, 2005).

Crop residues include crops like wheat, barley, teff and other straws.

4.2 Forage Conservation Techniques

Forage conservation is an avenue for ensuring continuity in ruminant feed availability. Although their nutritive qualities differ from those of fresh materials, adequate levels of nutrients are retained in the conserved feed to merit use in dry periods. Because rain-based pasture and fodder production is seasonal, there are times of plenty and times of scarcity. It is thus imperative to conserve the excess for use in times of dry season scarcity.

The aim of conservation is to harvest the maximum amount of dry matter from a given area and at an optimum stage for utilization by animals. It also allows for re-growth of the forage. The two main ways of conserving fodder are making hay or making silage.

4.2.1 Hay Making

Hay is fodder conserved by drying to reduce the water content so that it can be stored without rotting or becoming mouldy (reducing the moisture content slows down the rate of growth of spoilage microorganisms). The moisture content should be reduced to about 15%. Hay is mostly used to maintain feed supplies throughout the year.

The aim in hay making is to reduce the moisture content of the green crop to a level low enough to inhibit the action of plant and microbial enzymes. The moisture content of a green crop depends on many factors, but may range from 650 - 850 g/kg, tending to fall as the plant matures. In order that a green crop may be stored satisfactorily in a stack or bale, the moisture content must be reduced to 150 - 200 g/kg. The custom of cutting the crop in a mature state when the moisture content is at its lowest is clearly a sensible procedure for rapid drying and maximum yield, but unfortunately the more mature the herbage, the poorer the nutritive value.

Procedures/Steps in hay making

1) Harvest at an optimum stage of maturity

This will provide a maximum yield of nutrients per unit of land and the highest digestibility. Stage of maturity is the most important factor that influences chemical composition and quality. The more immature the plant at harvesting, the higher the quality of the hay will be. This is primarily due to the proportion of leaf to stem. Hay quality decreases with advancing maturity. Most forages should be mowed just after reaching an early bloom stage of maturity. Time of cutting is, therefore, a compromise between quality and quantity of the harvested forage. The first cut of hay from a hay crop is usually of better quality than subsequent cuttings. Big farms use mechanical harvesters that are tractor-mounted. Even small machinery is beyond the economic capacity of the subsistence farmer. The most widely used hand tool for harvesting is the sickle. The scythe is a more efficient hand tool for harvesting forage. The scythe can mow at about five times the speed of the sickle.

2) Proper drying

This is essential for the hay to store safely without heating excessively or becoming moldy. Maximum leafiness, green color, nutrient value and palatability can also be retained. The grass should be dried quickly and not unduly exposed to the sun to maintain these characteristics. Problems in hay making vary according to the type of hay crop and prevailing weather at harvest. Slowness of drying is the major problem in sub-humid and humid conditions.

Too rapid drying, shattering of the finer parts of the plant, and bleaching with consequent loss of carotene and vitamins are problems encountered under hot and dry conditions. For drying hay under small scale conditions, follow the following:

- Place the material into small heaps 20 - 30 cm high and turn the heap frequently in the sun to encourage quick drying.
- Turning should be completed before the forage is completely dry to avoid excessive shattering of leaves and overexposure to the sun.
- It is better if turning is done when wet with dew, especially when high leaf shattering is expected.
- If the weather is humid or rainy, facilitate drying by placing the cut material off the ground using a homemade tripod consisting of three poles. Freshly cut forage contains 75 - 80% moisture, whereas the maximum moisture content for safe hay storage is 25% for loose hay and 20% for baled hay.
- Hay of higher moisture content should not be stored because its nutritive value may

be greatly lowered.

3) Baling hay

Baling the hay allows more material to be stored in a given space. A good estimate of the amount stored makes feed budgeting easier. Baling can be manual or mechanized; manual baling is more economical for small-scale dairy farmers. Manual hay baling is done using a baling box with dimensions 85 cm long × 55 cm wide × 45 cm deep, open on both ends (Fig. 18). If the hay is well pressed, the box will produce an average bale of 20 kg. Baled hay is easier to handle, and baling reduces wastage. Baling also reduces transportation cost and storage space requirements (Lukuyu and Gachuiri, 2012).

4) Storing hay

Hay must be stored in a dry environment. Good quality hay should never be poorly stored. The type of storage may vary from area to area. A good stack of loose or baled hay will provide satisfactory storage in arid areas where there is little rainfall. More expensive shelters may be required for high rainfall areas. It is advisable to store hay by kinds and grades in case variable qualities are stored. Hay can also be stored by creating hay stacks. Stacks may be covered by plastic sheets to keep out rain.

The surface layer of a stack may also be "thatched", in the same manner as a thatched roof of a house. Hay stacks should be fairly compressed and loaf-shaped or conical shaped to shed rain water. It is advisable that the stack rests on a platform just above the ground to provide air circulation and prevents the hay becoming wet from below.

5) Feeding hay

Hay can be fed to cattle, including sheep and goats. If enough hay is available, the animals should be fed excess, i. e. provide them with a greater total volume than they will eat. The feed that they leave (the steamier part of the hay) can be fed to cattle. Hay is best fed on hay racks to reduce wastage.

Factors for the nutrient losses of dry forage (hay)

(1) Losses during curing of the hay

1) Shattering loss

Leaves contain 2 to 3 times as much protein as stems. Leaves are also richer in carotene, B-vitamins, minerals, and energy. Legume forages contain a larger proportion of leaves than grasses. The fine leaves dry more rapidly than the coarse stems to which they are attached. This results in considerable shattering loss unless great care is taken. In field cured hay, losses from a leaf shattering range from 2% to 5% for grass hay and 3% to 39% for legume hay.

In areas with a long, severe dry season, tree legumes may lose their leaves during the driest period.

2) Bleaching and fermentation loss

In general, the carotene or pro-vitamin A content is proportional to hay's greenness. With severe bleaching, more than 90% of the vitamin A potency may be destroyed. These losses will not be excessive with good weather and proper curing methods. Color loss is due to destruction of chlorophyll by sunlight.

3) Leaching loss

This is the loss due to washing of nutrients by rain. Repeated showers are more damaging than one heavy rain. Leaching may lower the feeding value of hay by one-fourth to one-third or even more with severe exposure.

4) Spontaneous combustion

Wet hay ferments and generates heat. This can result in spontaneous combustion and fire. This usually occurs a month to six weeks after storing. Indicators of potential spontaneous combustion are hay that feels hot to the hands, a strong burning odor, and visible vapor. Tropical pasture grasses generally take 50 – 55 hours of drying in good weather and 70 – 75 hours in dull weather.

There are two methods that can be used to determine when hay is dry enough for storage. These are the twist method and the scrape method.

- Twist method: Twist a wisp of the hay in the hands. If the stems are slightly brittle and there is no evidence of moisture on the twisted stems, the hay is dry enough for safe storage.
- Scrape method: Scrape the outside of the stems with the finger or thumbnail. If the epidermis can be peeled from the stem, the hay is not sufficiently cured. If the epidermis does not peel off, the hay is usually dry enough to store.

(2) Characteristics of good quality hay

- It is leafy, fine-stemmed, and adequate but not overly dry, thus, giving assurance of high protein content. The leaf-to-stem ratio should be high for two-thirds of the plant nutrients are in the leaves.
- It is bright green in color indicating proper curing, a high carotene or pro-vitamin A content and good palatability. Brown hay indicates loss of nutrients due to excess water or heat damage.
- It is free from foreign material, such as weeds, stubble, etc.
- It is free from mold and dust. Musty, moldy, or dusty hay is not only unpalatable but can contribute to respiratory diseases.
- It is fine stemmed and pliable—not coarse, stiff and woody.
- It has a pleasing, fragrant aroma; it “smells” good enough to eat (Table 3. 4).

Table 3. 4 Forages suitable for conservation as hay

Fodder type	Agro-ecological zone	Dry matter yield (kg/acre)	20 kg hay bales (No.)
Boma rhodes grass	Lower highland 3	4,868	243
Elmba rhodes grass	Lower highland 3	3,944	197
Lucerne	Lower highland 3	2,718	136
Vetch	Upper highland 1	1,432	72

4.2.2 Silage Making

Silage is high-moisture fodder preserved through fermentation in the absence of air. These are fodders that would deteriorate in quality if allowed to dry. An ideal crop for silage making should contain an adequate level of fermentable sugars in the form of water-soluble carbohydrates, have dry matter content in the fresh crop of above 20% and possess a physical structure that will allow it to compact readily in the silo after harvesting. Crops not fulfilling these requirements may require pre-treatment such as field wilting for reducing moisture content, fine chopping to a length of 2 - 2.5 cm to allow compaction and use of additives to increase the amount of soluble carbohydrates.

Silage making requires the use of an airtight and well-built container, usually designed and used for that purpose. These containers used for silage making are called "silos". Different types of silos are utilized for silage making. The choice of the type of silo to be used in a given farm depends upon different factors. The most important factors include the choice of the farmer, the availability of space, the availability different materials used for the construction of the silo and availability of capital necessary for construction of the selected type of silo. The different types of silo and their detailed discussion are given below.

Types of Silo

Silo is an airtight place or receptacle for preserving green feed for future feeding on the farm. Silos can be either underground or above ground. The qualification being that the silo must allow compaction and be airtight. Five types silo are described below, these are: tube, pit, above-ground, trench and tower.

- Tube or plastic sack: Silage can be made in large plastic sacks or tubes. The plastic must have no holes to ensure no air enters. This is popularly referred to as tube silage. Chop the wilted material to be ensiled into pieces not more than 2.5 cm long.
- Pit silo: Silage can also be made in pits that are dug vertically into the ground and then filled and compacted with the silage material.
- Above-ground silo: It is made on slightly sloping ground. The material is compacted and covered with a polythene sheet and a layer of soil is added at the top. When finished, it should be dome shaped so that it does not allow water to settle at the top, but rather collect on the sides and drain away down the slope.
- Trench silo: It is an adaptation of the pit silo, which has long been in use. It is much cheaper to construct than a pit silo. Construction is done on sloping land. A trench is dug and then filled with silage material. This method is ideal for large-scale farms where tractors are used. Drainage from rain is also controlled to avoid spoiling the silage.
- Tower silo: Cylindrical and made above-ground. They are 10 m or more in height and 3 m or more in diameter. Tower silos containing silage are usually unloaded from the top of the pile. An advantage of tower silos is that the silage tends to pack well due to

its own weight, except for the top few meters.

Material of silage making

Silage can be made from grass, fodder sorghum, green oats, green maize or Napier grass. Harvesting forage crops at appropriate stages:

- Napier grass must be harvested when it is about 1 m high and its protein content is about 10%.
- Maize and sorghum should be harvested at dough stage, that is, when the grain is milky. At this stage, maize and sorghum grains have enough water-soluble sugars so it is not necessary to add molasses when ensiling, however, when ensiling Napier grass, it is necessary to add molasses to increase the sugar content.
- To improve silage quality, poultry waste and legumes like lucerne and *Desmodium* may be mixed with the material being ensiled to increase the level of crude protein. However, since protein has a buffering effect that increases the amount of acid required to lower pH, poultry waste and legumes should be incorporated within limits. Poultry litter should not exceed 5% and legumes should not exceed 25% of the total material ensiled.

Procedures of silage making

- First cutting the material—the same considerations regarding when to cut, discussed above for hay, also apply to silage.
- The material then needs to be chopped into pieces no more than 2.5 cm long.
- Sprinkle the chopped material with a molasses and water mixture; for every sack use 1 liter of molasses mixed with 2 - 3 times as much water. This is especially for material like Napier grass that has low sugar content.
- Maize bran or cassava flour can be added to improve the carbohydrate (energy) content. It is then sprinkled with a molasses and water mixture.
- Place the chopped material, sprinkled with the molasses/water mixture, into the black plastic sack—this can be made from a length of specially made 1,000 - gauge plastic "tubing" that is 1.5 meters wide (available from agro-vet and hardware shops).
- Cut 2.5 meters length, tie off one end and then fill with the material, compressing it well and then tie off to seal.
- Stack the filled sacks until needed. Fermentation is usually complete after 21 days.

Losses during silage making

- Nutrient losses may occur during silage making.
- In the field during cutting, losses due to respiration during wilting will be about 2% per day.
- If it rains, leaching may cause some loss.

- Overheating due to poor sealing gives a brown product, which may smell like tobacco and result in severe damage to nutrients, e. g. proteins.
- Effluent losses of 2 - 10% that occur from moisture seepage contain soluble and highly digestible nutrients.
- Seepage should be avoided by wilting the herbage.

Characteristics of good quality silage

- Well-prepared silage is bright or light yellow-green, has a smell similar to vinegar and has a firm texture.
- The pH below 4. 2 (acid) for wet crops and below 4. 8 for wilted silages
- The moisture content should be 60 - 70% and acid lactic content between 3% and 13%.
- Acid butyric content is less than 0. 2%.
- Have the same nutrient content as the original fodder
- Liked by the animals (Fig. 19)

Natural microorganisms ferment the sugars in the plant material and in the added molasses into weak acids, which then act as a preservative. The result is a sweet-smelling, moist feed that cattle like to eat once they get used to it (Alyaa, 2012).

Characteristics of bad quality silage

- Bad silage tends to smell similar to rancid butter or ammonia (contained in household bleaches).
- There is growth of fungus.
- Have dark color such as dark brown
- Silage in a bad condition
- When the silage is held, there's an oily and watery texture.
- pH more than 6. 0

Additives used in silage making

During silage preparation, different types of additives can be incorporated to improve the quality. These include:

- Fermentation stimulants: Some crops may not contain the right type or the right number of lactic acid bacteria.
- Bacterial inoculants and enzymes can hasten and improve fermentation by converting carbohydrates to lactic acid.
- Most inoculants contain *Lactobacillus plantarum* fermentation inhibitors include acids such as propane, formic and sulfuric.
- Inorganic acids are more effective but are strongly corrosive thus not recommended.
- Of the organic acids, formic is more effective than propionic, lactic or acetic.
- Substrate or nutrient sources (grains, molasses, urea or ammonia) are used when there are

insufficient soluble carbohydrates in the material to be ensiled (e. g. legumes, Napier grass, crop residues). They are also used to increase the nutritive value of the silage.

- Molasses can be added at the rate of about 9 kg/t of silage.

Note: Use of additives is not a prerequisite for making good silage, but it is good for crops such Napier grass, lucerne and grasses such as *Cynodon dactylon* (star grass), *Brachiaria brizantha* (signal grass), and *Setaria sphacelata* (bristle grass) because it improves fermentation and nutritive value of the resultant product (Mwendia et al., 2012).

Storage and feeding of Silage

- Tube silage should be stored under shade, for example in a store.
- Rats and other rodents that could tear the tube need to be controlled.
- When feeding, open the tube and scoop a layer, and remember to re-tie without trapping air inside.
- When feeding from the pit, scoop in layers and cover after removing the day's ration, making sure the pit is airtight.
- Drainage from the top should be guided to avoid rainwater draining into the pit.
- When feeding from the above-ground method, open from the lower side of the slope, remove the amount you need for the day and re-cover it without trapping air inside.
- To avoid off-flavors in milk, feed silage to milking cows after milking, not before, or feed at least 2 hours before milking.

4.3 Storage of Feedstuffs

There are different ways to store forages for feeding when pasture production is limited. Hay making is the most popular method of feed conservation because it stores well over a long period of times, but sometimes silage is more suitable when curing the hay is difficult due to the type of forage being harvested or to weather conditions.

- Storing teff straw: Straw is stored in heaps of conical or pyramidal forms either at homesteads or in the fields. The heaps are stored usually in open areas, and rarely under shades. The heaps are made in such a way that they do not let rain water percolate the heaps. The heaps in the field are fenced with thorny bushes to protect them from animals. A properly made heap (compact and tilted slope from the top to the foot) can stay for several years with minimum quality deterioration. Large traders and commercial livestock producers store baled straw under shades.
- Storing hay: Hay is stored in heaps. Users usually store for up to a year. Hay is stored under shade and in the open air. Limited commercial hay producers mainly on the peripheries of Addis Ababa usually produce baled hay store hay for less than five months and sell during the dry season when feed demand is high. Hay can be stored for more than a year if it is under shelter, protected from moisture and rodents.
- Storing stover: Stover is usually first stored in heaps in the fields for a short time,

after which they are transported to homesteads and stored in heaps again. The duration of field storage varies from area to area. However, it could easily cause nutrient damage if exposed to rain, sun and dust kept for longer time. Usually it is the first to be consumed by animals after harvesting, keeping behind the other forages with very limited amounted.

- Storing wheat bran: Like other agro-industrial by-products wheat bran is sensitive to humidity. The shelf life of bran is lower than that of seed cakes. Sometimes bran can be stored for two months in the dry season while it can be stored only for one month during the wet season. Bran cannot be stored for long because it solidifies, and the shelf life of bran is also affected by type of storage. Bran stored in ventilated stores can have a longer shelf life than bran stored in non-ventilated stores.
- Storing Oilseed cakes: Seed cakes need care while stored. The cakes should be aerated, lest they would be heated and become too dry, which shortens their shelf life. The cakes should also be protected from moisture, while being transported or stored. Linseed, cotton and Noug seed cakes have a shelf life of 5－6, 2－3, and 3 months, respectively. Noug cake stored in piles over one another may be heated and its shelf life reduced.

4.4 Supplementary/Concentrate Feeds

Ration formulation is the art of mixing up of different feed ingredients so that to produce a ration that can meet the requirements of the animal. Such formulated feeds are sometimes termed as compound feeds. They are also termed as balanced feeds when they contain the amount of nutrients required by the animal in appropriate proportions. Formulated rations are largely prepared in relatively large feed processing factories to be sold to different animal farms. But some farms produce their own compound feeds by establishing the feed mixing facilities inside the farm.

The process of cattle feed manufacturing is comparatively simple, once the feed mixture is arrived at. The main aim of the process is to ensure that the final mixture contains uniformly the same proportion of the ingredients even in the smallest possible sample of, say, one gram as originally intended. Obviously, to attain uniformity in concentration, a mixture of ingredients is to be ground to fine mesh and mixed thoroughly. This mixed powder can be used directly for feeding. Alternatively, the mixed ground material can be palletized.

In dairy cattle production, the formulated feeds manufactured through ration formulation are mainly used for supplementation of the roughage feeding. This supplementation might not exceed 25 to 30 % the total amount of the daily feed of the milking cows. Currently, many dairy farms, especially in urban areas, provide higher proportions of concentrate supplementation to their animals. This practice is believed to have many problems than benefits.

In brief, various steps in the processing of cattle feed manufacturing can be enumerated as follows:

- Selecting, receiving and proportioning the feed ingredients
- Grinding the feed ingredients
- Molassifying and mixing the ingredients
- Pelletizing and cooling, as the requirement may be
- Packing

4.4.1 Types of Concentrates and/or Supplements

Of course, the type of concentrates required depends very much on the quality and quantity of roughage offered to the different classes of animals. In general, the following types of concentrates can be distinguished:

- Early weaner mixture: For young calves up to 5 - 6 months (minimum till one month after weaning), with a minimum DCP content of 18 - 20%
- Young stock mixture: For young stock from 6 months till about 1.5 years of age, with a minimum DCP content of 15 - 17%
- Medium protein mixture: For young stock till calving, dry animals and possibly animals in late lactation, with a DCP content of 12 - 14%
- High yielding mixture: For cows in early lactation (up to 12 to 15 weeks), with a minimum DCP content of 15 - 17%, to allow for high milk production from mobilized body reserves (milking from the back)
- Standard mixture: For high yielding cows in the second lactation stage, or medium yielding cows, with a DCP content of 12 - 14%

As concentrates are usually used as a supplement in balancing a roughage diet, the actual quality (expressed in DCP %) depends on the actual amount and quality of roughage offered. Although single ingredients can be used, mixtures of ingredients are more common. In supplementary concentrates, it is necessary to:

- Reach the proper ratio of protein to energy per category of animals.
- Supply deficiency quantity of protein and energy (production level); include possible deficiency of minerals and vitamins.
- Reach the most economical mixture (least cost calculation).
- Counter balance certain characteristics of individual ingredients, such as taste, fat content, certain substances like gossypol, and laxative aspects such as molasses.

In many cases, protein is the main lacking ingredient, especially in high potential yielding cows. Moreover, protein is usually the most expensive ingredient. Therefore, economics may not allow expressing the genetic potential of a high yielding cow.

4.4.2 Essentials of a Good Concentrate/Supplement

Essentials

1) Energy

Concentrate (mixtures) should have an energy content of over 1,000 milk feed unit (FUM) per kilogram of DM as to have a minimum possible production response of at least 2 kilograms of milk per kilogram of concentrates. If a concentrate contains less than 1,000 FUM per kilogram, it indicates a high CF or ash content. A high CF/ash content is not desirable when roughage is of medium/low quality. This generally is the case. CF reduces the feed density and digestibility following by an increasing heat production and consequently depresses appetite. Especially brans, hulls and chaff of cereal grains have a high CF content. A high CF content indicates adulteration (chaff, husks, sand, sawdust) as the result of poor or/and unhygienic handling. In order to increase energy value and/or to improve the taste, molasses could be added as to avoid diarrhoea (up to maximum 15% for adults and 5% for calves).

2) Fat

Good concentrate contains at least 3% crude fat, accomplished by the inclusion of byproducts from oilseeds. However, under warm and humid conditions the fat may quickly become rancid affecting the keeping quality. Rancidity affects the intake as the product becomes less palatable.

3) Crude Fiber (CF)

Concentrates for calves (early weaner mixture) should not contain more than 9 % CF, as the rumen of the young calve is not yet completely functioning. Cotton seed cake cannot be included in the concentrate for young calves, as the gossypol in the cake acts as a poisonous substance for calves.

4) Minerals

Inclusion of at least 3% mineral mix is desirable. The mineral mix should contain the major and trace elements. The recommended quantities depend on local conditions, such as soil type, type of forage, type of concentrates, production level. Furthermore, it is recommended to provide NaCl (common salt) *ad lib*, either as a "lick" or dissolved in water (2.5 % = 2.5 kg salt in 100 liters of water).

5) Vitamins

In indoor cattle keeping systems, vitamin D may have to be supplemented in the concentrates (calves, young stock). If rations are devoid of fresh, green, leafy materials, extra vitamin A may have to be supplied (calves, young stock). In general, inclusion of 0.1% vitamin AD3 preparation is recommended.

Consistency

To reduce losses and to stimulate quick intake, concentrates could be offered as a thick porridge. The porridge should not be prepared more than 4 hours before milking/feeding. Dry meal (without the availability of drinking water) increases losses and reduces the speed of intake and possibly the overall appetite. It is therefore not recommended.

Quality and maximum allowance of ingredients

Proper information should be available and obtained about the quality of ingredients as to allow composing a balanced mixture at least costs (Table 3. 5).

Table 3. 5 Some maximum allowances of ingredients in concentrates for rations in which concentrate has a maximum of up to 50% DM

Product	Maximum percentage in mixture
Maize by-products	40 - 50%
Wheat by-products	25%
Rice bran	0 - 20%
Malt germs	10%
Coconut products	50%
Groundnut cake	20%
Cottonseed cake	20% (not for calves)
Sunflower cake	10%
Rapeseed cake	10% (goitrogenic substances)
Molasses	15% (5% for calves)
Sugars	5%
Slaughter by-products	limited
Fish meal	5%

Source: Chiba, 2009.

In general, ingredients should be checked for mould, soil contamination and residues (e. g. sweet potatoes and cassava). But also, particular aspects have to be considered. Examples are shown in Table 3. 6:

Table 3. 6 Ingredients should be checked in feedstuff

Gossypol contents	In cottonseed products
Aflatoxin	In peanut products
Goitrogenic substances	In rapeseed
Hairy seeds	Like cottonseed, should be decorticated (hairs/fibers removed)
Sugars/molasses	Can only be used in limited quantities, or it may cause diarrhoea
Sugars	May depress utilization of the ration. It results in excessive amounts of methane gas. It either escapes unutilized and/or causes a decrease of the pH in the rumen, to fall below the level
Mouldy and/or sour products	Are to be taken care off
Maize and rice products	Should be limited in quantity in mixed feeds. It might result in the production of (very) soft butterfat, which turns rancid quickly
Soybeans	Undesirable due to their high fat content in large quantities
Sesame and sunflower products	Are less tasty. Ratio of some ingredients depends on the percentage of hulls and husks

4.4.3 Ration Formulation

Ration balancing and composition

While incorrect ration balancing is not a common cause of stress, farmers should be aware of a few points. Excess nitrogen, usually called crude protein (CP), can cause a variety of problems. Excess N can cause a drop in dry matter intake, can affect fertility, needs energy for its excretion, can interfere with vitamin A metabolism and, if the cows are hungry, can result in death through bloat. Therefore, rations must be properly balanced. The most obvious symptom of excess nitrogen is a distinct lack of appetite for well fertilized pasture and a tendency to select for the lowest and poorest parts of the pasture.

Excess energy in the form of readily fermentable carbohydrates can lead to acidosis and other metabolic problems. For this reason, diets should not contain less than 40% roughage and a minimum of rapidly fermentable constituents.

Mineral nutrition can be very important, especially in pasture systems where excess potassium can induce magnesium and calcium deficiencies. Deficiencies, imbalances or excesses may not be immediately apparent, but must be avoided as a range of problems from foot rot to infertility can be due to poor mineral nutrition.

A chronic imbalance between calcium and phosphorus manifests itself through cows walking with their hind feet tucked forward under the body. The cattle should be observed walking on a hard surface and any abnormalities noted. This sickle-hocked walk will persist even after the problem has been corrected and leads to injuries to the bulbs of the hoof and impairs the herd's ability to walk to and from pastures.

Factors considered in feed formulation

Before formulating a ration, the following information is needed:

- Nutrients required: Obtain information on nutrients from feed requirement tables developed by various bodies.
- Feedstuffs (ingredients): Prepare an inventory of all available feedstuffs. For home-made rations, use materials available at home as much as possible; commercial feeds may use a wider range (Lukuyu and Gachuiri, 2012).
- Nutrient composition: Nutrient composition of each feedstuff should be known. Analysis of the ingredients is most desirable, but if not possible, obtains estimates from textbooks. Book values, however, can at times be misleading, especially for by-products. Also consider the palatability of the ingredient and any limitations such as toxicity.
- Costs of the feeds: Always consider the cost of the ingredients. Least-cost formulations should be made to obtain the cheapest ration.
- Type of ration: Ration type may be complete (total mixed ration), concentrate mix or a nutrient supplement of protein, vitamin or mineral.

➢ Expected feed consumption: Rations should be formulated to ensure that the animal consumes the desired amount of nutrients in a day. For example, if a heifer requires 500 g of crude protein per day and consumes 5 kg of feed, the crude protein content should be 10%. If it consumes 4 kg per day the crude protein content should be 12.5%.

Mixing of concentrates

Mixing of concentrates can be done on the farm or in special plants. Mixing on the farm can be done simply by using a spade. It is best to start with the ingredient taking part in highest proportion. The remaining ingredients are added in order of decreasing proportion, minerals and vitamins last. Then the mixing can start, using the spade to make the mixture as homogeneous as possible. The incorporation of urea in these "home-made" mixtures is not recommended because of the risk of poisoning.

Calculating compositions

Nowadays, most mixed feeds are composed with the aid of computers, which are able to combine several ingredients in such a way, that the cheapest mixture with the desired feeding value is obtained with similar DM values.

1) Composing a ration from two ingredients

In composing simple rations from two ingredients, it is possible to use the "Pearson Square Method". An example of this method is presented, using DCP content. It is also possible to do this with energy (FUM). The procedure for calculation is as follows:

➢ Make a square and place the desired DCP % of the mixture in the center of the square.

➢ Place the DCP % of the 2 available ingredients at the upper (A) and lower (B) left-hand corners of the square.

➢ The difference between the figure in the left-hand corner and the desired DCP % is placed in the diagonal right-hand corner of the square. The figure at the upper right-hand corner is the number of parts of feed A that must be used and the figure at the lower right-hand corner the number of parts that must be used of feed B.

Example:

A mixed feed with 16% DCP in the DM is required. Available ingredients are: maize meal 7% DCP, 1210 FUM; and soybean cake 40% DCP, 1150 FUM. Calculate the amount of each feed ingredient to obtain the mixture.

To solve this problem and obtain the mixture with 16% DCP, the Pearson Square Method is used in the following way:

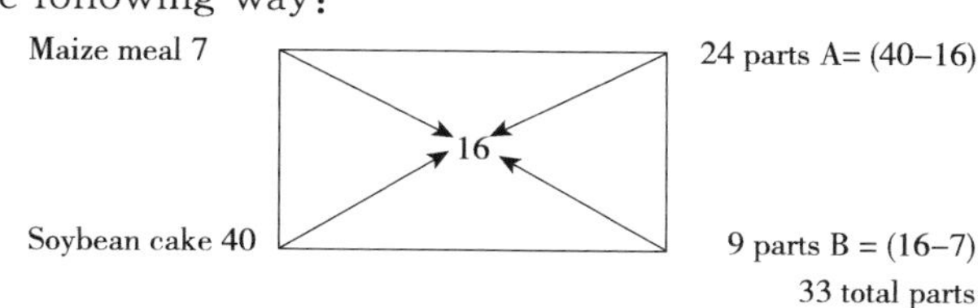

The mixture consists out of 24 parts maize meal and 9 parts soybean cake. Or, when expressed in percentages: maize meal 72.73% (24/33 × 100%) and soybean cake 27.27% (9/33 × 100%). The ratio of soybean cake and maize meal is: 27.27/72.73 = 1 kg : 2.67 kg.

To check the level of the DCP %:

Maize meal	24/33 × 7 % DCP = 5.09% DCP
Soybean cake	9/33 × 40 % DCP = 10.91% DCP
	Total DCP % = 16.00%

2) Composing a Ration from Three Ingredients

For reasons of economics and overall feeding value, it may be attractive to mix three available ingredients (with similar or converted DM contents) into a concentrate with a certain minimum required nutritive value.

Example:

A concentrate is required with a medium protein content of 130 g DCP/kg and the following ingredients are available:

Ingredients	FUM (per kg DM)	Gram DCP (per kg DM)
Wheat bran	818	125
Maize meal	1,210	65
Soybean cake	1,150	460

How to make the required mixture (13% DCP)?

First, the difference is calculated between the desired DCP and the content of each ingredient:

wheat bran	125−130=−5
maize meal	70−130=−65
soybean cake	460−130=+330

Two of the ingredients contain less DCP than the desired content and one ingredient contains more. These differences are to be combined in such a way that they add up to about zero. Then the mixture has its desired DCP content.

Ingredients	Difference	Multiplication	Balance
Wheat bran	−5	1	−5
Maize meal	−65	5	−325
Soybean cake	+330	1	+330
		7	0

The multiplication factors indicate which proportion each ingredient contributes to the desired concentrate (13% DCP in DM). The desired concentrate consists out of 1/7 wheat bran, 5/7 maize meal and 1/7 soybean cake. It is possible to check if the answer is correct. If 7 kg DM of the desired concentrate is made, then the total protein content is: 7 × 130 = 910 g DCP.

5 MONITORING AND EVALUATION OF FEED QUALITY

5.1 Feed Evaluation

Feed evaluation is the testing of feed quality, providing information on the composition of feed or feed ingredients as well as their suitability. The dairy cattle feed is made up of many ingredients, which are broadly grouped into providers of energy (fats, oils and carbohydrates), protein (amino acids), vitamins, minerals and product quality enhancement.

Feed evaluation is a key process in the dairy industry. Feed ingredients need to be tested in order to formulate the complete diet, and diets have to be evaluated to determine their suitability for dairy cattle. Evaluation provides different types of information, as required by nutritionists and farmers. In general, the range of tests that can now be performed is wide and it is now possible to obtain results rapidly.

Feeds and feed ingredients can be evaluated physically as well as chemically. The physical evaluation of feed mostly provides preliminary information on the quality of the material. It involves assessing physical qualities such as weight, color, smell and whether the material has suffered from any contamination by other materials. Chemically, feed is made up of water and dry matter. The dry matter contains organic and inorganic compounds (Korver, 2011). The organic part of the feed is made of mainly carbohydrates, proteins, vitamins and fats and oils. The inorganic part is made of mineral elements, also known as ash. Feed or feed ingredients can be analyzed to provide values of each of these components.

5.2 Method of Feed Evaluation/ Feed Analysis

Different chemical, physical and biological methods have been developed to evaluate the quality of animal feeds. Feeds and feed ingredients can be evaluated physically as well as chemically. One can obtain the information needed on the quality of feeds using these methods.

The physical evaluation of feed provides information on the quality of the material. It is a simple method of assessing the physical qualities of feed, such as weight, color, smell and information about the level of contamination by other materials.

The chemical method of evaluation, on the other hand, provides information about the chemical contents of a feed. The chemical compounds found in a given feed can be determined through the chemical method of evaluation also known as feed analysis.

5.2.1 Chemical Analysis of Feeds

Apart from obtaining values of the chemical composition, the extent of utilization of these chemical components by the animal, termed as nutritive value, is also measured in this system. The most important approach to feed analysis is one of practically possible techniques of analyzing the chemical composition of animal feeds. This technique determines the chemical composition of feeds through direct determinations of moisture, ether extract (fat), ash (mineral), nitrogen (crude protein), and fiber fraction (crude fiber) and the nitrogen free extracts (soluble sugars).

The most important precondition when analyzing feedstuffs and diets is the sampling procedure:

- The sample taken must be representative for the entire lot.
- Samples are to be taken randomly from several different points of the lot. Subsequently the samples are then mixed into a single blend to produce a collective sample, again is divided into several representative lab samples for analysis.
- Sampling equipment should not influence the sample taken via contamination or sedimentation.
- Feed samples need to be stored in a manner that the ingredients will not be altered (temperature, oxygen, sunlight etc.) before analysis.
- A sampling report should be prepared in order to assign the sample correctly after analysis.

Proximate or the Weende method of analysis

Proximate or the Weende analysis of feed is a quantitative method used to determine different macronutrients in a feed. Basically, it is the partitioning of feed compounds into six categories by means of their common chemical properties.

These six categories of nutrients in this method of analysis are moisturized (crude water), crude ash (CA), crude protein (CP), ether extracts (fats or lipids, EE), crude fiber (CF) and nitrogen-free extracts (NFE).

Procedures:

- The feed sample is initially dried at 103℃ for 4 hours.
- The weight loss of the sample is determined and the crude water or the moisture fraction is calculated and determined in percentage terms.
- Burning the sample at 550℃ for 4 hours removes the carbon from the sample, viz. all organic compounds are removed. The remaining part or the ash content determines the mineral content of the feed.
- Again, calculating the weight loss of the feed sample from the dry matter to crude ash (CA) content, mathematically determines the organic matter or the organic fraction of the sample.
- The nitrogen content of the feed is the basis for calculating the crude protein (CP) content of the feed.

- The method established by Kjeldahl converts the nitrogen present in the sample to ammonia, which is determined by titration.
- Assuming that the average nitrogen content of proteins is 16% multiplying the nitrogen content in percentage obtained via Kjeldahl analysis with 6.25 gives an approximate crude protein content of the sample.

In recent years, the over 100 years' old proximate system has been advanced and improved. Especially the imprecision of CA, CF and NFE as well as CP had been criticized. Modern methods to determine the exact composition of the CA fraction via atomic absorption spectroscopy and the CP fraction via amino acid analyzers, near infrared spectroscopy (NIRS) etc. have been established. Improving the information gained from analysis of feedstuffs and diets also involves the determination of sugars and starch (polarimetric methods) contained in the NFE fraction of the proximate analysis.

Dumas method of analysis

Alternatively, the Dumas method can be applied to measure the CP content. Fats and lipids are extracted continuously with ether; after evaporation of the solvent, the residue remaining is the ether extract (EE) fraction.

The carbohydrates in a feed sample are retrieved in two fractions (the CF and the NFE) of the proximate analysis. The fraction, which is not soluble in a defined concentration of alkalis and acids, is defined as crude fiber (CF). This fraction contains cellulose, hemicelluloses and lignin. Sugars, starch, pectin, hemicelluloses, etc. are defined as nitrogen-free extracts (NFE). This fraction again is not determined chemically, it is rather calculated by subtracting CP, EE and CF from the organic matter content of the sample.

Van Soest method of analysis

Van Soest developed a procedure to detect the different components of the cell wall. This helps specify the CF and NFE fraction. Thereby the complete amount of cell wall components is obtained by digesting (boiling) the feed sample in a so-called neutral detergent solution and results in the neutral detergent fiber fraction (NDF). The residue after digestion in a solution with sulfuric acid is called the acid detergent fiber (ADF) and contains mainly cellulose and lignin. Finally, the remaining sample is treated with a sulfuric acid with an even higher concentration resulting in a decomposition of cellulose leaving mainly lignin. This fraction is called acid detergent lignin (ADL) (Fig. 3.1).

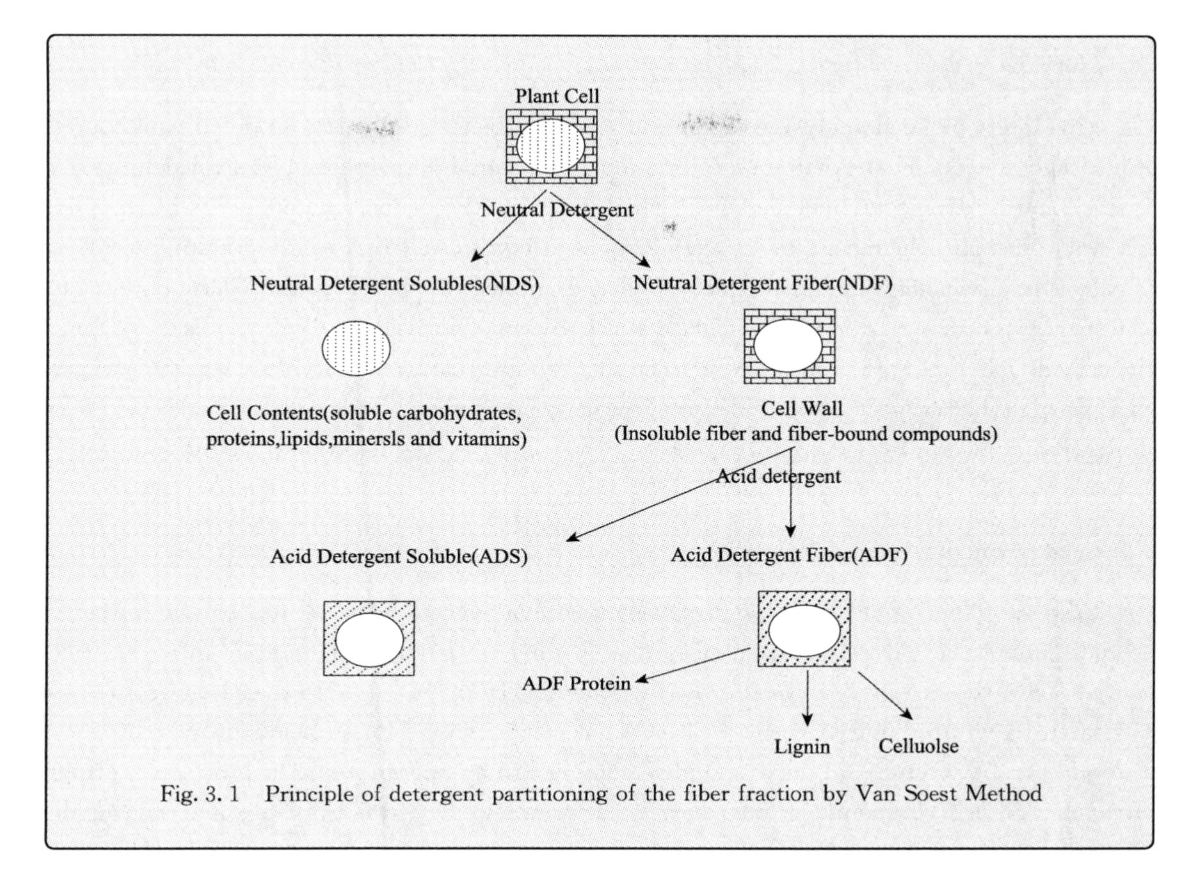

Fig. 3. 1 Principle of detergent partitioning of the fiber fraction by Van Soest Method

The combination of the proximate analysis with these modern methods allows for a detailed feedstuff analysis. Accordingly, the following fractions of the feed analysis will be apparent.

Dry Matter

Dry matter is defined as the non-moisture portion of a feed ingredient or diet. The sum of moisture and dry matter content of a feed on a percentage of the total will always equal to 100. The DM content contains the essential nutrients within a given feed ingredient or forage.

Feeds, and thereby diets, vary widely in their moisture content. Pastures and wet feeds have moisture content between 75% and 90% (10 – 25% DM). Dried feeds usually have <15% moisture (>85% DM). Moisture or dry matter content of a feed is determined by heating a weighed sample of feed in a convection drying oven until a constant weight reaches (24 – 48 hours). DM is expressed as a ratio of original sample weight (moisture+DM) or converted to a percentage. For example, a feed sample weighs 150 g wet and 50 g dry. The DM ratio would be 0. 33 (50/150) and %DM 33. 3% (50/150×100%). The moisture content of this feed would be 66. 7% [100 – 33. 3 or (150 – 50) /150×100%] .

Fiber

The detergent feed analysis system is used to characterize fiber or total cell wall content of a forage or feed. That portion of a forage or feed sample insoluble in neutral detergent is termed neutral detergent fiber (NDF), which contains the primary components of the plant cell wall, namely, hemicelluloses, cellulose, and lignin. As cell wall production increases, as occurs in advancing plant maturity, NDF content will increase. As the NDF content of a feed increases, dry matter intake will decrease and chewing activity will increase. Within a given feed, NDF is a good measure of feed quality and plant maturity. For legume forages, NDF content below 40% would be considered good quality, while above 50% would be considered poor. For grass forages, NDF$<$50% would be considered high quality and $>$60% as low quality.

Crude Protein

Feed protein content is often considered a good determinant of quality. In actuality, protein cannot be directly measured, it is estimated from feed sample nitrogen (N) content. On average all biological proteins contain 16% N, therefore protein content is estimated by multiplying N% by 6.25 ($6.25 = 1/0.16$). Thus, crude protein does not differentiate between N in feed samples coming from true protein or other non-protein nitrogen (NPN) compounds, nor does it differentiate between available and unavailable protein.

Energy

The energy content is often used to compare feeds and evaluate quality. Feed energy content is not directly measured like other nutrients but derived through regression equations. Traditionally ADF alone or with CP was used to predict the energy value of various feeds. Most laboratories report feed energy values based on cattle equations, reporting total digestible nutrients (TDN) and net energy (NE) values. Cattle TDN values are the best estimate we have and should reasonably reflect feed energy for llamas and alpacas given the similarity in digestive function.

Other feed fractions

Additional analyses may be completed on a feed sample, including fat content (ether extract) and mineral analysis. Ether extract is a chemical method by which all lipid (fat) soluble compounds are extracted by being dissolved in ether. This technique is of little value in evaluating feed quality except in the cases of comparing feeds with high fat content.

Total feed mineral content can be measured by a procedure where the feed sample is completely combusted or burned into ash. This does not separate out any individual minerals and does not separate macro-and micro-minerals of interest from silica and other less important minerals. Selected macro-minerals (calcium, phosphorus, magnesium, potassium, sodium, and sulfur) and micro-minerals (iron, copper, zinc, manganese, and molybdenum) can be determined using sophisticated wet chemistry atomic absorption spectroscopy.

5.2.2 Visual Assessment of Forage Quality

Although forage testing through the use of the chemical analysis techniques is the most definitive method of determining forage quality, it often is not complete. Associated costs, lack of laboratory availability or constant forage turnover are the most often reasons people cite for not testing their forages. The first two reasons are not good excuses; however, the third is an issue on many farms that purchase small lots of hay often. One can use their various senses to evaluate small amounts of forages, though sensory evaluation, even if it does not provide any sense of nutrient content (Table 3.7).

Table 3.7 Visual and chemical analysis of forages for assessing quality

Testing Method		Description/Comment
Sensory evaluation		
Visual	Stage of maturity	Look for the presence of seed heads (grass forages) or flowers or seed pods (legumes), indicating more mature forages
	Leaf to stem ratio	Look at forage and determine whether the stems or leaves are more obvious; good-quality legume forages will have a high proportion of leaves, and stems will be less obvious and fine
	Color	Color is not a good indicator of nutrient content, but bright green color suggests minimal oxidation; yellow hay indicates oxidation and bleaching from the sun, and hay will have lower vitamin A and vitamin E content
	Foreign objects	Look for the presence and amount of inanimate objects (twine, wire, cans, etc.), weeds, mould, or poisonous plants
Touch		Feel stiffness or coarseness of leaves and stems; see if alfalfa stems wrap around your finger without breaking; good-quality hay will feel soft and have fine, pliable stems
Smell		Good quality hay will have a fresh mowed grass odor; no musty or mouldy odors
Chemical testing		
Dry Matter		Measures amount of moisture in forage; moisture content will determine how well the forage will store without mould; goal for any hay <15% moisture (>85% dry matter)
Neutral Detergent Fiber		Measures total cell wall content of plant and indicates maturity; the higher the value, the more mature and lower quality the forage; goal < 40% alfalfa and < 55% grasses
Acid Detergent Fiber		Measures the more indigestible portion of the cell wall and reflects the degree of lignification; higher values indicate more mature, lower quality forages; goal: <35% alfalfa and < 35% grasses
Crude Protein		Crude protein content reflects the maturity of forage as well as fertilization amount; good-quality forages generally will have higher protein content; goal >9% grasses and > 15% alfalfa

5.3 Marketing of Dairy Feeds

An efficient feed marketing system is rewarding both for marketing agents and livestock producers. Feed shortage in quantity and quality has been a critical problem in the Ethiopian livestock production system. The single largest expense in animal production is feed cost and it dictates the feasibility of livestock enterprise. Overall, the poor feed marketing system characterized by poor market information and localized thin markets and limited premium price for quality is among the dominant factors contributing to the feed shortage both in terms of quantity and quality.

The feed market is basically informal. Formal market institutions are rare, indicating the low attention paid to the feed production and marketing system. In some parts of Ethiopia, roughage feed sale is done mainly during market days or religious holidays and weekends. Usually roughage (mainly crop residues) feed sellers roam around the towns to sell their feed. Lack of the available feed resources processing, improving/treating using different methods, packaging and marketing is the main bottleneck in the livestock production system and the absence of even simple market places with available resources also contributes to inefficiency of feed marketing and utilization system. Hence, development of marketplaces for feeds is an important step in improving the operation of the feed market. In the past, farmers used to lease in land for *in situ* grazing for their preferred animals. Prices of feeds are usually higher during market days, where many buyers and sellers meet.

Crop residues and hay are the major marketable roughage feeds in most areas. The type of crop residue marketed in a particular area is mainly determined by the type of crop grown as influenced by the agro-ecology. Among crop residues, teff, barley, wheat, and millet straws, and sorghum Stover are the most marketed in the most areas of the country.

Among these crop residues, teff straw is the most marketed, perhaps because of its quality, as perceived by farmers. While sorghum Stover is marketed mainly in the lowland sorghum growing areas; teff, wheat, barley and millet straws are marketed mainly in the highland and mid-highlands. The crop residues of teff, wheat and barley straws are used as feed, and for house construction and mattress making. In some areas, mud house construction during the wet season raises the price of teff straw significantly. Hay is used as feed and for roofing while sorghum Stover is used mainly as feed, but also for fuel and construction. In general, teff straw is preferred for draught oxen and fattening, while barley and wheat straws and hay are preferred for dairy.

The most commonly marketed agro-industrial byproducts are wheat bran, wheat-short, and oil-seed-cakes, such as linseed, cotton and Noug seed cakes. Their availability is, mainly, limited to urban and semi-urban areas of big cities like Addis Ababa.

Dairy farmers purchase concentrate feeds, either directly from the feed processing plants or from retailers. The emerging of dairy cooperatives are becoming the main sources of concentrate feed mainly to their members.

>>> SELF-CHECK QUESTIONS

Part 1. Choose the best answer(s) from the given alternatives.

1. Feeds are processed
   A. To facilitate handling　　B. To increase feeding value
   C. To facilitate pelleting　　D. All
2. Which of the following is the most important agro-industrial byproduct feed?
   A. Oilseed meal　　B. Cotton seed meal
   C. Wheat pollard　　D. All
3. For grazing herd, the amount of grazed forage consumed can vary greatly depending on
   A. Herd size　　B. Pasture acreage
   C. Grazing condition　　D. All
4. Which of the following criteria could be used in renovating a pasture?
   A. If the pasture contains 75% or more of the desirable species, then consider not renovating
   B. If the pasture contains 40 - 75% of the desirable species, then consider over-seed
   C. If the pasture contains <40% of the desirable species, then consider reestablishment.
   D. All
5. Which of the following is false about the nutritional requirement during peak production?
   A. Energy and protein needs are greater than what she can consume in her feed
   B. In early lactation cows mobilize body energy stores like fat and other tissue to meet needs.
   C. Feeding enough nutrients at peak production is not critical to minimize stress.
   D. All

Part 2. Matching.

A	B
1. A receptacle for preserving green feed	A. Silage
2. High moisture preserved fodder	B. Hay
3. Feed with large quantities of starch and sugar	C. Silo
4. Reduce particle size and nutritional value	D. Cottonseed meal
5. Contain gossypol	E. Cereal grain
	F. Grinding

Part 3. True or false.

1. Planning for successful pasture establishment or renovation should begin month in advance.

2. A good set of records will tell the dairy farmer what has happened in the business.

3. Proper seeding rate of the forage seeds depends on the forage species and seeding method.

4. Efficient feed marketing system is rewarding for marketing agents and livestock producers.

5. Hay is the most popular storage method because it stores well over a long period of time.

6. Silage is high-moisture fodder preserved through fermentation in the absence of air.

7. The ideal seeding method depends on the type of equipment available and whether you plant on a no till or a conventional seedbed.

8. Renovating a pasture should not be based on existing percentages of the desirable species present in the pasture.

9. Lactating dairy cows have the most complex nutritional requirements and their requirements change through early, mid, and late lactation.

10. The high growth rates of tropical grasses are associated with greater stem development and hence lower leaf to stem ratios than in temperate grass species.

>>> REFERENCES

Blanchet K, Moechnig H, DeJong-Hughes J, 2009. Grazing Systems Planning Guide [E] . http: //www. extension. umn. edu/agriculture/dairy/grazing-systems/grazing-systems-handbook. pdf

Hoffman K, 2013. Forage utilization Managing Dairy Nutrition for the Organic Herd: Nutritional Requirements & Rations July [E] . http: //articles. extension. org/pages/68574/managing-dairy-nutrition-for-the-organic-herd: -nutritional-requirements-and-rations

Humphreys L R, 1991. Tropical Pasture Utilization [M] . Cambridge, UK: Cambridge University Press.

Lemus R, 2009. Guidelines for pasture establishment [E] . http: //extension. msstate. edu/sites/default/files/publications/publications/p2541. pdf

Lukuyu B, Gachuiri C, 2012. Feeding dairy cattle in East Africa [E] . https: //cgspace. cgiar. org/bitstream/handle/10568/16873/EADDDairyManual. pdf

Moranm J, 2005. Tropical Dairy Farming, Feeding Management for Small Holder Dairy Farmers in the Humid Tropics [M] . CSIRO Publishing.

National Research Council, 2001. Nutrient Requirements of Dairy Cattle [E] . 7th ed. Washington, DC: The National Academies Press.

Payne, W. J. A. ; Wilson, R. T. (1999). An Introduction to Animal Husbandry in the Tropics [M] . 5th ed. Blackwell Publishing Ltd.

Walker J. 2013. Alkaline Treatment of Low Quality Forages (Corn Stovers) [E] . http: //igrow. org/livestock/beef/alkaline-treatment-of-low-quality-forages-corn-stovers/

MODULE 4: PERFORMING FEEDING AND MANAGEMENT OF DAIRY CATTLE

>>> INTRODUCTION

Productivity of dairy cattle is to a large extent dependent on how well it is fed. Dairy cattle are highly sensitive to changes in feeding regimes, and production can fall dramatically with small variations on a day-to-day basis. A good farmer should set a good feeding schedule and as much as possible adhere to it. Milk yield per cow and the cost of feed to produce milk have by far the greatest influence on profitability in a dairy operation. Feed costs for the dairy cattle herd represent 50% to 60% of the total cost associated with the production of milk. In addition, properly implemented dairy cattle nutrition programs can improve milk production, health, and reproductive performance of dairy cows (DCNMDDC, 2016).

If a dairy farm is to be successful, the dairymen must continually strive to adopt practices that allow the greatest output of milk at the most economical cost. Successful dairying in the future will depend on high levels of milk production, culling of low producing cows, controlling feed costs, and using good replacement stocks. Average production per cow increases due to better nutrition and feeding, overall better management practices and the genetic improvement of the herd.

For proper living and production, a cow needs energy, protein, minerals, vitamins and water. Healthy cows will also make the transition from dry to peak easier. Health, feeding and reproduction management for efficient milk production requires specific knowledge, skills and good management (Delaval, 2001).

1 IDENTIFYING THE DIGESTIVE SYSTEM OF CATTLE

Digestion refers to the breaking down of larger feed particles into smaller particles in such a way that they can be taken up by the bloodstream and transported to the places in the body where they are needed. The process of digestion in the digestive systems of animals takes place under the action of an enzyme. Enzymes are chemical substances that stimulate the action of digestion without being changed into other substances (Parish, 2011). The processes of digestion taking place in the digestive organs of an animal as discussed below.

1.1 Anatomy of the Digestive System

Digestion of feeds takes place inside the digestive system. The part of the body that starts at the mouth and ends at the anus is called the alimentary canal or the digestive tract. It is a continuous tube that has special parts or organs with special functions along its length and has a number of accessory organs and different glands that secrete chemical substances, called glands, which help it work.

The alimentary canal (digestive tract) is divided into different parts. These parts include mouth, esophagus, stomach (rumen, reticulum, omasum and abomasum), small and large intestines. The detailed discussion of these different parts is given below.

Mouth

This is a well-designed structure for ripping pasture and grinding feed. It also contains different parts necessary to support the digestive process.

Tongue

It is extremely movable has a covering of sharp, backward pointing projections that help direct food towards the throat.

Teeth

These are designed for grinding and chopping. There are no upper front teeth (incisors). Instead, there is a dental plate which the lower incisors come into contact with.

Salivary gland

As much as 50 to 80 quarts of saliva can be produced by salivary glands and added to the rumen each day. Saliva provides liquid for the microbial population; recirculates nitrogen and minerals and buffers the rumen. Saliva is the major buffer for helping to maintain a rumen pH between 6.2 and 6.8 for optimum digestion of forages and feedstuffs. There are six glands, which produce saliva to aid the initial digestion of food (salivary glands). Two glands are situated below the ear (parotid); two near the back of the throat (mandible) and two under the tongue (sublingual). These glands produce 70 - 180 liters of alkaline (pH 7.7 - 8.7) saliva each day. The volume of saliva produced depends on the diet.

Pharynx

It is the cavity between the mouth and the opening of the esophagus. It is a common route for the passage of both food to the esophagus and air to the lung.

Esophagus

It is the muscular tube from the pharynx to the stomach. Feed can be moved either up or down the esophagus by muscular contraction. The feed of ruminant animals consists of fibrous nutrients that cannot be utilized by the animals unless they are broken down into smaller components with the help of enzymes.

Stomach

True ruminants such as cattle have a stomach with four compartments. The ruminant stomach occupies almost 75% of the abdominal cavity, filling nearly all of the left side and extending significantly into the right side. The four compartments of the ruminant stomach include rumen, reticulum, omasum and abomasum (Fig. 4. 1).

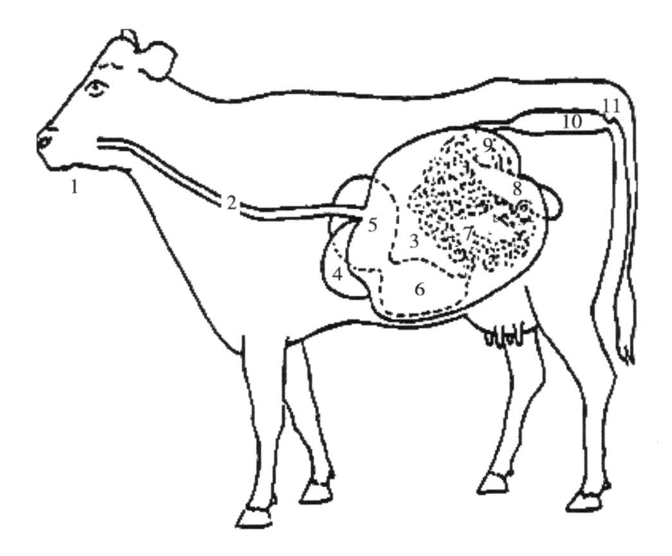

Fig. 4. 1 The digestive system of cattle

1. mouth 2. oesophagus 3. rumen 4. reticulum 5. omasum 6. abomasum
7. small intestine 8. caecum 9. large intestine 10. rectum 11. anus

1) Rumen

The rumen is sometimes called the "paunch". It is lined with papillae for nutrient absorption and divided by muscular pillars into sacs. The rumen acts as a fermentation vat by hosting microbial fermentation. Fifty percent to sixty-five percent of the starch and soluble sugar consumed is digested in the rumen. Rumen microorganisms (primarily bacteria) digest cellulose from plant cell walls.

Large quantities of gas (eructation), mostly carbon dioxide and methane, are produced in the rumen.

Production amounts to 30 – 50 quarts per hour and must be removed; otherwise bloating occurs. Under normal conditions, distension from gas formation causes the cow to belch and eliminate the gas.

The rumen is always contracting and moving. Healthy cows will have one to two rumen contractions per minute. The contractions mix the rumen contents, bring microbes in contact with new feedstuffs, reduce flotation of solids, and move materials out of the rumen. Lack of or a decrease in frequency of rumen movements is one way of diagnosing sick animals.

2) Reticulum

The reticulum is called the "honeycomb" because of the honeycomb appearance of its lining. It sits underneath and towards the front of the rumen lying against the diaphragm. Ingesta flow freely between the reticulum and rumen. It is also responsible for:

- Collects smaller digested particles and move them into the omasum while the larger particles remain in the rumen for further digestion.
- Traps and collects heavy/dense objects consumed by the animal. When a nail, wire, or other sharp, heavy object is consumed by a ruminant, it is very likely that the object will be caught in the reticulum.
- During normal digestive tract contractions this object can penetrate the reticulum wall, and make its way to the heart, where it can lead to hardware disease.
- The reticulum is sometimes referred to as the "hardware stomach" .

3) Omasum

The omasum is spherical in shape and connected to the reticulum by a short tunnel. It is called the "many piles" or the "butcher's bible" in reference to the many folds or leaves which resemble the pages of a book. These folds increase the surface area, which increases the area that absorbs nutrients from the feed and water. Water absorption occurs in the omasum. Cattle have a highly developed, large omasum.

4) Abomasum

The abomasum is the "true stomach" of a ruminant. It is the compartment that is most similar to a stomach in a non-ruminant. The abomasum produces hydrochloric acid and digestive enzymes such as pepsin (breaks down proteins) and receives digestive enzymes secreted from the pancreas such as pancreatic lipase (breaks down fats).

Small intestine

The small and large intestines follow the abomasum as further sites of nutrient absorption. The small intestine is a tube up to 150ft long with a 20 - gallon capacity in a mature cow. Digesta entering the small intestine mix with secretions from the pancreas and liver, which elevate the pH from 2.5 to between 7 and 8. This higher pH is needed for enzymes in the small intestine to work properly. Bile from the gallbladder is secreted into the first section of the small intestine to aid in digestion. Active nutrient absorption occurs throughout the small intestine, including rumen bypass protein absorption. The intestinal wall contains numerous "finger-like" projections called villi that increase intestinal surface area to aid in nutrient absorption. Muscular contractions aid in mixing digesta and moving it to the next section.

Most of the digestion and absorption of fat occurs in the small intestine. Rumen micro-organisms change unsaturated fatty acids to saturated acids through the addition of hydrogen molecules. Thus, more saturated fat is absorbed by cows than by simple-stomach animals. Feeding large quantities of unsaturated fatty acids can be toxic to rumen bacteria, depress fiber digestion, and lower rumen pH.

Large intestine

The large intestine functions to absorb water from material passing through it and then to excrete the remaining material as feces from the rectum.

Gall bladder and pancreas

They are two glands along this part of the digestive tract; they provide Secretions that help digestion in this section of the gut.

Caecum

It is the large area located at the junction of the small and large intestine, where some previously undigested fiber may be broken down. The exact significance of the caecum has not been established.

1.2 Digestive Process and Digestion of Feed Nutrients in Cattle

Dairy cattle are ruminants that have compartmentalized stomach. The ruminant digestive system uniquely qualifies to efficiently utilize high quantities of roughage feeds. On high-forage diets ruminants often ruminate or regurgitate ingested forage. This allows them to chew their "cud" to reduce the particle size and improve digestibility. As ruminants are transitioned to higher concentrate (grain-based) diets, they ruminate less. Once inside the reticulorumen, forage is exposed to a unique population of microbes that begin to ferment and digest the plant cell wall components and break these components down into carbohydrates and sugars.

1.2.1 Carbohydrate Digestion

Rumen microbes use carbohydrates along with ammonia and amino acids to grow. The microbes ferment sugars to produce volatile fatty acids (VFAs), methane, hydrogen sulfide, and carbon dioxide. The VFAs (acetic acid, propionic acid, and butyric acid) are then absorbed across the rumen wall, where they go to the liver. Once in the liver, the VFAs are converted to glucose via gluconeogenesis. Because plant cell walls are slow to digest, this acid production is very slow. Coupled with routine rumination (chewing and rechewing of the cud) that increases salivary flow, this makes for a rather stable pH environment of the rumen (around 6.0).

1.2.2 Protein Digestion

Two sources of protein are available for the ruminant to use: protein from feed and microbial protein from the microbes that inhabit its rumen. A ruminant is unique in that it has a symbiotic relationship with these microbes. Like other living creatures, these microbes have requirements for protein and energy to facilitate growth and reproduction.

During digestive contractions, some of these microorganisms are "washed" out of the rumen into the abomasum where they are digested like other proteins, thereby creating a source of protein for the animal.

All crude protein the animal ingests is divided into two fractions, degradable intake protein (DIP) and undegradable intake protein (UIP, also called "rumen bypass protein"). Each feedstuff (such as cottonseed meal, soybean hulls, and annual rye-grass forage) has different proportions of each protein type (FDH, 2016).

Rumen microbes break down the DIP into ammonia (NH_3) amino acids, and peptides, which are used by the microbes along with energy from carbohydrate digestion for growth and reproduction. Excess ammonia is absorbed via the rumen wall and converted into urea in the liver, where it returns in the blood to the saliva or is excreted by the body. Urea toxicity comes from overfeeding urea to ruminants. Ingested urea is immediately degraded to ammonia in the rumen. When more ammonia than energy is available for building protein from the nitrogen supplied by urea, the excess ammonia is absorbed through the rumen wall. Toxicity occurs when the excess ammonia overwhelms the liver's ability to detoxify it into urea. This can kill the animal. However, with sufficient energy, microbes use ammonia and amino acids to grow and reproduce.

The rumen does not degrade the UIP component of feedstuffs. The UIP "bypasses" the rumen and makes its way from the omasum to the abomasum. In the abomasum, the ruminant uses UIP along with microorganisms washed out of the rumen as a protein source.

Vitamin synthesis

The rumen micro-organisms manufacture all of the B vitamins and vitamin K. Vitamin synthesis in the rumen is sufficient for growth and maintenance. Under most conditions, cattle with functioning rumen do not require supplemental B vitamins or vitamin K in the diet. Niacin (vitamin B_3) and thiamine (vitamin B_1) may be needed under stress conditions.

1.2.3 Fat digestion

Most of the digestion and absorption of fat occurs in the small intestine. Rumen micro-organisms change unsaturated fatty acids to saturated acids through the addition of hydrogen molecules. Thus, more saturated fat is absorbed by cows than by simple-stomach animals. Feeding large quantities of unsaturated fatty acids can be toxic to rumen bacteria, depress fiber digestion, and lower rumen pH.

2 FEEDING THE DAIRY CATTLE

A high-producing dairy cow has been called a "genetic monster" meaning that she is capable of producing far more milk than ever would be required by her calf. In order for her to express her full genetic potential, she must be kept as contented and comfortable as possible so that she experiences as little stress as possible.

The dairy animal's environment has profound effects on her productivity through its effects on her growth rate while young, and on her reproduction and milk production as a cow. Environmental factors can be conveniently separated into three distinct subdivisions. First, the way in which the cow interacts with her herd mates; second, the interactions with the farmer and his labor, which can be referred to as stockmanship; and third, the physical environment, which includes the farm's facilities, climate and weather.

2.1 Feeding Dry Cows

2.1.1 Reasons for Drying off

Involution of the udder

The principal reason for the dry period is to allow the secretary tissue of the udder to involute. During this period, the secretory cells actually break down and are resorbed, and a new set of secretory cells is formed. This cell renewal process takes approximately six weeks. If a cow is allowed no dry period at all, a loss of milk of at least 30% in the subsequent lactation will be the result.

Fetal development

During the last eight weeks before calving, the fetus gains almost 60% of its birth weight, an overall rate of gain for the cow of about 0.75 kg/day. At the very least, the cow must be fed for fetal growth. Mineral nutrition, including the trace elements, especially iodine must not be neglected.

Replenishment of body reserves

If a cow is thin at drying off, some replenishment of body reserves can be achieved, but a gain of more than 20 to 25 kg is not desirable. If an attempt is made to feed for more than half a condition score, in addition to feeding for fetal growth, there will be a marked increase in dystocia and metabolic problems at calving. Feed conversion is also significantly less efficient when a cow is dry. The conversion of feed energy into body reserves is about 62% efficient in the lactating cow and only 48% in the dry cow. It is simply good economics to feed for condition before drying off.

Mineral reserves are an entirely different matter to energy reserves. The high-producing cow will have severely depleted her body reserves of minerals, especially calcium and phosphorus, during her lactation. These reserves can only be completely replenished when the cow is dry. Adequate mineral nutrition during the dry period is very important.

2.1.2 Length of the Dry Period

Many studies have shown that dry periods of less than 45 days will markedly reduce the milk yields in the subsequent lactation, primarily because of the time needed for involution and regeneration of secretory cells. There is no advantage in dry periods of more than 60 days. Long dry periods occurring after an abortion, are distinctly disadvantageous, since cows can easily become too fat, leading to metabolic problems and to disappointing milk yields after calving. The exact date of calving is never known, and these cows will often calve a week or more before the estimated date.

Given these facts, it is wise to aim for a dry period of eight weeks (56 days). Do not make a practice of drying off for two calendar months before the due date; this merely wastes milk for four or more days, which could otherwise have been sold.

2.1.3 Management at Drying off

When to dry off

Dry off either when the cow is within 56 days of recalving or when the cow's milk yield drops to a level sufficiently low be cost effective (uneconomic), or too much trouble to warrant continued milking. Uneconomic milk production, usually considered to be about 5 L per day, should be a rare event. This level may vary seasonally depending on the need to build quotas and the cost of feeding. On well-managed farms nearly all cows will be dried off eight weeks before calving.

The actual dry-off date can be calculated as 224 days from the last service date. Do not rely on a cow calendar because the magnets can easily be accidentally misplaced; rather mark the dry-off date clearly in the milk recording book once the cow has been confirmed incalf or use a computer-produced action list as a guide.

How to dry off the cow

The first step is to cut out all concentrates, and even keep the cow on poor roughage, for three or four days before and after the dry-off date. Cows do not like to be separated from their herd mates and such separations, combined with the low plane of nutrition, will assist rapid cessation of milk secretion. Cows, however, should not be separated if there is any chance that they will damage themselves in attempting to break through fences to rejoin the herd. Having made them as "unhappy" as possible, most cows can then be dried off simply by ceasing to milk them.

However, cows are increasingly being dried off at yields over 20 kg per day. Under these circumstances, undesirable swelling of the udder can occur. Excessive udder enlargement can lead to udder damage. When enlargement occurs, the cow can be left unmilked for two or at most three milkings, then milked out for the last time. Cessation of milking for two milkings will cause an immediate suppression of milk secretion through damage to the secretory cells, and the single milking out will help relieve pressure in the udder and reduce the chances of udder injury.

In consultation with the local veterinarian, a decision will have been made on dry cow therapy. In general, a recognized "dry cow treatment" appropriate for the most prevalent mastitis organisms should be used immediately after the last milking. Disinfect the ends of the teats well; insert the dry cow remedy into each quarter and then teat dip. Keep the cow on a low plane of nutrition for another 3 or 4 days. By now, a week will have passed since drying off, and for the remainder of the dry period the cow should be fed to meet her nutritional requirements only.

Mastitis and drying off

Cows exhibiting clinical signs of mastitis should not be dried off until the mastitis is clear. Any cow with a history of mastitis during her lactation should definitely have dry cow treatment. All recently dried off cows must be very closely inspected for at least the first week after drying off and the cause of any excessive swelling or redness ascertained and, if necessary, treated. Maintain strict hygiene when handling the teats of any cows, but especially dry cows and pregnant heifers.

2. 1. 4 Management during the Dry Period

Nutrition of the dry cow

Once the cow is dry, usually from about a week after the dry-off date, she must be fed correctly for the growth of the fetus and the replenishment of her body reserves. Many dairy textbooks also recommend that the cow be given a chance to adapt to the nutrition she will be getting after calving. A common recommendation is to feed the same dairy meal that the lactating cows receive, for a week or two before calving. This practice is probably overrated, except where there is a very marked difference between the diets of the lactating and dry cows. It can also increase the incidence of parturient paresis. If roughages are good (ME greater than 9. 6 MJ/kg) and similar to those which the cows in milk are eating (the same silage, hay and/or pasture), then the microbial population will adapt rapidly enough to the concentrates in the first week or so of lactation.

Actual amounts and kinds of supplements that need to be fed will depend on the roughages and on the condition of the individual cows. Dry cows should not normally be fed more concentrates than about 0. 5% of their body weight per day. If calculations show that a great deal more is needed then, either the cows were far too thin at drying off, or the roughage quality is hopelessly inadequate. Most dry cows will need nothing but mineral supplementation if roughage quality is good (ME greater than 9. 6 MJ/kg) and they have been dried off at target condition. As a general rule, dry cows should not be fed diets which are high in calcium or phosphorus. Therefore, no legumes, especially not lucerne hay, should be fed. Well-fertilized temperate pastures sometimes contain more minerals than required by the dry cow. Tropical pastures are unlikely to be a problem and can even be advantageous. Mineral nutrition must not be neglected, but excess calcium contributes to milk fever.

Nutrition at calving

If the dry cows are on dry feeds and their dung is firm, then a slightly laxative ration should be fed in the few days prior to calving. Feeding about two kilograms of wheaten bran, in lieu of concentrates, will usually loosen up the feces and facilitate calving.

The cow's appetite is depressed at calving and on no account should the cow be fed a high level of concentrates, because acidosis, laminitis and edematous udders are all aggravated by excess concentrates at calving. A practical strategy is simply to increase concentrates, at the rate of a kilogram a day, from the day of calving until the desired maximum rate has been reached. This strategy also gives the rumen's microbial population time to adjust to the change in diet. If a high level of minerals was fed in the few days immediately before and after calving, this should be reduced to normal levels after the first 2 to 3 days (Wilson, 1990).

2.1.5 Management at Calving

Parturition is the time in the animal's life when she is more sensitive to environmental stimuli. It is, therefore, very important that the cow has good experiences at calving. Ideally the cow should calve down in a clean pasture or maternity stall, with a minimum of other animals to disturb her. In particular, there should be no cows in milk, which may try to steal the calf, or horses and dogs which may interfere with the calving. The calf must get colostrum, by stomach tube if necessary, within a few hours of birth. The calf should be removed from its mother within the first 24 hours. This is far less traumatic for both cow and calf than allowing them to form a firm bond over several days. The only advantage of leaving a calf with its mother is that suckling promotes the production of the hormone oxytocin. This hormone causes contractions of the uterus and assists to expel the afterbirth.

Heifers need especially careful handling at parturition and should have been thoroughly accustomed to the milking parlor in the weeks before calving. If possible, bring the first-calver's calf with her into the parlor for the first couple of milkings. This will significantly relax the mother and make training her to the milking machine a good deal easier. The better the "vibes" at calving, the better the milk production for her whole lactation.

The most common mistake at calving is to assist too soon. There is no substitute for experience, but the study of a good veterinary textbook will help. If it is necessary to assist at the birth (in the case of Holstein-Friesians, this could be at more than 10% of calvings), do not chase the mother around. When you have ascertained that she needs help, say after being in labor for three hours and beginning to look tired, bring the cow very quietly into a hospital pen where she can be easily restrained with a head clamp at a manger. Tame cows (they should all be tame!) can often be handled in the field.

2.2 Feeding the Milking Cows

Developing diets and feeding strategies for the cowherd is facilitated by a basic understanding of the production cycle of the cow and her changing nutrient requirements. By knowing and anticipating the changing nutritional needs of the cow, producers can plan their feeding programs and lower feed costs. Cows use the nutrients provided to them for bodily processes in the following order:

- Maintenance: keep alive and moving
- Lactation: providing milk for the calf
- Growth: including weight gain
- Reproduction

A lactating cow requires nutrients for maintenance, growth if she is young (<30 months), and growth of the unborn calf if she is pregnant, milk production.

Feeding during the early lactation phase

Is the period from 0 to 70 days after calving? Milk production increases rapidly during this period, peaking at 6 to 8 weeks after calving. Increasing concentrates by about 0.5 kg per day after calving will increase nutrient intake while minimizing off-feed problems, e.g. lack of appetite and acidosis (rumen acids increase, which stops normal function and digestion). Feed intake can be increased, and rumination stimulated by chopping the forage to small pieces. Chopped forages are better consumed so that cows increase milk production, and also reduce forage wastage.

Feeding during mid-lactation phase

Mid-lactation period is the period from day 100 to day 200 after calving. By the beginning of this phase, cows will have achieved peak production (8 – 10 weeks after calving). Peak dry matter intake has also occurred with no more weight losses. Cows should reach the maximum dry matter intake no later than 10 weeks after calving.

At this point, cows should be eating at least 4% of their body weight. The cow should be fed a ration that will maintain peak production as long as possible. For every 2 kg of expected milk production, large-breed cows should eat at least 1 kg of dry matter. For each extra kilogram of milk at peak production, the average cow will produce 200 – 225 kg more milk for the entire lactation. Thus, the key strategy during mid-lactation is to maximize dry matter intake. During this period, the cow should be fed high quality forage (minimum 40% to 45% of the ration dry matter) and the level of effective fiber should be maintained at a level similar to that of early lactation.

Concentrates should not exceed 2.3% of body weight and sources of non-forage fibers such as beet pulp, distillers' grains and cereal bran can replace part of the starch in the ration to maintain a healthy rumen environment. Protein requirements during mid-lactation are lower than in early lactation. Therefore, rations for dairy cows in mid-lactation should contain 15 – 17% crude protein. During this period the cow should be bred to initiate a new pregnancy (60 – 70 days after calving).

Feeding during late lactation

This phase may begin 200 days after calving and end when the cow dries off. During this period, milk yield continues to decline and so does feed intake. However, the intake easily matches milk yield. The cow also gains weight during this period to replenish the adipose tissue lost during early lactation. However, as lactation approaches an end, more of the increase in body weight is due to the increased size of the growing fetus. Sources of protein and energy are not very critical during this period. Cheap rations can be formulated with non-protein nitrogen and a source of readily fermentable carbohydrates such as molasses (NRC, 2001).

Management of milking cow

Feeding and management programs for dairy cows in the first 2 to 4 weeks after calving, have a direct and long-term impact on their health, milk production, and reproductive performance, and thus a dairy farm's potential profitability. When managing these recently fresh dairy cows, two key underlining concepts impact the design and implementation of their feeding and management programs.

First and foremost, these cows need to be managed and fed diets that maximize dry matter intake as quickly as possible to minimize the amount of time and the degree of negative energy balance that occurs during early lactation. By maximizing dry matter intake during this time frame, less body fat stores are mobilized, resulting in lower blood concentrations of non-esterified fatty acids (NEFA) and ketones (e. g. beta-hydroxybutyrate; BHBA) and a liver that is more capable of making glucose to support milk production.

Second, early disease diagnosis and intervention help to minimize the detrimental effects on dry matter intake and may lower the risk of culling a cow from the dairy herd. As we learn more about this critical time period, we can refine management and feeding practices for these dairy cows to improve profit. Outlined are areas that impact milk production, health, and reproductive performance of fresh dairy cows (Donna, 2014).

When possible, housing dairy cows for the first 2 to 4 weeks after calving in a separate group allows one to target management time, labor, facilities, feed resources, and other financial resources for this group of cows. Changes in diet and housing create additional stress besides the stress associated with calving. With additional stress, blood concentrations of cortisol and other stress hormones increase, and dry matter intake may decrease. Thus, consistent and specialized management for fresh dairy cows is critical to reduce stress and includes the following:

Fresh cows should be housed in facilities that provide at least 30 inches of bunk space, adequate fans and sprinklers to reduce heat stress, and adequate resting space with at least one free stall or $100ft^2$ of resting space per cow. Concrete alleys should be grooved to prevent cows from slipping. Time spent in the holding pen should be kept to a minimum, but no longer than 2 to 2.5 hours daily. If possible, fresh first-calf heifers should be housed separately from mature cows. If adequate facilities are not available for a separate fresh group of cows, these cows are best put directly into the milking herd.

Keep daily observation and taking a cow's body temperature for the first 10 days post-calving can help in the early diagnosis of diseases. Cows with metritis will have elevated body temperatures and should be treated according to procedures developed after consulting with your local veterinarian. Cows should be observed to make sure they are eating and ruminating (chewing their cuds).

2.3 Feeding the Dairy Bull

Proper management and nutrition of bulls is essential to ensure cow/calf producers maximize reproductive efficiency and genetic improvement of the calf crop. In addition, the herd bull influences overall herd fertility more than any other single animal, and loss of fertility by a bull can cause substantial loss to a potential calf crop. Each cow produces one calf per year; however, bulls should contribute to the calf crop by 25 – 60 times via siring 25 – 60 calves. Additionally, bulls influence their daughters' production in the cow herd. Therefore, bull selection can be the most powerful method of genetic improvement in the herd, but bulls with low fertility, structural problems and low *libido* reduce the percent calf crop weaned. Nutrition and management of dairy calves and bulls is discussed in the headings below:

Pre-weaning nutrition

Nutrition and management of dairy calves (including bull calves) are to maintain them on their dam until normal weaning at 6 – 9 months old. Under normal environments, the plane of nutrition from dam's milk and forage should be adequate for normal growth rates of bull calves. It is important that adequate nutrition is available to the dam. Additional management options such as early weaning or creep feeding can be considered when the calves' plane of nutrition is less than desired; however, the cost should be considered when making your decision.

Post-weaning nutrition

During the post-weaning period, both under-nutrition and over-nutrition can have negative impacts. Under-nutrition results in delayed puberty and over-nutrition can reduce semen production and quality. Diets should be balanced to meet the nutrient requirements for the desired animal performance and body condition score should be monitored to ensure that the bulls are not being under-or over nourished. Typically, many of the diets to develop dairy bulls contain from 40 – 60% concentrate.

Breeding Season Nutrition

There is limited opportunity to manage bull nutrition during the breeding season. They are basically on the same plane of nutrition as the cows. However, you should assess the body condition score of bulls during the breeding season as well as observe bulls' ability to service the cows. Bulls often lose from 100 to 200 pounds during the breeding season. If bull (s) get extremely thin during the breeding season, you may want to replace him because his ability to service the cows will probably be reduced.

Post-breeding season

Nutritional management post breeding is influenced by both age of bulls and amount of weight loss during the course of the breeding season. Once the breeding season is over, producers usually turn bulls out to a separate pasture to regain lost weight and prepare them for the next breeding season. Mature bulls in fairly good condition after the breeding season can be managed on pasture or an all roughage diet without supplements during the winter.

Hay quality should be 8% to 10% crude protein and fed at 2% of body weight. Rations should be modified based on available feed ingredients and to manage the bulls to maintain moderate body condition. Young bulls are still growing so the ration should be formulated to gain 1.5 to 2 pounds per day depending on the magnitude of weight loss during breeding (Julie et al., 2001).

General Management of Bull

- For 305 days lactation period. The production record for the daughters of the bull being used compares with a production record of daughters of other bulls within the same herd and within the same year.
- The bull's pedigree "parentage of the bull", should also be considered when selecting a bull. The bull's parentage i. e. dam should be known to calve with ease and to be resistant to diseases such as mastitis. Only proven bulls should be used by a dairy farmer.
- Bulls should be handled with care from the calf hood until they reach maturity.
- A bull should be dehorned as bulls can be dangerous. The bull should be exercised regularly to keep it in shape.
- A young bull can only be used to serve from 18 months old. Mating should be increased gradually to three times a week because more than this can exhaust and shorten the reproductive life of a bull.
- A bull should be kept in its own paddock and lead to female cows for maturing only during a planned period.
- Bulls that are allowed to roam with the female cows on heat are served without the

farmer's knowledge.

- Notwithstanding that, record keeping becomes virtually impossible.
- Inbreeding is bound to take place if proper management is not done. The bull is likely to serve young heifers that are not fully developed.
- The disadvantage of using bulls is that sterility goes undetected, as the bull could be seen servicing cows yet no calves at the end of the year. Wasted time as no cows conceive.
- If the bull was not selected properly the progeny would be of poor quality.
- To avoid inbreeding bulls have to be changed every two years and are very expensive.
- Heavy bulls should not be allowed to service young heifers for fear of injury. Bulls infected with the reproductive disease spread the disease quickly.

3 UNDERTAKING ROUTINE DAIRY HUSBANDRY PRACTICES

3.1 Definition and Use of the Common Terms in the Dairy Farm

Culling rate (S)

The percentage of cows culled per year. This is usually 25% to 30% of the herd. In other words, in a herd of 100 cows only 70 to 75 of the cows will remain in the herd after one year. This implies that at least 25% to 30% heifers have to be brought into a herd every year to maintain herd size.

Replacement rate (R)

The percentage of cows brought into the herd per year, expressed as a percentage of the number of cows in the herd at the start of the year.

Rate of increase (E)

Subtracting culling rate from replacement rate equals rate of increase. For example, if there were 100 cows at the start of the year, 25 cows were culled, and 33 heifers brought in, then culling rate = 25%, replacement rate = 33%, and rate of increase = 8%.

Age at first calving (AFC)

The time from birth to calving, i.e. when a heifer becomes a cow, is called the age at first calving. The mean age at first calving is the generation interval in the herd. The younger the AFC, the greater the potential rate of increase and the fewer the heifers which have to be fed and cared for. It should not normally be less than two years of age or greater than three and heifers should have achieved about 90% of their average mature mass when weighed just before calving.

Note the following three points very well. First, if a farmer wishes to expand his herd, then the younger the AFC, the greater the potential rate of expansion. Second, the younger the AFC, the sooner the heifer becomes a cow and therefore, the fewer heifers of all ages in the herd. Third, in a stable herd, AFC has no effect on the number of surplus heifers for sale; AFC only affects the total number of heifers of all ages.

Number surplus heifers (V)

Either the sale of surplus heifers, or herd expansion by retaining all the heifers are very important parts of dairy farm income. The number of surplus heifers is affected by the average inter-calving period. The shorter the inter-calving period, the more calves that will be born and the greater the probability that more heifers will be born.

Total number of heifers (H)

The maximum number of heifers is determined by both the AFC and the inter-calving period. The minimum number which must be kept on the farm for herd replacements depends on the age at AFC and the culling rate plus an allowance for infertility and mortality.

Inter-calving period (ICP)

The period in a cow's life that is between the birth of one calf and her next calf. The ICP is so important that it measures the fertility and the breeding superiority of a given cow.

The recommended targets for ICP are:

- A mean of 365 days
- No cows should have an ICP less than 330 days.
- 90% from 330 to 400 days, and less than 10% an ICP of more than 400 days

The ICP is best understood if broken up into its component periods and events as illustrated in Fig. 4. 2. The reader is advised to commit this diagram to memory because successful dairy farming is inextricably linked to a thorough understanding of the implications of the factors affecting the inter-calving period and their management.

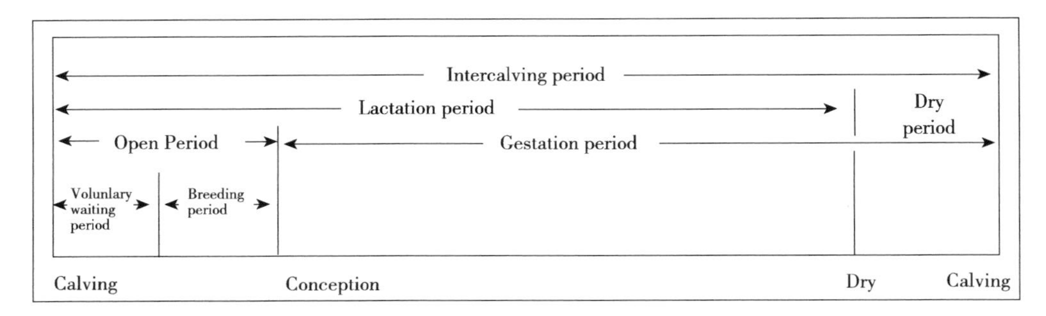

Fig. 4. 2 Diagrammatic representation of the periods and events in the inter-calving period of the dairy cow

Total services to conception and services per conception (TSPC & SPC)

The total number of services per conception (TSPC) is the number of inseminations in the entire herd divided by the number of confirmed pregnancies. In practice this is not easy to calculate, and the usual figure quoted is services per conception (SPC) for pregnant cows only. This is an important distinction because it could happen that cows which fall pregnant do so readily, and the use of ICP could be masking a real problem amongst those which do not. Note, too, that the SPC and TSPC figures are influenced by the rate of culling. Ruthless culling of apparently infertile cows may result in a good SPC figure but hide a real problem. The figure for percentage take to first A. I. helps in interpretation: targets: SPC 1. 3 to 1. 6, TSPC<2.

Take to first insemination

It is usually calculated as the number of cows pregnant to one insemination divided by the number of first inseminations. For example, 70 first inseminations and 49 pregnancies would be $49/70 \times 100\% = 70\%$ take to first A. I. ; target: >60%.

Heats spotted or heats missed

The number of heats spotted divided by the total possible heats. The total possible heats are usually calculated by assuming that all open cows (non-pregnant) cycle every 21 days. Due to cystic ovaries, and many other causes, not all cows cycle and the length of the estrus cycle is very variable. Therefore, it is impossible to achieve 100 % success.

Target: >75% of theoretical possible number of heats.

Lactation Period

The time from calving until the cow is dried off, i. e. the time during which the cow is producing milk.

➢ Target: 300 to 305 days (43 weeks)

➢ Range: 265 to 340 days (38 to 49 weeks)

The length of the lactation period depends on the open period and the level of production.

3. 2 Dairy Herd Structure/Dairy Herd Dynamics

Lactation can be initiated with drugs, but, in general, if a cow does not produce a calf, she will not produce milk. Calves are also necessary to provide replacements for the dairy herd, and surplus heifers can form a very important part of the dairy farmer's income. Therefore, reproduction management is vital to successful dairying.

The cow's early life, from birth until she is mated for the first time, is dealt with in detail. It is important to understand that a heifer must not be mated at a too light mass. From 65% to 70% of her expected mature mass is normally considered to be a desirable mating mass. After mating, the heifer must continue to grow so that she is as close to her mature mass as possible at first calving. The age at which the heifer becomes a cow, that is her age at first calving, depends on the rearing system (she must meet the target weights mentioned above) and the desired generation interval. The sooner a heifer calves down, the sooner she contributes to the economy of the farm in the form of milk and calves. Therefore, the farmer must weigh up the possible extra rearing costs of, say, calving heifers at close to two years of age, which is the earliest practicable age, against the loss of income from milk and calves by delaying calving to a later age. If a farmer wants to expand his herd, then the lower the age at first calving, the faster is the rate of increase in the herd. Interestingly, in a stable herd, the age at first calving has no effect on the number of surplus animals available for sale each year. A greater age at first calving merely means that there are more heifers on the farm. A final consideration is that the older the heifer, the lower her fertility. In other words, the later a heifer is mated, the more difficult it is to get her into calf. For this reason, heifers should not calve down later than three years of age.

Once a heifer has calved for the first time, she becomes a cow, and takes her place in the milking herd. It can be shown that the farmer should aim for a calf per cow per year. Therefore, the cow must begot back into calf so that she re-calves as close to one year after her previous calf as possible. In practice, cows not pregnant after a reasonable period are culled for infertility. Although a cow may not be truly infertile at culling, the farmer may decide that the extra cost of keeping her and continuing to try to get her into calf is not justified. As a generalization, it can be said that the average dairy cow in Ethiopia produces three calves in her lifetime. This average is calculated from a large population of cows, some of which have fewer than three lactations and others which may go on for as many as ten lactations. Expected age structures in expanding and stable herds are shown in Table 4. 1. The number of heifers is dependent on the age at first calving (AFC) and the age when surplus heifers are sold.

Table 4. 1 Typical age structure of expanding and stable dairy herds

Lactation number	Percent of cows in herd in each lactation	
	Expanding herd	Stable herd
1	33	25
2	22	20
3	16	15
4	11	12
5	8	9
6	5	6
7	4	5
8	0	4
9	1	3
10	0	1
Average lactations	2. 8	3. 4

As mentioned above, some cows are culled for fertility, and some may be culled for other reasons, e. g. mastitis. Also, some cows die through accident or disease. The annual culling rate (including mortality) is typically from 20% to 30%. This means that to maintain a herd of 100 cows, 20 to 30 replacement heifers must calve into the herd each year.

3.3 Practical Feeding of the Producing Herd

During the formulation of rations for lactating dairy cows, the quality of the ration should be commensurate with the requirements of the cow. The requirement is directly related to the milk yield, which is in turn dependent on the stage of lactation. As such cows in early lactation will require more nutrients compared to those in late lactation. Since it is not practically possible to formulate a separate ration for each cow, the cows should be fed in groups (strings) with common nutrient requirements. Cows in the same stage of lactation will have almost similar requirements. Therefore, the rations can be formulated according to the phase (stage) of lactation.

3.3.1 Phase Feeding 1 (1 - 70 days)

The first phase lasts from calving to peak milk production, which occurs at about 70 days. During this phase, milk production increases more rapidly than feed intake, resulting in higher energy demand than intake, leading to a negative energy balance. This results in mobilization and use of body reserves and loss in body weight (negative energy). The energy is mobilized from fat reserves, protein from muscle and calcium and phosphorus from bones. However, energy is most limiting. The health and nutrition of the cow during this phase is critical and affects the entire lactation performance. The cow is expected to achieve peak production during this phase, failure to which the lactation milk yield is reduced. Excessive weight loss may be detrimental to cow's health and reproductive performance (cow may not come on heat at the optimum time), leading to long calving intervals.

Concentrates should be added to the basal diet to increase the energy and protein content as forage alone will not be sufficient. Cows that are poorly fed during this early phase do not attain peak yield and milk production drops from week 1. If excessive concentrates are added too rapidly (non-accustomed cows) to the ration, they can lead to digestive disturbances (rumen acidosis, loss of appetite, reduced milk production, low milk fat content). It is therefore recommended that concentrates should be limited to 50 - 60% of diet dry matter, the rest being foraging to ensure rumination (proper function of the rumen).

If high amounts of concentrate are fed during this time, buffers (chemicals that reduce the acid in the rumen and available commercially) can be helpful. At this stage, high protein content is important since the body cannot mobilize all the needed protein and bacteria protein (synthesized in the rumen by bacteria) can only partially meet requirements. A ration with a protein content of 18% CP is recommended for high yielding cows. If the cow is underfed during this stage, milk production cannot recover even when balanced rations are fed at later stages. This is attributed to the fact that cows in later stages of lactation, use energy more efficiently to restore body reserves than for milk synthesis. It should be noted that cows come on first heat during this phase and regaining a positive energy balance is critical in achieving this.

3. 3. 2 Phase Feeding 2 (70 - 150 days)

The second phase lasts from peak lactation to mid-lactation. The voluntary dry matter intake is adequate to support milk production and either maintain or slightly increase body weight. The aim should be to maintain peak milk production for as long as possible with milk yield declining at the rate of 8 - 10% per month. The forage quality should be high; 15 - 18% whole ration crude protein content is recommended. Concentrates high in digestible fiber (e. g. wheat or maize bran rather than starch) can be used as an energy source.

3. 3. 3 Phase Feeding 3 (151 - 305 days)

The third phase lasts from mid-to end-lactation, during which the decline in milk production continues. The voluntary feed intake meets energy requirements for milk production and the body weight increase. The increase is because body reserves are being replenished, and towards the end of lactation, it is because of increased growth of the fetus. It is more efficient to replenish body weight during late lactation than during the dry period. Animals can be fed on lower-quality roughage and more limited amounts of concentrate than during the earlier two phases.

3. 3. 4 Phase Feeding 4 (305 - 365 days, Dry period)

This phase lasts from the time the cow is dried to the start of the next lactation. The cow continues to gain weight primarily due to the weight of the fetus. Proper feeding of the cow during this stage will help realize the cow's potential during the next lactation and minimize health problems at calving time (e. g. ketosis, milk fever and dehydration, dystocia).

At the time of drying, the cow should be fed a ration that caters for maintenance and pregnancy, but 2 weeks before calving the cow should be fed on high-level concentrates in preparation for the next lactation. This extra concentrate (steaming) enables the cow to store reserves to be used in early lactation. To avoid over-conditioning, cows should not be fed large amounts of concentrate. The aim is to achieve a body condition score of 3. 5 - 4. If the diet is rich in energy, limit the intake of concentrates. Feeding bulky roughages can help increase the rumen size to accommodate more feed at parturition (birth). Before calving, feed concentrate progressively to adapt the rumen microbial population. This will minimize digestive disturbances in early lactation when the diet changes to high concentrate (Lukuyu, 2012).

3. 4 Condition Scoring of the Dairy Cows

In early lactation, high potential dairy cows frequently produce far more milk than can be supported by feed intake alone. They do this by drawing on body reserves that were built up before calving. This phenomenon is shown in Fig. 4. 3 where the condition score decreases due to the withdrawal of body reserves.

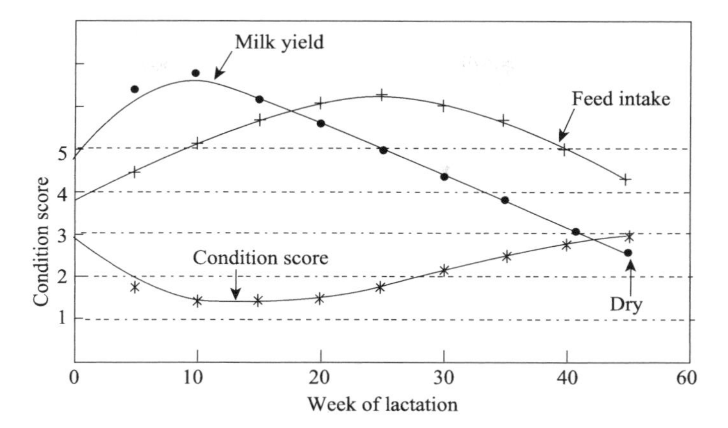

Fig. 4. 3 Relative changes in milk yield, feed intake and condition score over the lactation

Few farmers weigh their cows at regular intervals, and, even if they do, live-mass alone is not a good indicator of body reserves. Cows of similar mass could be small and fat, or large and thin. Similarly, cows could have the same body reserves and yet have very different masses. Live-mass is also affected by gut fill and by pregnancy. Body condition scoring is a technique for quickly and reliably estimating the body reserves of cows. These scores can then be used in making management decisions.

3. 4. 1 Technique

The local condition scoring technique is based on the method developed at the British National Institute of Research in Dairying. Two score areas are involved, namely the loin area and the tail head area subjectively, by feeling with the hand the amount of fat cover over the transverse processes (horizontal projections) of the lumbar vertebrae, and around the tail head.

The cow is awarded a condition score on a scale of 0 (very poor) to 5 (grossly fat), with half scores to give an11 - point scale. In most cases, the tail head score is used, but this may be adjusted by half a point if it differs greatly from the loin score.

Scoring Method

- Score the tail head area by feeling the amount of fatness. This gives a better estimate than a visual inspection alone because of the set of tail head and thickness of coat.
- Score the loin area in a similar way, using the same hand, when the cow is relaxed.
- Assess the scores to the nearest half point. Cows must be handled for accurate assessment of half points.
- If the tail head score differs from the loin score by one point or more, adjust the tail head score accordingly by no more than half a point as shown in Table 4. 2. The adjusted tail head score is then used as the condition score (Fig. 4. 4).

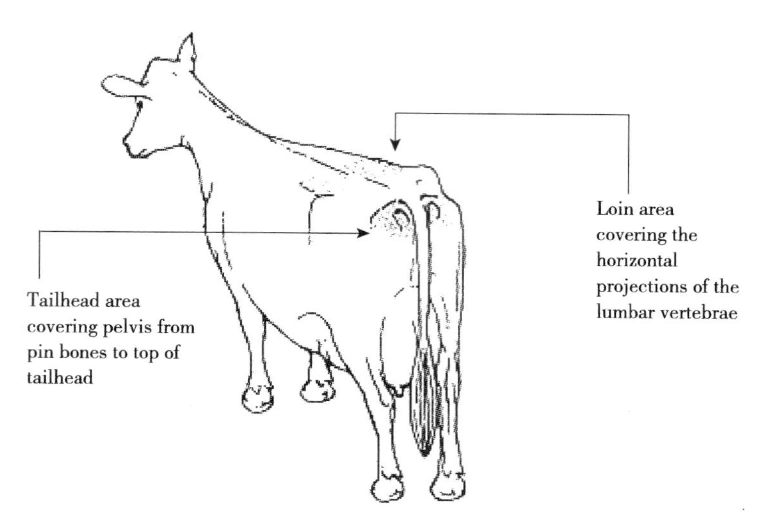

Fig. 4. 4 The two score areas

Table 4. 2 Adjustment of tail head score according to loin score

Tail head score	Loin	Difference	Adjustment	Adjusted tail head score
4. 0	2. 5	1. 5	−0. 5	3. 5
1. 5	2. 5	1. 0	+0. 5	2. 0
3. 0	2. 5	0. 5	none	3. 0

Get together at regular intervals, at least twice a year, with other condition scorers to revise technique and scores. Good stockmen who like their cows tend to score over. Experience has shown that people who work in isolation with only one herd tend to drift away from the definitions. For example, if that herd is generally a bit thin, and has few or no cows with condition scores greater than 2. 5, the 2. 5′s will inevitably become 3′s and so on. Regular comparison with friends and colleagues is essential.

3. 4. 2 Condition Score and Live-mass Change

It is apparent that there is a relationship between condition score and live-mass change. This is not a straightforward relationship, however, as it is complicated by stage of pregnancy and age of the cow. It is normal for high yielding adult cows to lose 30 to 60 kg body mass during the first 80 days of lactation, representing a drop of 1 to 1. 5 points in condition score. In an investigation on the influence of condition score on live-mass at the Cedara College of Agriculture, it was found that one condition score point represented, for Holstein-Friesland, a change in live-mass of 44 kg for cows with the same frame size, determined by body measurements.

3. 4. 3 Condition Score and Maiden Heifers

Work at the NIRD indicates that maiden heifers at service should be at condition score 2. 5 to 3. 5 to give the best conception rates. The conception rate of heifers below score of 2 will respond to extra feeding both before and after service time (e. g. an extra 2 kg of concentrate 6 weeks before to 6 weeks after service). Do not allow heifers to reach a score of 4 or more, because conception rates will be reduced; restrict feed if necessary. Heifers should be fed to grow steadily from birth to first calving. They should not be suddenly pushed if they are too light or too thin at any time. Rather accept a slightly older age at first calving than try to compensate completely for past mistakes.

3.4.4 Condition Score and Milk Yield

The cow's milk production in any lactation is the result of interactions between genetic potential, nutritional and environmental factors, including, specifically, body condition at calving and the system of management. The most important score is the score at calving. If this score is correct, it is possible to ensure that scores at other stages will also be correct. The condition at calving is dependent on the condition at the start of the dry period, and the level of feeding prior to calving. It is generally agreed that cows should have an average condition score of between 2.5 and 3.5 at calving. They therefore have the opportunity to mobilize reserves in early lactation and to "milk off their backs". One kilogram in live-mass can be converted to 5 to 8 kg milk, clearly illustrating the importance of cows being good condition at calving. Supporting evidence for this target condition score at calving on milk yield is presented in Table 4.3. Cows were classified into groups according to their score at calving, and actual performance related to predicted performance for the different groups.

Table 4.3 Effect of condition score at calving on milk yield

Score at calving	Number of cows	Difference in milk yield per day (kg)	Difference in total yield (0 to 84 days) (kg)
0.5 to 1.5	283	−1.8	−150
2	159	0	0
2.5 to 3.5	213	+1.1	+95
4	8	−1.8	−150

Source: Haresign, 1981.

Taking the yield of cows calving at a condition score of 2 as average (Table 4.3), those in poor condition and those at a condition score of 4 had markedly lower yields than average. It would be expected that these results would translate into losses of over 1 000 kg over the whole lactation for the very thin and very fat cows. The highest yields were obtained from cows calving at condition score 2.5 to 3.5. Cows at condition score 3.5 produced 182 kg more milk during the first 84 days of lactation than predicted. Cows which calve at a condition score of greater than 3.5 are more susceptible to severe metabolic and physical problems. Excess fat leads to fat infiltration of the liver (fatty liver). These cows have difficulty in establishing their normal feed intakes after calving, and this reduced intake, coupled with heavy lactation, leads to a rapid mass and condition loss. The result of this excessive fat mobilization, caused by the sudden energy crisis, is ketosis. Other conditions to which obese cows are more susceptible include milk fever, digestive disorders (displaced abomasum), infectious diseases (such as mastitis), and reproductive disorders (such as retained placenta and metritis).

Several studies have shown differences in the response of milk yield to condition score at calving. These differences can be explained largely by differences in the composition of the rations fed to cows, and the feeding regime. Research indicates that there is a complex interaction between the cow's genetic potential and her level of body energy reserves on the one hand, and diet composition (fiber content and energy concentration) on the other hand. If a cow that is thin at calving is to achieve her potential milk yield, she must be given the opportunity of consuming sufficient dietary energy and protein to meet her requirements. Dietary factors such as grazing intake could prevent her from reaching a desired intake to meet her requirements. When diets with a high fiber content or low energy concentration are offered to the cows, intake is more limited by the physical capacity of the rumen than by physiological mechanisms. The well-documented observation that fat cows eat less than thin cows is an indication of these physiological feedback mechanisms. Good stockmen should try to balance the various dietary factors and cow factors which physically and physiologically influence intake.

Cows that are thin at calving can only achieve intakes of energy that are high enough to meet their requirements for milk production when offered diets that will not lead to substantial restriction by physical capacity. If restricted rumen capacity or bulky diets are a problem, the logical approach would be to have these cows in a better body condition at calving but keeping in mind that over-conditioned cows have a physiological mechanism which can also impair intake. From examination of intakes of cows on pasture systems, it is generally accepted that their intakes are lower than when the same cows are offered a high energy total mixed diet. In early lactation, it is therefore unlikely that the dietary intake of high-producing cows grazing pasture or consuming long hay will supply the total energy demand, and body reserves will be needed.

The whole process of cyclic changes in mass and condition score, with cows losing condition in early lactation and gaining condition again in late lactation or the dry period, is considered to be a biologically inefficient process. The efficiency of conversion of dietary energy to milk energy is 0.62, but when this energy is stored in body fat, and later used for milk production, the efficiency factor is 0.51. The indirect route is still less efficient if body reserves are laid down during the dry period. The dry period is not the time to condition cows. Body reserve recovery should take place from mid to late lactation, while the cow is still in milk. Cows in milk are about 25% more efficient at feed conversion than are dry cows, therefore it is more cost effective to regain lost condition during lactation. Whether or not the differences in biological efficiency, shown above, for energy utilization *via* the direct or the indirect route, can be directly related to economic efficiency, will depend on the relative costs of feeds used to increase condition before calving, and for milk production in early lactation. British results have shown that it is only economically more efficient to have cows in good condition at calving when the extra condition was produced from well-managed grassland.

3.4.5 Condition Score and Reproduction

A relationship between body condition and fertility has not been fully established. Despite this, there is evidence which suggests that conception rates are impaired if cows are served at condition scores below 2. When condition scores drop below 2, fewer cows come into estrus, and conception rate in those that do is usually low. Work done at NIRD illustrates this trend (Table 4.4).

Table 4.4 Effect of condition score on conception rates

Condition score	Conception rate (%)
Below 1.5	52
1.5	56
2.0	68
Above 2.0	72

Source: MAFF, 1978.

Research has also shown that cows, with a condition score of above 3.5 at calving, lost more condition in early lactation, had longer intervals to first estrus and conception, and required more services per conception. It is evident that the rate and extent of live-mass loss and mobilization of body reserves resulting from a negative energy balance are associated with a reduction in reproductive efficiency. Cows of high potential milk yield have an even greater chance of their level of energy intake being inadequate for normal fertility. The cow should be in a stable or improving condition (and generally, that means gaining mass) at the time of service. There is evidence that it is important to stem the tide of mass loss after calving as quickly as possible, and to stop the absolute loss from going too far. This "trough mass" should be reached at about 35 days after calving. Research has shown that the fertility of replacement heifers is also affected by level of nutrition, and change of condition score, during the mating period. The condition score at service, the rate of loss of condition score during the mating period, and nutrition before and around the time of service, are important factors which should be taken into account when attempting to improve reproductive efficiency. The emphasis should be on maintaining or improving the cow's condition during the mating period.

3.4.6 Condition Scoring and Dairy Cow Management

The scoring method described above is simple and can be carried out quickly. With a little practice it provides consistent scores. Cows can be scored when standing in abreast parlors, cubicles, crushes, feeding sheds, insemination stalls, or on cattle scales. The same scoring system can be used for dairy heifers and for different breeds.

With any recording system, there are two problems to overcome. The first is to ensure that the information is recorded; the second is to ensure that it can be recovered and used. Monthly, or preferably fortnightly, recording of all cows is the simplest way of ensuring that the information is recorded. The simplest place to record condition scores, and for that matter, live-mass, is to use extra columns/rows in the milk recording book or cow byre sheet. This is especially appropriate for feeding programs which require body score, live-mass and milk yield. The ideal cow byre sheet would have columns for milk yield, condition score, live-mass and butterfat and protein tests.

With a little practice, and with an assistant to do the writing, it is possible to weigh and score at the rate of three cows per minute. A major advantage of monthly recording is that a change in the average herd score, if any, could be an indication of feed availability to the whole herd and so ensure that supplementary feeding is appropriate. Fortnightly recording is desirable for users of computerized feeding programs or any rationing system which uses condition as a factor in determining feed requirements.

If monthly scoring is still too much trouble, then, provided that care is taken to ensure that the information is recorded conscientiously, cows can be scored at calving, when they will be isolated and handled anyway, and again at first service, pregnancy diagnosis, and drying off.

3.5 General Management and Routine Husbandry Practices

Appropriate housing and husbandry are essential for the health and well-being of dairy cattle. Animal handling is a key factor in successful dairy farming. Employers have an obligation to properly train employees about proper animal handling practices. Most husbandry systems impose restrictions on some freedoms of cattle. However, modern dairy farming should not cause unnecessary discomfort or distress to the animals. Producers meet the needs of their cattle under a variety of husbandry and management systems (NFACC, 2009).

3.5.1 Identification

Individual cow identification is playing an important role in day-to-day herd management decisions. Proper identification becomes more difficult as herd sizes increase. The overall objective of proper identification is to ensure that each animal can be followed from the time it is born to when it is sold or slaughtered. In this way, herd managers can monitor daily, visually or electronically, all parameters about individual cow.

It is essential that calves are identified as soon as possible after birth and that the registration numbers are kept as a unique lifetime number for all cattle improvement purposes, such as herd book registration, type classification, milk-recording, artificial insemination and general health. In smaller herds, herdsmen are able to identify animals more easily, but in larger herds an accurate identification system is an absolute must, following a method facilitating quick on-the-spot identification with a minimum of time and effort at all times.

Routine management activities including the details of identifications and removal of extra teats are discussed in module 6.

3.5.2 Pedometers

The pedometers are used mainly for monitoring the activity level of cows coming into estrus. The pedometer, which is attached to a cow's leg, utilizes a microprocessor with a built-in movement sensor to monitor and record the cow's typical 24 - hour activity. It interprets this information to help the herd manager determine when this activity occurred, thus facilitating optimal insemination timing and also monitoring the health of individual cow. Be sure that there are no obstacles where the ankle bands may get stuck and eventually lost, e.g. with mushroom washers in the holding area before the cows are entering the milking parlor.

3.5.3 Inspection

The inspection will concentrate on key areas as listed below and an overall assessment of the hygienic conditions and management practices at the premises. Generally, routine inspections are undertaken on:

Animal health and cleanliness

The milking animals should be inspected to ensure that there are no obvious signs of ill health. Check that the producer makes sure that milk from any animal showing individually a positive reaction to test for tuberculosis or brucellosis is not used for human consumption. Establish that there are effective isolation facilities available for animals infected or suspected of being infected with any diseases that could have any adverse effect on other animals and milk. Check that veterinary products are stored in a lockable storage or container and that this is locked when not attended.

The milking animals should be inspected for cleanliness. Heavily soiled animals, which are incapable of being adequately cleaned prior to milking, should be kept separate and milked last. The milk from those animals should not be sold for human consumption, as there is potentially a high risk of contamination.

Proper milking procedures/ milking premises/ milking equipment

The milking premises should be assessed in relation to construction and location. Walls, floors and any fixtures should be kept clean, be freely draining and in a good state of repair. Surfaces of equipment that are intended to come into contact with milk or are used for milking should be examined to ensure that they are maintained in a sound condition. A representative proportion of the milking equipment including but not limited to, claw pieces, liners and shells, pipe work / joints, milk flow sensors, mastitis detectors, receiver & recorder jar (s), balance tank & transfer pumps, should normally be examined internally and externally for cleanliness and physical condition. At a milking time inspection, what can be examined will be more limited. Depending on the time of the inspection, the cleaning and disinfection routine will be observed, monitored and/or discussed.

3.5.4 Hoof Trimming

Hoof trimming is an important tool to prevent and treat lameness and should form part of an overall hoof-health program. Each hoof must be trimmed to its own "normal" structure in order to prevent hoof disease. Over-trimming is a common error that can cause lameness. Only skilled individuals should trim hoof on cattle (NFACC, 2009).

Condition of the feet and legs of dairy cattle shouldn't lightly be taken. A cow with sore-feet may realize losses in milk production, diminished breeding efficiency and decreased salvage value in the case of severe lameness. As the number of cows in confinement increases, periodic hoof trimming is necessary for cows to reach their full genetic potential. Hoof trimming is very labor intensive, which is why many times it is neglected. Before you can begin hoof trimming, you must understand:

The correct hoof shape:

- Notice the 45° angle of the hoof to give the greatest amount of shock absorption through the pastern yet provide plenty of heel depth.
- Another way to get an idea of a correctly shaped heel is to look at a young calf's hoof (1 - 2 months old).

Toes on each hoof should be about equal length, with all four feet approximately the same shape.

The hind feet are likely to get longer on the toes than the front feet and may need trimming more often.

Trim the feet of cows that show excessive hoof growth or signs of lameness.

It is best to trim hooves when cows are in the latter part of lactation, so you don't disturb milk production.

3.5.5 Dehorning

Disbudding and dehorning are done for the safety of cattle and their caregivers. Disbudding refers to removal of the horn bud prior to three weeks of age. Removal of the horn after this age is referred to as dehorning. Disbudding is recommended over dehorning because it is less invasive. All calves should be disbudded to avoid injuries and behavioral problems associated with horns in later life. It is also important that the job of disbudding be done correctly to avoid the re-growth of horn in the future.

Pain control reduces animal discomfort during disbudding and dehorning. The most popular local anesthetic, lidocaine, is effective for two to three hours after administration. The use of analgesics in addition to a local anesthetic can minimize pain and stress in the hours that follow dehorning.

3.5.6 Grooming

To optimize health and maximize production, cows need to live and act like cows. Grooming is one behavior they seek to perform. In all different types of housing situations, they can be seen rubbing their faces and sides on barn walls, water troughs, or low hanging branches. Grooming is a behavior that appears to promote cow health, calm, comfort, and overall performance. The brush, which resembles something from a car wash, is hung so that when the cow starts to rub against it, begins to rotate. (Fig. 20)

Within the first day of access to the mechanical brush more than half the cows were using it. After a week most of the cows were getting themselves brushed. The grooming device helps to satisfy the animals' need for grooming, while at the same time improving their cleanliness.

>>> SELF-CHECK QUESTIONS

Part 1. Choose the best answer (s) from the given alternatives.

1. Which of the following is not the main function of the digestive system of the dairy cow?
   A. intake of food
   B. Breakdown of food into compounds
   C. Transport of the compounds into the body
   D. All
2. Which of the following choice is false about concentrate feeds?
   A. Concentrates are containing crude fiber <18%
   B. Concentrates are containing crude protein >18%
   C. Concentrates rich in nutrient required for production.
   D. None
3. Which of the following is true about the energy portion of the feed?
   A. The energy portion of the feed fuels all body functions
   B. Enabling the animal to undertake various activities including milk synthesis
   C. Energy requirement of a lactating cow increases with milk production and butterfat content
   D. All
4. Which of the following is not a good source of protein for dairy cows?
   A. Oilseeds and oilseed cakes
   B. Cotton seed meal or cake
   C. Sunflower meal or cake
   D. Wheat bran
5. A lactating cow requires nutrients for all except?
   A. Maintenance
   B. Growth if she is young (<30 months)
   C. Growth of the unborn calf if she is pregnant
   D. Milk production

E. None

6. Nutritive value of feeds is determined by a number of factors except?

A. Composition B. Odor

C. Texture D. Non

7. Which of the following is not true about the conventional digestion trial?

A. Measuring a feed's digestibility

B. Somewhat time consuming

C. Not tedious and costly

D. All

8. Herd composition is the result of a number of interrelated management decisions, except

A. Rate of replacement & short-term goal

B. Culling policy

C. Rate of reproductive success

D. All

9. Which of the following is not true about body condition scores of dairy cattle?

A. A herd of cattle in good body condition will produce more, and will be highly susceptible to, disease problems.

B. It is a method of evaluating fatness or thinness in cows.

C. As the production level of a herd increases, body condition of scoring is more important.

D. A routine program for body condition scoring can help detect potential health problems.

E. None

10. Which of the following is true about practical feeding of the producing herd?

A. During the formulation of rations for lactating dairy cows, the quality of the ration should be commensurate with the requirements of the cow.

B. The requirement is directly related to the milk yield, which is in turn dependent on the stage of lactation.

C. Cows in early lactation will require more nutrients compared to those in late lactation.

D. All

Part 2. Fill in the blank space.

1. Cows consuming too much energy become too fat, resulting in low conception rates

2. A protein is a major component of products such as milk and meat, but lack of protein can't adversely affect milk production.

3. The rumen micro-organisms manufacture all of the B vitamins and vitamin K.

4. Proper management and nutrition of the dry cow are critical for obtaining maximum dry matter intake, but not good health.

5. Mid-lactation period is the period from day 100 to day 200 after calving.

6. Restoring body energy and nutrient reserves is more efficient if accomplished during late lactation rather than the dry period.

7. Beginning at a fairly early age, the bull calf has a higher growth rate than the heifer, but *not* requires more energy and other essential nutrients.

8. Lactating cows need larger proportions of water relative to body weight than do most livestock species since 87% of milk is water.

>>> REFERENCES

Amaral-Phillips D M, 2014. Management of Fresh Dairy Cows Critical for a Dairy's Profitability [E]. http://articles.extension.org/pages/69058/management-of-fresh-dairy-cows-critical-for-a-dairys-profitability

Amaral-Phillips D M, Scharko P B, Johns T J, et al., 2006. Feeding and Managing Baby Calves from Birth to 3 Months of Age [E]. https://afs.ca.uky.edu/files/feeding_and_managing_baby_calves_from_birth_to_3_months_of_age.pdf

Gillespie J R, Flanders F B, 2010. Modern Livestock and Poultry Production [M]. 8th ed. Delmar Cengage Learning.

Heinrichs A J, Ishler V A, 1994. Body Condition Scoring as a Tool for Dairy Herd Management [E]. http://extension.psu.edu/animals/dairy/nutrition/nutrition-and-feeding/body-condition-scoring/body-condition-scoring-as-a-tool-for-dairy-herd-management/pdf_factsheet.

Heinrichs A J, Ishler V A, Adams R S, 1996. Feeding and managing dry cows [E]. http://extension.psu.edu/animals/dairy/nutrition/nutrition-and-feeding/dry-cow-nutrition/feeding-and-managing-dry-cows.

Lukuyu B, Gachuiri C, 2012. Feeding dairy cattle in East Africa [E]. https://cgspace.cgiar.org/bitstream/handle/10568/16873/EADDDairyManual.pdf

Moranm J, 2005. Tropical Dairy Farming, Feeding Management for Small Holder Dairy Farmers in the Humid Tropics [M]. CSIRO Publishing.

National Research Council, 2001. Nutrient Requirements of Dairy Cattle [E]. 7th ed. Washington, DC: The National Academies Press.

Pond W G, Church D C, Pond K R, 1995. Basic Animal Nutrition and Feeding [M]. 4th ed. John Wiley and Sons, Inc.

Wagner J, Stanton T L, 2014. Formulating Rations with the Pearson Square [E]. http://extension.colostate.edu/docs/pubs/livestk/01618.pdf

MODULE 5: MANAGING REPRODUCTION AND BREEIDING IN DAIRY CATTLE

>>> INTRODUCTION

Reproduction forms the basis of livestock improvement as it allows the transfer of genetic material from one generation to the next and can greatly influence genetic gain. Reproduction is probably the single most important factor affecting the economics and profitability of dairy cattle operations. Cows must have efficient reproductive performance. This is essential for the production of the main commodity of interest; milk, as well as to provide replacement animals. For bulls, reproduction is all about the capacity and ability to sire a large number of viable offspring in each mating. No dairy production system is sustainable without an acceptable level of reproduction. From the viewpoint of reproduction, the main factors which contribute to economic losses are delayed puberty, long calving intervals, short productive life (due to culling for infertility or sterility) and high calf mortality.

This module generally, discusses aspects of the reproductive processes and provides practical guidelines for management of dairy animals for optimum reproductive efficiency. Specifically, this module enables to identify and describe the male and female reproductive organs, comprehend reproductive process in dairy cattle, describe the function of the endocrine glands and hormones in reproduction, describe estrus induction and synchronization techniques and appropriate breeding system, artificial insemination, pregnancy diagnosis, evaluate breeding efficiency and describe reproductive problems and their corrective measures.

1 IDENTIFYING MAJOR REPRODUCTIVE ORGANS OF DAIRY CATTLE

Successful reproduction on modern dairy farms requires an understanding of reproductive processes of the dairy cow and bull. A sound knowledge of the location, structure and function of the reproductive organs is essential for efficient insemination and for avoiding damage to cows. Thus, this knowledge can be useful in identifying and correcting many situations leading to poor reproductive efficiency. This section discusses the basics of the reproductive tract and how this knowledge can be used to increase reproductive success.

1.1 Female Reproductive Organs

The cow reproductive tract or organs includes vulva, clitoris, vagina, cervix, uterus, oviducts, ovaries, follicle and corpus-luteum.

Vulva: The vulva is the external opening of the reproductive and urinary systems. The exterior and visible part of the vulva consists of two folds called the labia majora. The labia minora are two folds located just inside the labia majora.

Clitoris: It is the sensory and the erectile organ of the female, it is located just inside the vulva. The clitoris develops from the same embryonic tissue as the penis in the male and produces sexual stimulation during copulation.

Vagina: The vagina is serving as an unrestrictive passageway for the calf at the time of birth; it is also the location of semen deposition during natural service. In addition, an important function of the vagina is a line of defense against the invasion of bacteria into the remainder of the reproductive tract. The vagina is the passage between the cervix and the vulva. It is 24 to 30 cm long with muscular walls. About 10 cm inside the vagina along its floor is the passage from the bladder, the urethra. It enters the vagina near a blind pouch, the urethral diverticulum. The lining of the vagina is moist during estrus and dry when the animal is not in estrus (Dunn et al., 2009).

Cervix: The cervix consists of dense connective tissue and muscle that connects the vagina to the uterus. The cervix is 3 - 10 cm long and from 1.5 - 6 cm in diameter. The walls of the cervix are very thick, and the body of the cervix contains three or four rings called annular folds, easily distinguished by rectal palpation around the pelvic floor (Gillespie and Flanders, 2010).

A narrow canal passes through the center of the cervix. It is spiral in form and tightly closed. The start of this canal is called *Os* which extends into the vagina, forming a blind pocket the *fornix*. The cervix relatively relaxed during estrus to allow the passage of sperm into the uterus; during pregnancy, it remains tightly closed to block the entrance of any foreign matter into the uterus and enlarges greatly during calving.

The cervix is larger in *Bos indicus* cattle (Zebu cattle) than *Bos taurus* cattle and increases in size with age. Furthermore, the position of the cervix will vary with the age of the cow and the stage of pregnancy. In non-pregnant cows, most operators should not have to go in beyond elbow length to locate the cervix and the rest of the reproductive tract. In heifers, the cervix should be picked up at wrist depth. It may be pulled out of reach by the weight of the developing fetus in a pregnant animal (Dunn et al., 2009).

Uterus: The uterus of a cow is a Y-shaped structure consisting of the uterus body and two uterine horns. Pregnancy normally occurs in the uterine horns and the fetus grows within the uterus, where it remains until parturition. The walls of the uterus are comprised of three layers: the serosa, myometrium and endometrium. The endometrium is the innermost layer and aids in the transportation of sperm to the oviduct. The myometrium is the muscle layer, which aids in the expulsion of the calf at the time of birth. Additionally, the uterus provides nourishment and protection to the developing fetus (Harms, 2001; Gillespie and Flanders, 2010).

- *Uterus body*: It is 2 - 3 cm long and separates into the horns. The body feels longer on palpation than 2 - 3 cm, because the horns are joined together by a ligament for about 12 cm beyond the point of division.
- *Uterine horns*: The two uterine horns are 35 - 40 cm long and 2 cm or more in diameter. They have a thick elastic wall and a rich blood supply to nourish the developing calf.

Oviducts: The oviducts are two tubes that carry the ova from the ovaries to the uterus. The oviducts are also called the Fallopian tubes. It is 20 - 25 cm long and 1 - 2 mm in diameter. It runs from the tip of the horn to the *infundibulum* (or funnel) which surrounds the ovary. The oviducts are close, but not attached, to the ovaries. The infundibulum is funnel-shaped that catches the egg as it is released from the ovary at ovulation and carries it to the *ampulla*, the enlarged upper end of the oviduct (Harms, 2001).

Ovaries: The cows have two ovaries that produce the ova and hormones (estrogen and progesterone) involved in regulating the estrous cycle and pregnancy. Ovaries can normally be felt alongside the uterine horns. They are oval in shape with size depending on age and breed (usually the size of an almond nut or slightly larger). In a dairy cow, each ovary is approximately 3. 8 cm long and 1. 9 cm in diameter. The ovaries are suspended from the broad ligament near the end of the oviduct and lie near the tips of the curved uterine horns (Hafez, 1993; Taylor and Field, 1998).

The female ovaries produce ova in the process of oogenesis. The development of the ova begins before the female is even born. Cells called oogonia develop in the ovaries of a fetus. By the time of birth, these oogonia have matured into oocytes. There are thousands of oocytes at the time of birth; however, only a small proportion of these develop into ova or reach ovulation.

Follicle: The process of oogenesis occurs within a follicle, either on a continual cycle or seasonally. Within the containment of a follicle, an oocyte will begin cellular divisions and other physical changes in response to the female sex hormone, estrogen. The follicle appears as a clear blister on the surface of the ovary.

The function of the follicle is to hold the growing ovum and to produce and store the hormone estrogen. Estrogen is secreted from the follicle as a signal to the remainder of the reproductive anatomy to prepare for the ovulation of an ovum. The follicle remains relatively hard throughout the development of the ovum, but it becomes very soft, ruptures, and expels the ovum at the time of ovulation. The ovum enters the infundibulum and then the oviduct to await fertilization (Ball and Peters, 2004).

Corpus-luteum (CL): after ovulation, the ruptured follicle collapses and a small hemorrhage occur. This blood-clotted area is called a corpus hemorrhagicum and only lasts two to three days. This area begins to be filled with a yellow mass of cells. This yellow body is called the CL. Its cells have the primary purpose of producing the female sex hormone, progesterone.

Production of progesterone prepares the female reproductive anatomy for pregnancy and lasts approximately twelve days unless the ovum is fertilized (in which case the CL remains until parturition). A degenerating CL becomes covered by connective tissue and is called a corpus albicans. The function of the corpus albicans is to remove the yellow cells of the CL and return the ovary to its normal shape and function (Harms, 2001) (Fig. 5. 1 and Fig. 5. 2).

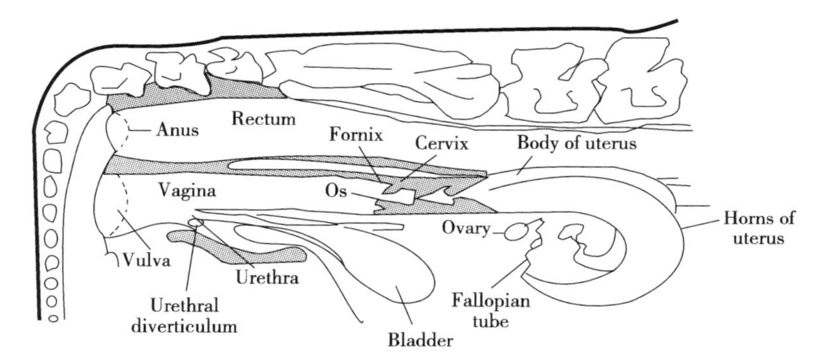

Fig. 5. 1 Reproductive tract of a cow

Source: Dunn et al., 2009.

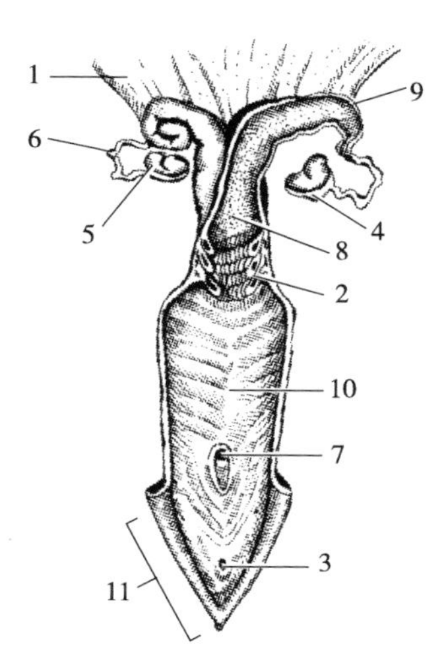

Fig. 5. 2 Top view of reproductive tract of a cow

1. Broad ligament 2. Cervix 3. Clitoris 4. Infundibulum 5. Ovary 6. Oviduct 7. Urethral opening 8. Uterine horn 9. Uterine body 10. Vagina 11. Vulva

Source: Harms, 2001.

1.2 Male Reproductive Organs

Good reproductive performance of a bull is necessary to obtain a high percentage calf crop. A bull must be fertile and capable of servicing a large number of cows for optimum production. A basic knowledge of the reproductive system will also help the producer to understand fertility examinations, reproductive problems and breeding impairments.

The major function of the male reproductive system is the production, storage, and deposition of sperm cells, producing male sex hormones, and serves as a passageway for expelling from the urinary bladder. The male reproductive tract is made up of several organs, glands, and muscles, each having a specific function. The reproductive tract of the bull consists of the testicles, scrotum, epididymis, vas deferens, urethra and penis, and three accessory sex glands the seminal vesicles, prostate and Cowper's gland (Turman and Rich, 2010). This basic anatomy is illustrated in Fig. 5.3.

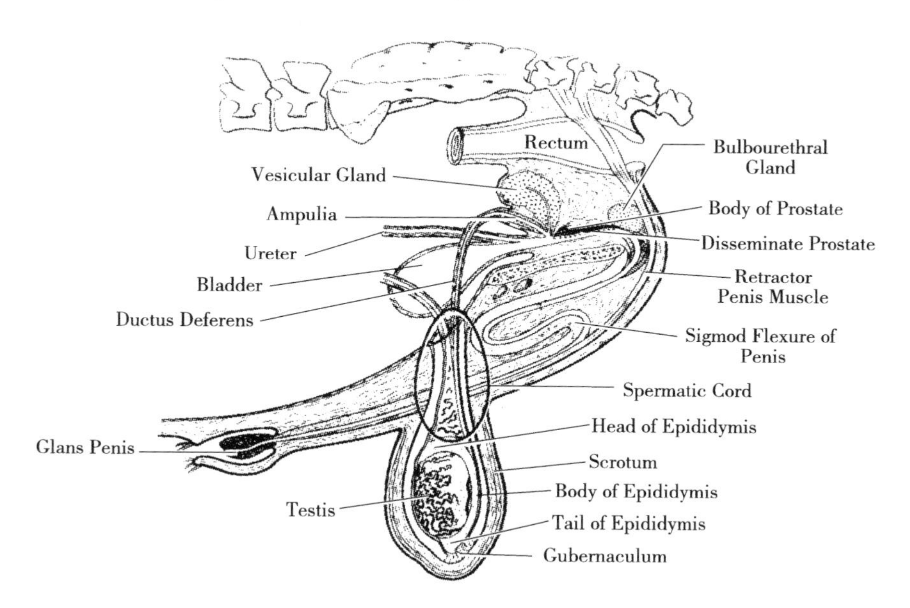

Fig. 5.3 Reproductive tract of the bull

Source: Harms, 2001; Dunn et al., 2009.

Testes/testicles: The testes/testicles are paired, ovoid shaped organs that produce sperm cells and the male sex hormone testosterone. Testosterone causes the development of secondary male characteristics and sex behavior (*libido*). The presence of testosterone maintains the masculine appearance of the animal. A male that is castrated at an early age does not develop the typical masculine appearance of the species and the reproductive organs do not continue to develop. When an adult male is castrated, the reproductive organs diminish in size and lose some of their functions (Gillespie and Flanders, 2010).

The testes are suspended from the body by the spermatic cord. The spermatic cord is a protective fibrous sheath consisting of smooth muscles, blood vessels, and nerves. It extends through the inguinal ring and attaches to the testes to suspend them within the scrotum. The scrotum protects and supports the testes.

Within the testes are microscopic cellular parts that function in the production of sperm cells and the male hormone testosterone. The development of sperm cells, called spermatogenesis, is a process of cell division and maturation that begins with stationary cells called spermatogonium and ends with motile spermatozoa.

➢ *Seminiferous tubules*: the seminiferous tubules are tubular structures that coil throughout the testes. The process of spermatogenesis takes place within the seminiferous tubules. The resulting spermatozoa, or sperm cells, are motile and tadpole-like. Once the maturation process has completed, the sperm cells proceed to the epididymis where they are stored until ejaculation or are absorbed by the body. Unusual climatic conditions (extremely high temperatures) or stress on a male can temporarily halt sperm cell production causing reproductive failure upon breeding.

➢ *Interstitial cells*: between the seminiferous tubules are groups of interstitial cells (cells of Leydig) that function in the production of the male sex hormone, testosterone. Initially, testosterone is produced in response to an interstitial cell stimulating hormone that is produced in the anterior pituitary of the brain and is transported in the bloodstream to its target, the interstitial cells. Testosterone is an androgen hormone that directs the development of masculine traits. This is observed in the development of secondary male characteristics such as coarse hair, horns that are long and large at the base, a deep voice, and pronounced muscle development. A constant level of androgens, especially testosterone, has a major influence on an animal's *libido*, as well as spermatogenesis (Harms, 2001).

Scrotum: It is the saclike part of the male reproductive system outside the body cavity that contains the testicles and the epididymis. The testicles are held in the abdominal cavity of the fetus. After the animal is born the testicles descend into the scrotum.

The primary muscle supporting the testes and coursing the length of the spermatic cord is the cremaster muscle. The cremaster muscle, spermatic cord, and tunica dartos muscle in the wall of the scrotum rise (during cold weather) or lower (during hot weather) the testes to maintain a constant testicular temperature of 4 – 6℃ below body temperature. This is because sperm must develop (spermatogenesis) under conditions cooler than body temperature.

If the testicles of an animal are held in the body cavity, the animal is sterile, or cannot produce live sperm. A ridgeling or ridgel is a male in which one or both testicles are held in the body cavity. This is also called cryptorchidism and is an inherited trait. The animal is usually sterile if both testicles are in the body cavity. If one testicle is retained in the body cavity and the other descends into the scrotum, the animal may be fertile. An animal with cryptorchidism (one or both testicles retained) should not be used for breeding (Harms, 2001).

Epididymis: This is a coiled tube connected to each testis and is responsible for the maturation, storage, and transportation of sperm cells. Sperm cells that are not moved out of the epididymis by ejaculation during copulation are eventually reabsorbed by the body. The epididymis is divided into three parts: the head, body and tail. The head is on the top part of the testicle and is very closely attached to it. The body extends down the outside of the testicle. The tail is the enlarged distal part which may be palpated at the base of the testicle.

Vas deferens: It is a tube that connects the epididymis and the urethra. It conducts sperm cells to the urethra. Its end is enlarged as the ampulla that opens directly into the urethra.

Urethra: It is the tube that carries urine from the bladder. This tube is found in both male and female mammals. In the male animal, both semen and urine move through the urethra to the end of the penis. The semen contains the sperm and other fluids that come from accessory glands. The urethra is the passageway or tube that extends from the bladder to the end of the penis and serves as the transportation route for semen and urine.

Penis: The penis deposits the semen within the female reproductive system. The urethra in the penis is surrounded by spongy tissue that fills with blood when the male is sexually aroused. This causes an erection that is necessary for copulation to occur.

The *sigmoid flexure* (found in bulls, rams, and boars) and the *retractor muscle* extend the penis from the *sheath*, a tubular fold of skin. After copulation, the blood pressure in the penis subsides and the retractor muscle helps draw the penis back into the sheath. Horses and other mammals do not have a sigmoid flexure. Erection is caused by the blood that fills the spongy tissue when sexual arousal occurs (Turman and Rich, 2010).

The *prepuce* is a double invagination of skin, which contains and covers the free, or pre-scrotal, the portion of the penis when not erect and covers part of the body of the penis behind the glans when the penis is erect. The external opening is called the preputial orifice. The lining of the prepuce is a freely movable membrane or modified skin that is attached firmly only at the glans penis and at the preputial orifice.

The *sheath* is the layer of skin that houses and protects the penis when it is not extended. In bulls the prepuce is 35 – 40 cm long and 3 cm in diameter. The preputial orifice is about 5 cm behind the umbilicus. The lining membrane forms longitudinal folds (Ball and Peters, 2004).

Accessory glands

Accessory glands are responsible for the production of secretions that contribute to the liquid non-cellular portion of semen, known as the seminal plasma. Semen and ejaculate are terms given to the sperm plus the added accessory fluids. The three accessory glands are the seminal vesicles, prostate gland, and Cowper's gland.

Seminal vesicles: Seminal vesicles, also called the vesicular glands, are paired accessory glands that secrete seminal fluid that adds fructose and citric acid to nourish the sperm and functions as a protection and transportation medium for sperm upon ejaculation. The seminal vesicles open into the urethra.

Prostate gland: The prostate gland is near the urethra and the bladder. The prostate gland secretes a thick, milky fluid that mixes with the seminal fluid and also provides nutrition and substance to the ejaculate.

Cowper's gland: It is also called the bulbourethral glands. It produces a fluid that moves down the urethra ahead of the seminal fluid. Just prior to ejaculation, Cowper's glands secrete a fluid similar to the seminal fluid, which cleanses and neutralizes the urethra from urine residue that can kill sperm cells. This helps protect the sperm as they move through the urethra. The mixture of the seminal and prostate fluid and the sperm is called *semen*.

2 REPRODUCTIVE PROCESS IN DAIRY CATTLE

During the lifetime of an individual female, higher reproductive efficiency yields more calves for use as replacements or for sale in the herd, as well as more lactations and therefore more milk. These considerations become more important as rearing conditions become more intensive, where expenses for labor and feed have to be compensated for by higher income.

In order to fulfill its role as an economically useful animal to its owner, dairy females must perform the following functions: grow rapidly from birth until puberty, attain puberty at an early age, conceive readily to a fertile mating, produce a viable calf, produce adequate milk for the calf and extra for sale, return to estrus early during the postpartum period and conceive again, continue producing calves and milk at regular intervals till the end of its productive life. The ability of an animal to meet these requirements will depend on many factors as outlined below.

2.1 Puberty and Sexual Maturity in Dairy Cattle

Puberty is defined as the process/time in which the young female become sexually maturated and capable of reproduction. In case of cattle, the onset of the first behavioral estrus accompanied by ovulation and development of a normal corpus-luteum (CL) in the ovary is considered as the time of puberty.

Puberty is determined by several factors, which are endogenous, for example, genotype, growth and body weight, as well as exogenous, for example, year or season of birth, rainfall, nutrition, thermal environment, photoperiod, rearing method and diseases (Perera, 1999). Generally, heifers attain puberty when they reach 55 - 60% of their adult body weight (Graves, 2009). Too early (young) mating and pregnancy will cause dystocia at the time of delivery, because of the narrowness of the birth canal. The recommended standard age and body weight for the first insemination of different breed are shown in Table 5.1:

Table 5.1 Recommended age and body weight of heifers for first breeding

Heifer / Breeds	Age in months	Body weight in kilogram
Holstein Friesian & Brown Swiss	13–15	340
Crossbred and Jersey	13–15	250
Zebu dairy type (Sahiwal, Red Sindhi)	15–20	220
Ayrshire and Guernsey	15–18	300

Source: Graves, 2009.

Generally, the age at which they attain puberty can be highly variable, ranging from 12–40 months. Thus, growth rate and body weight are more important determinants of puberty than age. Under optimum conditions, *B. taurus* cattle and their crosses attain puberty earlier than zebu cattle. However, zebu cattle generally have a longer productive life than *B. taurus* cattle and some of the disadvantages of late puberty are compensated for by their longevity (Perera, 1999).

2.2 Estrus Cycle in Cows

Estrous cycle is the regular sequence of stages which the cow undergoes from one heat to the next. Estrous cycles give females repeated opportunities to become pregnant throughout their productive lifetime. This cycle involves a sequence of events in preparation for mating, conception and pregnancy. The cycle repeats in preparation for a new mating cycle if pregnancy does not occur. The estrus cycle has an average length of 21 days. Any period between 18 and 24 days is considered normal (Connor, 1993; Ball and Peters, 2004).

2.2.1 Regulation of Estrus Cycle

The estrus cycle is regulated by the hypothalamic-pituitary-gonadal axis, which produces hormones that dictate reproductive events. The reproductive axis is composed of the hypothalamus, pituitary, the ovary and uterus. Fig. 5.4 illustrates the interaction of the structures and hormones that control the estrus cycle (Rasby, 2015).

HYPOTHALAMIC-PITUITARY GONADAL AXIS

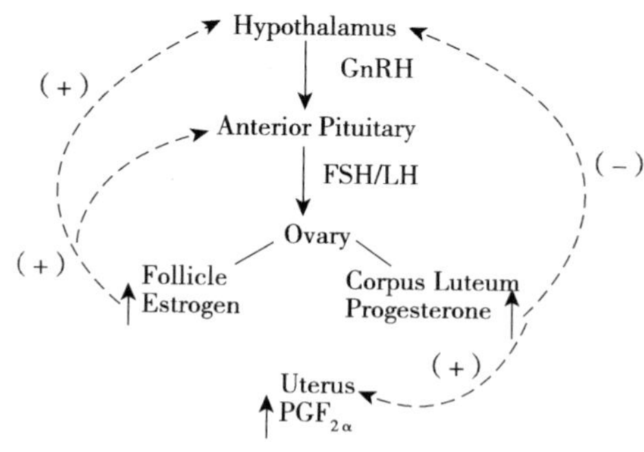

Fig. 5.4 the interaction of the structures and hormones that control the estrous cycle (the estrus cycle) changes during estrus cycle

Source: Rasby, 2015.

Hypothalamus is a specialized portion of the ventral brain. Its primary function is to produce gonadotropin-releasing hormone (GnRH) in response to circulating estrogen, or to cease GnRH production in response to progesterone.

Pituitary is composed of the anterior and posterior lobes. The anterior lobe is related to reproduction in dairy females. It produces the gonadotropins follicle-stimulating hormone (FSH) and luteinizing hormone (LH) in response to GnRH and estrogen. FSH and LH production is inhibited by progesterone.

Ovary is the third portion of the reproductive axis.

➢ *Follicles* contain ova and produce estrogen. Follicles range in size and maturity at different stages of the cycle, but usually only one is selected to ovulate.

➢ *CL* is a structure that forms from the previous cycle's ovulation point. The CL is responsible for progesterone production. Both estrogen and progesterone are produced following FSH and LH stimulation of the ovary.

Uterus contributes to reproductive control, as it produces prostaglandin $F_{2\alpha}$ ($PGF_{2\alpha}$). It functions in destroying the CL.

2.2.2 Stages of Estrus Cycle

Each estrus cycle consists of a long luteal phase (days 1 - 17) when the cycle is under the influence of progesterone and a shorter follicular phase (days 18 - 21) when the cycle is under the influence of estrogen. The estrous cycle is divisible into four stages: 1) estrus/heat, 2) met-estrus, 3) di-estrus, and 4) pro-estrus. Anestrus is the non-breeding period, the period when there is no evidence of estrus occurrence (Dunn et al., 2009).

Stage 1: Pro-estrus (before heat)

Days 19 - 21 of the cycle are the period of preparation for sexual activity. The sequence of hormonal release essentially begins with the synthesis and release of GnRH from the hypothalamus. GnRH functions to stimulate the anterior pituitary to produce and release FSH and LH. FSH and LH are transported to the ovaries, where they initiate a series of morphological changes that lead to ovulation and pregnancy if fertilization occurs. FSH acts upon the germinal epithelium of the ovary to cause the formation of the follicle (the "Graafian follicle") containing the ovum.

Stage 2: Estrus (standing heat, sexual desire)

As the follicle grows, more estrogen is produced. As increasing amounts of estrogen are released into the bloodstream and travel to the anterior pituitary, it acts in a positive feedback fashion, stimulating pulsatile LH release. It also affects the nervous system of the cow, to modify the cow's behavior and cause her to display estrus 'heat' such as restlessness, mounting, and most importantly, causing the willingness to be mounted by other animals (standing heat). The vulva swells and becomes distended while the cervix and uterus produce copious quantities of clear stringy mucus. This time of peak estrogen secretion can last from 6 to 24 hours, with ovulation occurring 24 to 32 hours after the beginning of estrus.

Stage 3: Metestrus (after estrus)

The female will not permit mounting in days 1 – 5 of the cycle. About 50% of cows and 90% of the heifer show metestral bleeding. Metestral bleeding does not indicate that the cow has conceived or not conceived. During this time LH is released from the anterior pituitary after estrus. Peak LH production is reached 10 to 12 hours after the end of standing heat. LH causes ovulation (release of the ovum from the follicle) and growth of the CL at the ovulation site. Thus, ovulation occurs during this time and the CL begins to develop.

Stage 4: Di-estrus

Days 5 – 19 of the cycle are characterized by absence of sexual desire. If the cow conceives, she passes from di-estrus into a state of anestrus (absence of cycling).

Anestrus (due to pregnancy)

The presence of a fetus at day-19 inhibits the release of prostaglandins from the uterine wall. The CL develops and after 3 to 5 days produces progesterone, the pregnancy hormone, which acts on the brain, inhibiting LH to release and sexual activity. It also prepares the reproductive tract for pregnancy and in prolonged concentrations causes udder development. Pregnancy is maintained by progesterone. At approximately six months into the pregnancy the CL starts to decline in activity as the cotyledons become the major source of progesterone. If by day 19 of the cycle the cow is not pregnant, her uterus releases hormone called prostaglandins that degenerate the CL and a new cycle begins.

2.2.3 Estrus Detection Procedures

Estrus or heat is the period in the reproductive cycle when the cow will permit mating. Accurate detection of estrus is the most important part of an artificial breeding program. Irrespective of the competence of the inseminator, the facilities available and the condition of the cows, the artificial breeding program will fail if cows are inseminated at the wrong time. Personal observation is the most accurate and reliable method of heat detection available (Dunn et al., 2009).

(1) Signs of estrus

It is essential to understand the primary and secondary signs of heat in order to achieve accurate and efficient heat detection.

1) Standing heat

It is the most sexually intensive period of the estrous cycle. A cow standing to be mounted is the most accurate sign of estrus. During this period, cows stand to be mounted by other cows or move forward slightly with the weight of the mounting cow. Standing to be mounted by a bull or another cow/heifer is the only conclusive sign that an animal is in standing estrus and ready to be inseminated (Fig. 21). The period of standing estrus usually lasts about 15 hours but can range from less than 6 hours to close to 24 hours.

Cows that move away quickly when mounting is attempted are not in true estrus. In order for standing behavior to be expressed, cattle obviously must be allowed to interact

(Connor, 1993; Perry, 2004). Table 5.2 indicates the signs to look for before, during, and after standing estrus.

Table 5.2 Signs to look for before, during, and after standing estrus and mounting activities

Before standing estrus (6 – 10 hours before)	Standing estrus (6 – 24 hours)	After standing estrus	
		Up to 10 hours later	1 – 3 days later
Will not stand to be mounted Vocal Nervous & restless Smells other cows Attempts to ride other cows Vulva moist, red, and slightly swollen	Stands to be mounted Nervous and restless Congregates Rides other cows in standing estrus Vulva moist, red, and slightly swollen Clear mucous discharge	Clear mucous discharge	Bloody mucous discharge

Source: Perry, 2004.

2) Secondary signs of heat

Because it is difficult to monitor animals 24 hours a day, you can check for secondary signs of standing estrus. Secondary signs can indicate that a cow is approaching standing estrus, and that she should be monitored closely over the next 48 hours. Or secondary signs may indicate she has recently been in standing estrus and should be monitored closely again in 17 – 25 days. Secondary signs can be very helpful, but used alone, they do not allow for a confident determination of standing estrus. These secondary signs are:

- Congregating: Cattle that are in standing estrus naturally seek out other animals in estrus and form a small group, referred to as the sexually active group. They make physical contact with each other, standing head to tail, circling, butting heads, and resting their chins on the back or hip of other cows/heifers. Whenever a small group of animals gathers together, it should be watched closely for animals in standing estrus.
- Mounting other animals: Animals in standing estrus or approaching this stage usually try to mount other animals. The cow that is doing the riding may or may not be in standing estrus. A cow/heifer that mounts another animal a few times and then leaves the sexually active group is likely not in standing estrus, but an animal that consistently mounts other animals may be and thus requires further observation.
- Clear mucus discharge from vagina: Estrogen causes thick, clear mucus to be released from within the cow's cervix. When a cow or heifer mounts another animal, stands to be ridden, or when the reproductive tract is stimulated during artificial insemination (AI), mucus may be expelled from the vulva. Strings of mucus hanging from the vulva or smeared on the tail and buttocks are a good sign that the cow/heifer is in or approaching standing estrus.

➢ Nervousness or restlessness: This may be excessive walking and bawling. Watch for any animal that is moving when other animals are relatively stationary. She might be walking a fence line in search of a bull.

➢ Roughed up tail head: Normally, the hair on the tail head lies down and points toward the tail, but the hair on the tail head of an animal that has been ridden may be roughed up to the point where it sticks almost straight up. A cow/heifer that has been ridden hard may sometimes have the hair rubbed off of her tail head. In muddy areas, mud will often be plastered on both flanks and sometimes up along the back and ribs. However, be careful when using roughed up hair or mud as an indicator of standing estrus. Both-sides of the animal must show signs of being ridden, since an animal cannot be ridden and marked on only one side.

➢ Swollen vulva: A moist, red, and swollen vulva is often associated with standing estrus. However, this can be difficult to determine and may be of limited value.

➢ Bloody mucus discharge from vagina: 2 – 3 days after standing estrus a bloody discharge from the vulva may be observed. This is normal and only means that the animal was in standing estrus earlier. It is too late to inseminate the animal at this time.

(2) Estrus detection procedures/techniques

The estrus detection methods include: visual estrus observation and mechanical methods of detection (detection aids) such as record systems, heat mount detector, tail paint and teasers. It is a tremendous task to detect standing estrus in a cow herd, and nothing can substitute for visually observing the cattle.

1) Detecting estrus by observation

To maximize detection of standing estrus, it is extremely important to monitor cows/heifers as closely as possible, these are: early in the morning and late at night as well as during the middle of the day. Checking for standing estrus at 6 a. m., noon, and 6 p. m. increased the estrous detection rate by 10% compared to detecting estrus at 6 a. m. and 6 p. m. alone. When cows were checked for standing estrus every 6 hours (6 a. m., noon, 6 p. m., and midnight), the estrous detection rate increased by 19% compared to checking at 6 a. m. and 6 p. m. alone (Perry, 2004).

2) Records systems

Good record keeping is an essential part of good herd management and is a very economical method in aiding estrous detection.

By accurately recording the animal number, dates in standing estrus, and dates inseminated (naturally or artificially), you can anticipate the next standing estrus for each individual (Connor, 1993).

Breeding wheel or herdex record system is wall-mounted reproductive record system that uses color-coded pins or markings to indicate reproductive events for each cow as shown on Fig. 22. By either turning a transparent plastic dial or sliding the plastic cover on a daily basis, future heats and reproductive events can be anticipated.

3) Tail paint

The use of oil-or water-based paints applied to the back of a cow's spine at the point most often rubbed by the brisket of the mounting companion cow (Fig. 5.5), was first promoted as an effective aid to estrus detection. The advantage of tail paint is that it is inexpensive. The tail-painting technique was shown to identify accurately 99% of cyclic cows.

Fig. 5.5 Tail paint on cows
Source: Dunn et al., 2009.

Tail paint is applied approximately 20 cm long and 3 – 5 cm wide over the butt of the tail. This can best be done by painting both against and with the 'grain' of the hair on the tail. A thick paint is needed. This paint should last for just over a week under most conditions, unless disturbed by the mounting action of another animal. If the cow is mounted, the paint will turn to powder and disappear, leaving only a ring of colored hair (Moran, 2012; Kerr and McCaughey, 1984).

4) Teasers

Teasers are animals (usually male) used to detect heat in the females of the herd. They should be sterilized or incapable of service and are mostly used in conjunction with other methods. Teasers should be used only where absolutely necessary. Breeders of *Bos indicus* cattle generally use teasers more than breeders of *Bos taurus* cattle, as *Bos indicus* cows do not display obvious signs of heat as readily. Teasers are unproductive animals utilizing feed, water and space that could be used by a breeding animal. To maintain peak efficiency, teasers must not be overused; they should be rotated frequently. There are various types of teasers:

- Vasectomized bulls: By removal of part of the vas deferens (vasectomy) a bull is rendered sterile.
- Cryptorchid bulls: A cryptorchid (or rig) is a bull in which one or both testes are retained in the abdomen. The retained testis cannot produce viable sperm but produces the male hormone testosterone which maintains *libido*. Any descended testicle must

be removed (castrated) to ensure sterility.

- Testosterone injected steers: *Libido* is induced in castrated males by testosterone injections. Repeated injections are needed to maintain hormone levels. This can be expensive.
- Nymphomaniac cows: Cows with cystic ovaries will actively seek out all cows on heat. These cows are generally a nuisance in most herds, creating management issues.

5) Heat mount detector

A heat mount detector (HMD) is a device which can be glued on the midline of the cow's back between the hip bones (Fig. 5. 6).

Fig. 5. 6 Panel reveal style of heat mount detector
Source: Dunn et al., 2009.

There are two main types of devices that are commonly used. The first type releases a colored dye and the second device reveals on colored panel similar to that of a 'scratch it' lottery ticket. Both systems rely on the sustained pressure and rubbing from the brisket of an animal which is mounting the standing cow to activate the device. The positioning of the HMD pads will depend on the size of the cow to which it is fitted and the size of her herd mates who will mount her to trigger it. Bigger cows need the device placed near to the tail head (Removal of the pad must be treated as a positive sign, because they are often lost with sufficient mounts).

The advantage of HMDs is that they are visible from a distance and do not require interpretation for results as do tail paint or chin-ball harness marks. Their disadvantages are that: Pads may dislodge, especially in areas where cows rub on trees due to irritation caused by fly or ticks. The adhesive may cause irritation and dermatitis and there may be false positives. Application of the pads requires handling and labor and extremes of weather may cause faulty operation.

2.3 Estrus Induction and Synchronization

Estrus synchronization involves the treatment of cows so that all or most will display estrus and ovulate within a very short time of each other.

Advantages

- Synchronization of estrus contributes to optimizing the use of time, labor, and financial resources by shortening the calving season, in addition to increasing the uniformity of the calf crop.
- Successful methods of estrous cycle control would facilitate the use of artificial insemination (AI) thereby allowing the greater exploitation of genetically superior sires.
- The AI could be carried out at a prearranged or fixed time if the control method was sufficiently reliable. This could potentially remove the need for the detection of estrus.
- It serves to improve the rate of estrus detection by concentrating the occurrence of induced heat into a shorter finite period of time.
- Synchronization may shorten the calving to conception interval, this is essential to the maintenance of a short (12 - 13 months) calving interval and is a primary factor affecting the profitability of any cattle breeding enterprise.
- This procedure is also effective for the induction of ovulation in acyclic cows (Jarnette, 2004).

Common products available for use in cattle estrus synchronization systems are given in Table 5.3. There are three main approaches for estrus induction and ovulation synchronization:

- The artificial induction of premature luteolysis using luteolytic agents such as prostaglandin $F_{2\alpha}$ ($PGF_{2\alpha}$). This will obviously only be effective in cycling cows with an active corpus luteum.
- Prostaglandin-induced luteolysis in association with gonadotropin releasing hormone to manipulate follicular and luteal function. This procedure could potentially be used for the induction of ovulation in acyclic as well as cyclic cows.
- The simulation of corpus luteum functions by administration of progesterone (or one of its synthetic derivatives) for a number of days, followed by abrupt withdrawal. The progesterone suppresses gonadotropin release, and hence follicular maturation, until it is withdrawn.

Table 5.3 Products, commercial names, and doses for synchronization procedures

Product	Commercial name	Administration	Dose
Prostaglandins	Lutalyse	i. m. injection	5 mL
	Estrumate	i. m. injection	2 mL
	In-Synch	i. m. injection	
	Prostamate	i. m. injection	
Progestins (Progesterone)	Melengestrol Acetate	Feed	0.5 mg/ (head · d)
	CIDR	Vaginal implant	1 implant (1.38 g)
Gonadotropin Releasing Hormone	Cystorelin	i. m. injection	2 mL
	Factrel	i. m. injection	2 mL
	Fertagyl	i. m. injection	2 mL
	Receptal	i. m. injection	2 mL

Source: Lamb, 2004.

2.3.1 Requirements of Estrus Synchronization

The single greatest reason for the failure of a synchronization program, estrus detection and breeding activities is due to poor management. There are several management requirements to consider before deciding whether synchronization and breeding will work in your operation (Lamb, 2004).

- Synchronization time/period: To achieve optimal pregnancy rates with any synchronization protocol, treatments should be initiated only when cows are at least 50 days postpartum and heifers should be at least 60% of mature weight.
- Nutrition: It is the single most important factor that could dictate the success or failure of that reproductive program. The body condition that cows calve at determines the rate at which those cows initiate their estrous cycles after calving. Therefore, it is essential to ensure that cows should have a body condition score of 3.
- Record Keeping: Maintaining a sound recording keeping system is a key to success in any reproductive management system. For synchronization to work, producers need to know when their cows calved, whether the cow had a difficult birth, and what the birth weights of all calves were. Without accurate records, these decisions can be extremely subjective.
- Facilities: Working facilities need to be able to accommodate the extra work. Not only should you consider reliable holding and sorting pens, but also a good solid alley and chute system. Anticipating an increase in facility use will certainly ensure a successful synchronization program.
- Labor: Reliable labor is an issue that many people neglect to consider when planning a synchronization program. Detecting when cows are in heat is important for the success of a synchronization program. Any labor associated with this process needs to know exactly how cows act when they are in heat.

2.3.2 Prostaglandin Synchronization Procedures

Prostaglandin ($PGF_{2\alpha}$) is a naturally occurring hormone. During the normal estrus cycle of a non-pregnant animal, $PGF_{2\alpha}$ is released from the uterus 16 - 18 days after the animal was in heat. This release of $PGF_{2\alpha}$ functions to destroy the CL. The release of $PGF_{2\alpha}$ from the uterus is the triggering mechanism that results in the animal returning to estrus every 21 days.

The most potent luteolytic agents commercially available are derivatives of $PGF_{2\alpha}$ (Lutalyse, Estrumate, ProstaMate) gives the herd owner the ability to simultaneously remove the CL from all cycling animals at a predetermined time that is convenient for heat detection and breeding. Prostaglandins act on the corpus-luteum, thus they can only be effective in cycling cattle (Jarnette, 2004). There are two treatment procedures using Prostaglandin ($PGF_{2\alpha}$): single dose $PGF_{2\alpha}$, injection, and double dose $PGF_{2\alpha}$ injection (Stevenson et al., 1999).

1) Single prostaglandin synchronization procedure

Single administration is used in cows with functional corpus luteum present on the ovary. Functional corpus luteum can be detected per rectum, by progesterone test in milk/blood, using heat records. If the animal is between days 6 and 16 of her cycle, she will generally come into heat 36 - 72 hours after injection of the drug and then insemination at detected heat (Fig. 5.7).

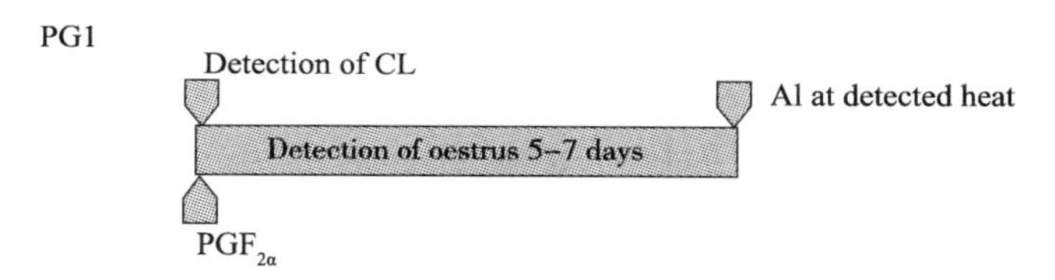

Fig. 5.7 Single prostaglandin synchronization procedure

Procedures:

- Day 0=detect CL, inject cows with $PGF_{2\alpha}$ (Estrumate 2 mL/cow through I. M.).
- Day 5 - 7=heat detection and provide AI for cows showing standing heat 8 - 12 hours later.

2) Double prostaglandin synchronization procedure

Two injections of $PGF_{2\alpha}$ with at 14 - day interval result in luteolysis in high percentage of animals with active ovaries (cycling). Cows showing heat after the first injection can be serviced or the whole group can be managed simultaneously after the second injection. Although timed AI is possible, the best results are obtained with AI at detected heat.

Procedure:

- Day 0=detect CL, inject cows with $PGF_{2\alpha}$ (Estrumate 2 mL/cow through I. M.).
- Day 5 - 7=heat detection and provide AI for cows showing standing heat 8 - 12 hours

later.

➢ Day 14 = Animals not detected in estrus after the first injection are re-injected 14 days later and bred over the next 5 - 7 - day period (Fig. 5. 8).

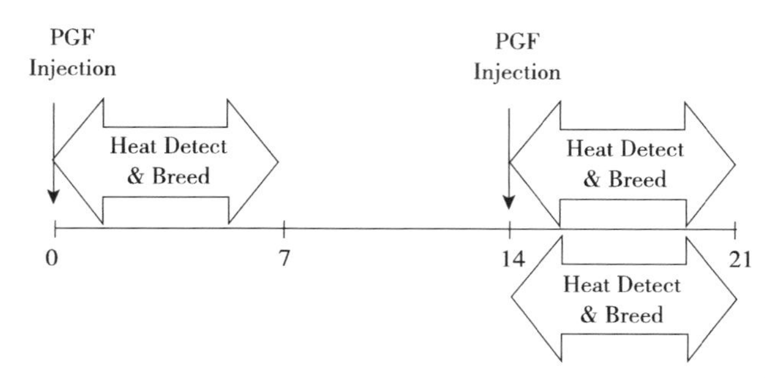

Fig. 5. 8 Double PFG synchronization procedure

Limitation of PGF

The major limitation of PGF is that it is not effective on animals that do not possess a CL. This includes animals within 6 - 7 days of a previous heat, pre-pubertal heifers and postpartum anestrous cows. Despite these limitations, prostaglandins are the simplest method to synchronize estrus in cattle.

Animals that are in anestrus use of PGF in combination with GnRH and/or a progestin source are much more effective options. Estrus and ovulation is highly variable due to differences between cows in the stage of follicular development at the time of PGF injection.

2. 3. 3 GnRH-PGF Synchronization Procedures

New synchronization protocols currently recommended for cows use GnRH in conjunction with PGF. A naturally occurring hormone, GnRH is more popularly known by the commercial brand names of Cystorelin, Factrel, and Fertagyl.

The use of GnRH in conjunction with prostaglandin is called Ovsynch regimen. GnRH is injected on day 0, followed by a prostaglandin on day 7 and a further GnRH injection on day 9 - 10 (Ball and Peters, 2004).

Benefits of Ovsynch: Synchronization is used in cows at all stages of estrus. Ovsynch ensures a homogenous ovarian follicular status at induction of luteolysis. As a result, the precision of estrus after prostaglandin-induced luteolysis and the synchrony of the LH surge are improved. Both follicular development and regression of the CL can thus be synchronized.

The GnRH-$PFG_{2\alpha}$ estrus induction and synchronization procedures include: Select-Synch and Ovsynch. The general procedure is basically the same, the way animals are subsequently handled for heat detection and breeding is the difference between the two methods.

1) Select synch procedure

- At day 0 cows are injected with GnRH then 7 days later they are injected with PGF.
- Heat detection begins 24 – 48 hours before the PGF injection and continues for the next 5 – 7 days.
- The PGF injection is excluded for cows detected in estrus on day 6 or 7.
- Animals are inseminated 8 – 12 hours after observed in standing estrus.
- Alternatively, heat detect and AI until 48 – 60 hours after PGF and then mass-AI the rest of the herd at 72 hours and give GnRH to those cows that have not exhibited estrus (Fig. 5. 9).

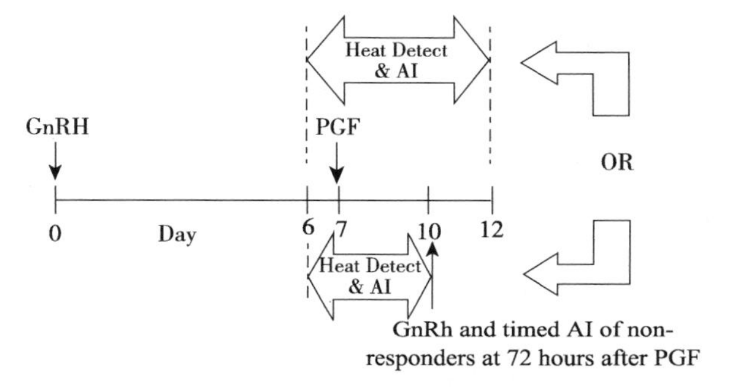

Fig. 5. 9 Select synch procedure

2) Ovsynch procedure

Ovsynch is a fixed-time AI synchronization protocol that has been developed, tested and used extensively in dairy cattle. The protocol builds on the basic GnRH-PGF format by adding second GnRH injection 48 hours after the PGF injection (Fig. 5. 10). This second GnRH injection induces ovulation of the dominant follicle recruited after the first GnRH injection. All cows are mass inseminated without estrous detection at 8 to 18 hours after the second GnRH injection.

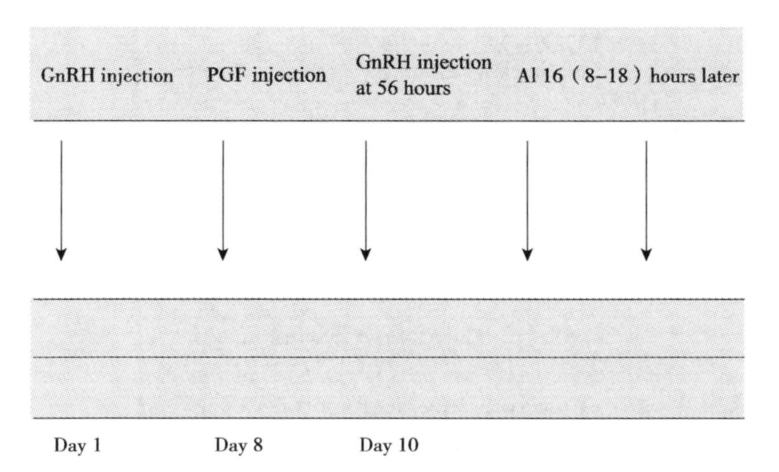

Fig. 5. 10 Ovsynch synchronization

Two injections of GnRH, 7 days before and 2 days after the prostaglandin ($PGF_{2\alpha}$), will effectively synchronize ovulation in more than 90% of lactating cows treated. Giving the second GnRH at 56 hours is very beneficial. Time of ovulation occurs 24 – 32 hours after the second injection of GnRH (Graves, 2009).

2.3.4 Progestin Synchronization Procedures

If a group of cows is treated with progesterone and then it is withdrawn from all cows simultaneously, this will theoretically synchronize ovulation in the group.

Implants are the most suitable method of administration of progesterone since withdrawal can then be precisely controlled by implant removal. The most common methods are MGA-PGF system and CIDR (Lamb, 2004).

1) Melengestrol acetate (MGA) -$PGF_{2\alpha}$ procedure

The MGA-PGF system (Fig. 5.11) is a proven method for synchronizing estrus in beef and dairy heifers. Melengestrol Acetate (MGA) is a synthetic form of the naturally occurring hormone, progesterone. MGA/PGF is best for synchronization of heifers. The advantages of feeding MGA are cost, kick-starts cows and heifers, whereas, the disadvantages are feeding the MGA, length of the system, not great for timed-AI.

Procedure:

- Mix MGA with 1.5 – 2.5 kg of concentrate supplement and feed at a rate of 0.5 mg/(head • day) for 14 days.
- Within 3 – 5 days after MGA feeding, most heifers will display standing heat.
- Do not breed at this heat as conception rates are reduced.
- Wait 17 – 19 days after the last day of MGA feeding, and inject all heifers with a single dose of PGF.
- For the next 5 – 7 days detect heat and inseminate animals 8 – 12 hours after detected estrus.

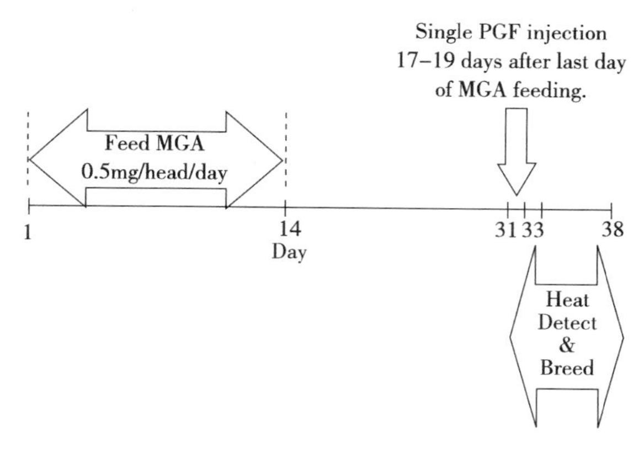

Fig. 5.11 The MGA-PGF synchronization

2) Intra-vaginal progesterone-releasing device procedure

A product called CIDR (controlled internal drug release) is available as an intra-vaginal progesterone-releasing device. CIDRs are T-shaped inserts and are placed into the vagina with

an applicator that collapses the wings for insertion. CIDRs have a nylon case with a silicone rubber cover and are designed to deliver natural progesterone (1.55 mg progesterone is impregnated) slowly over a 7 - day period to prevent heat expression.

The advantages of using a CIDR are: works for timed-AI, easy to apply, kick-starts cows, whereas, the disadvantage includes: cost, need facilities, vaginal infections, not great for heifers.

An injection of prostaglandin can then be used to bring animals in heat before removing the inserts, as shown below. CIDRs are easy to apply and remove and have excellent retention rates. The most successful recommended procedure is as follows (Fig. 5.12):

Procedure:

- Day 1 = administer CIDR to the cows by insertion into the vagina, with careful attention to hygiene.
- Day 7 = administer by injecting a dose of prostaglandins.
- Day 8 = remove CIDR inserted.

Start heat detection 36 hours after removal. Detect heat for three days. Inseminate normally, following the a. m. to p. m. rule (Dunn et al., 2009).

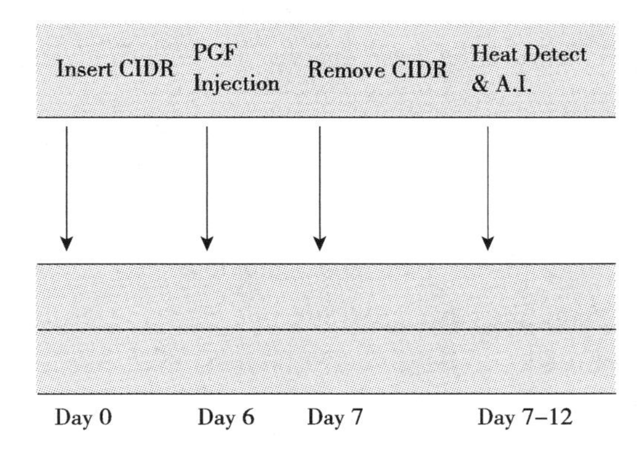

Fig. 5.12 Intra-vaginal progesterone-releasing device procedures

3 METHODS OF BREEDING IN DAIRY CATTLE

Once the cow/heifer has been identified as being in estrus/heat breeding can be achieved through natural service or artificial insemination, and irrespective of the method, the aim should be to achieve increased chances of conception. Generally, there are two methods of dairy cattle breeding/mating these are: natural mating/breeding and artificial insemination/breeding.

3.1 Natural Breeding

Natural mating/breeding is where the cow is taken to a bull and left for some time for the bull to serve. The use of bulls for natural service remains widespread even in areas where artificial insemination has proven to be very efficient. Many farmers believe that pregnancy rates are higher when a bull is used. The use of natural service may be indicated when personnel are inefficient to perform the tasks associated with heat detection and the techniques of AI, when the long term genetic gain is of minor importance and when local conditions do not provide the infrastructure necessary for successful AI (Moran, 2012).

Advantages of natural mating/breeding method

- The cow has an opportunity to be served more than once; this increase the chance of conception.
- The semen is fresh and of good quality since there is no handling.
- Where the farmer does not own a bull, cost of service is lower compared to AI.

Disadvantages of natural mating/breeding

- Rearing a bull is not economical, especially to a small holder farmer
- There is a risk of spreading breeding diseases.
- There is risk of inbreeding if the bull is not changed frequently
- There is no opportunity to select the type of bull the farmer wants.
- Some danger on the cow or bull, while mounting.

3. 1. 1 Managing a Breeding Bull

In undertaking natural mating, the bull must be morphologically and functionally sound, see section 3. 4 of Module 1 for a detailed description on bull selection for breeding.

- Appropriate attention should be given to factors affecting fertility of bulls that include unbalanced nutrition (under feeding and over feeding) and diseases. Adequate nutrition is vital, since it hastens puberty and body development.
- It is also important to keep in mind that overfeeding can lead to reduced *libido*.
- Bulls should also be tested for venereal diseases like Brucellosis, Trichomoniasis and Vibriosis that affect fertility and culled if they react positively.
- In addition, prevention measures like vaccination and deworming should be undertaken against other diseases that influence fertility indirectly.
- It is important to undertake a regular fertility test for bulls to verify their external physical soundness, reproductive health (congenital and inflammatory problems), scrotal circumference and semen quality.
- To avoid undesired mating, farmers should castrate all other bulls they have and prevent outside bulls from coming into the herd.
- Another factor to be considered is the temperament of breeding bulls which is genetically determined and is also affected by management. Viciousness makes bulls

unmanageable and it is often the result of confinement and ill treatment. Proper daily handling, firm training and exercise make most bulls fairly tractable and easily controlled.

3. 1. 2 Natural Mating Methods and Procedures

Natural mating can be done in two ways.

Free/pasture mating: This method of mating is practiced by farmers who own bulls which run full time with the cows. One bull can serve 30 – 40 cows. It has the advantage that no heat detection required and disadvantage of lack of accurate records and the possibility of transmission of reproductive diseases e. g. brucellosis.

Hand mating: The bull is enclosed in its pen and the cows are brought in when they show signs of heat. Most small-scale farmers will practice this method since bulls are owned by a few farmers and others bring their cows for service at an agreed fee. The recommended bull-cow ratio is 1 : (50 – 60). The advantage is keeping accurate records while the disadvantage is the farmer has to detect heat (Moran, 2012).

3. 1. 3 Improving Conception Rate by Natural Mating

The following guidelines have recommended increasing the chances of conception through natural service (Kelay, 2002):

- It is recommended to use bulls between 15 months and four years of age.
- Breeding bulls should be replaced to avoid inbreeding. Use a bull for two to four years in a herd without replacement or switching with bulls from other sites. Cows in heat must be served at the right time.
- Cows seen in estrus in the morning should be served that same day while those showing estrus later in the day should be served the next morning. Take the cow to the bull heat and leave it for at least twelve hours.
- Young inexperienced heifers should be mated with older, experienced bulls.
- Young inexperienced bulls should be given to older, experienced cows.
- The bull should be kept fit and in good health, particularly the legs and feet.

3. 2 Artificial Breeding

Artificial insemination also known as artificial breeding is a technique in which semen is collected from the male, evaluated, processed, preserved, stored and inseminated to females (Karuna et al. , 2012).

Artificial insemination as breeding methods has contributed to the development of the dairy sector in the last 60 years worldwide. The process of artificial insemination starts with a healthy bull, that is, disease free and producing ample quantities of high quality semen. The fertility of the cow is also important. The role of man who stands between the cow and bull play an important role. The efficiency of the person handling the semen and depositing in the cow will influence to a great extent on the conception rate in AI program (Moran, 2012). The advantages and disadvantages of AI are described below (Sinishaw, 2004; Gebremedhin, 2005):

Advantages of AI

- AI offers the best genetic improvement possible since AI bulls are derived from very top and most heavily researched and selected herds.
- AI is delivered to eliminate the transmission of venereal diseases and genetic defects, since we can have a control over the quality and safety of semen before it is applied into the cow.
- Less number of bulls are kept for a very large number of cows (Kelay, 2002) and can be carried out in the absence of bulls, enables accurate recording of information such as breeding dates, pregnancy rates, inter-estrus intervals and days to first service that can be used to monitor fertility.
- Small scale farmers through AI can access good bulls cheaply.
- It is cost effective since the farmer does not have to rear a bull.

Disadvantages of AI

- It requires very accurate heat detection and proper timing of insemination for greater chances of conception. Heat detection depends on farmers, which may not have the skill to do so.
- Require a well-organized infrastructure, including all weather roads, telephone, electricity, etc.
- High costs related to the production (collection and processing), storage and transport of semen.
- High cost of training and transportation of technicians, inseminators inefficiency due to inadequate training of technicians or lack of experience of technicians,
- Difficulty in assessing the conception rate of cows when farmers are not cooperative in bringing cows after AI failure.

3.2.1 Semen Collection, Processing and Storage

Most artificial insemination centers use an artificial vagina to collect semen. With this device, temperature, pressure and friction are used to stimulate ejaculation when the donor bull mounts a teaser animal. Cows may be used as teasers but have disadvantages. Other bulls or steers are satisfactory. In some instances, mechanical decoys, called dummy bull are used.

When bulls are unable to mount because of injury or some other factor, electro-ejaculation can be used that it employs passage of a fluctuating current between electrodes on a probe placed in the bull's rectum. This current stimulates accessory sex glands and the muscles of ejaculation. Electro-ejaculation is usually regarded as a last resort for semen-collection. It is time consuming, requires expensive equipment and risks injury to the donor animal. Semen quality is not as good as that collected by the artificial vagina (AV). Demand for a bull's semen determines the frequency of collection. Two ejaculates twice a week (e. g. 2 on Monday and 2 on Thursday), will allow an average bull to produce 30,000 doses or more in a year (Dunn et al., 2009).

(1) Semen collection procedure using artificial vagina

Semen donor bulls, on top of qualifying as a semen donor on different assessment and evaluation criteria (genetic superiority, health aspect, semen quality, etc.) should be trained and well prepared for semen collection. On the other hand, the semen collecting personnel should also be well trained on the procedures and techniques of semen collection and handling. The collection of semen is performed in a specially prepared place called the *manege*. The room must meet appropriate sanitary and health conditions because it is considered a laboratory room. The floor has to be flexible and anti-slip. The semen donor bulls properly washed before entering the semen collection area. In the first stage of the preparation the bull makes a few rounds in the manege and is allowed to sniff around and to jump other bulls and to perform empty jumps or make false jumps on a dummy cow or a dummy bull. A well aroused male with distinctive sexual impulses is brought to the dummy cow or the teaser in order to produce semen (Fig. 23).

The semen is collected in an artificial vagina (Fig. 24) which is composed of: 1) a glass container for semen with a water jacket and a volume scale, 2) a thin, flexible latex sleeve, 3) a latex cone joining the end of the vagina with the collection tube, 4) a rubber cylindrical casing with a valve for pouring water and blowing air, 5) a bag-a thermal protector and a mechanical container.

When the artificial vagina is installed attention has to be paid that:

- Every ejaculate is collected in a separate vagina.
- The latex sleeve has to be installed in such a way that there are no folds inside the casing.
- The sleeve should be moderately stretched in order to pour water and blow air.
- The water temperature in the vagina should be 40 - 42℃.
- The temperature of the container should vary between 35 - 37℃.
- The inlet to the vagina should be lubricated such as vaseline.

When collecting the semen, attention must be paid that the bull does not touch the rump of the teaser with its penis. The person collecting the semen should take the bull's penis by the prepuce and direct it to the inlet of the vagina. The artificial vagina should never be put over the penis. Characteristically for its species, the bull performs a single copulation push. Next, the collected semen is evaluated (Barszcz and Wiesetek, 2012).

(2) Semen processing

About ten million normal, actively moving sperm are required to produce conception using AI. They must be placed at the correct site in the reproductive tract of a cow which is at the correct stage of the breeding cycle.

When processing a dose of frozen semen, the aim is to ensure that sufficient live sperm are present for conception when the dose is thawed to body temperature. Processing includes: evaluation, dilution, cooling, packing, freezing, quality control (evaluation after freezing) and storage.

(3) Semen evaluation

After examining a drop of raw semen under low power on a microscope to obtain an estimate of sperm concentration and activity, samples are taken for detailed assessment. One drop is placed in a test tube with a special stain to determine the percentage of the spermatozoa which were alive and of normal conformation at the time of staining.

The quality of semen collected should be evaluated. Evaluation parameters and described by Barszcz and Wiesetek (2012) as follows:

- Volume of the ejaculate 2 - 8 ml or over
- Color: White with a characteristic cream hue
- Texture: Milky, milk with cream, creamy
- Smell similar to cow's milk
- Spermatozoa concentration: 0.6 - 1.5 million per milliliter.
- pH: 6.2 - 6.8
- Motility: More than 60 - 80% of sperm cells should have forward motility
- Morphology: More than 80% of spermatozoa with normal morphology

(4) Semen dilution/extension

An average ejaculate may contain 5,000 million sperm in 5 mL of raw semen. Super-fertile mature bulls may yield up to 12,000 million sperm per ejaculate. With natural service this could only produce one calf. By diluting the ejaculate, the raw semen can be extended to give 200 or more individual 0.25 mL doses each containing 25 million live sperm. While only 10 million sperm are required for conception, more than double this number are placed in the straw to allow for losses during freezing. Glycerol is the chemical added to protect the sperm during the freezing process. Diluents may be based on skim milk, egg yolk citrate or specially prepared chemical diluents such as TRIS with egg yolk. Antibiotics are added as a precautionary measure and most diluents have added fructose (fruit sugar) to supply energy to the sperm (Dunn et al., 2009).

(5) Packing

The common semen packing material is the Cassou or French straw system. There are two types of straw: the medium straw (0.5 mL volume) and the mini-straw (0.25 mL volume). They are plugged at one end (the double plug end) with a sealing powder which is retained between two cotton plugs. The automatic filling and sealing machine and printing machines are used to fill and seal the semen at a temperature of 4℃. All straws are marked with the bull's name (i.e. ID. No, breed and blood level of the bull), the center of origin and the batch number before filling and sealing the semen. Batch numbers specify the year and day of collection for each straw of semen (e.g. 04/10/15). Records of each day's collections and the number of straws processed are kept by the center of origin.

(6) Cooling and freezing

At body temperature sperm swim very rapidly, exhaust their energy reserves in a relatively short time, and die. The rate of a chemical reaction depends on the temperature at which the process is taking-place. If the temperature is reduced, the rate of reaction is also reduced. Cooling semen to near freezing point slows the sperm down, by slowing their internal reactions, and extends their life for several days if they are protected by suitable chemicals. By further cooling, activity can be effectively stopped to give the sperm an almost indefinite life. Some of the sperm are killed in the freezing process, but with correct processing, many survive and will revive on thawing. Liquid nitrogen boils at a temperature of minus 196℃, so it is the most suitable refrigerant available. Straws of semen are frozen by suspending them in the vapor above the surface of the liquid nitrogen, to give a controlled rate of cooling and freezing. They are then stored in the liquid nitrogen until use.

(7) Semen storage

For ease of handling and to minimize the risk of damage to the sperm through exposure, straws are always packed in plastic goblets. Semen should only be transferred to and from liquid nitrogen containers in goblets, because there is no safe exposure time for individual straws, i. e. transfer should be very abruptly to avoid exposure of the straw and death of the sperm cell. All frozen semen is transported and stored under liquid nitrogen, which is by far the most satisfactory refrigerant in terms of operator safety, ease of handling and availability. It also has a wide margin of safety in the temperature range for storage. The liquid has a boiling point of minus 196℃. Both the liquid and gaseous forms are relatively inert chemically and will not burn, support combustion or react in any way with other materials. These properties make it a very suitable refrigerant.

Safety precautions

Liquid nitrogen can be dangerous if not handled correctly because of its extremely low temperature. The following points must be given strict attention.

- Avoid contact with the liquid. Prolonged contact with the skin or contact with wet skin may result in severe burns. Contact with eyes (which are continually moistened by tears) may severely affect eyesight.
- Use metal forceps when removing straws from liquid nitrogen containers. Skin, if in contact with the cold metal of the buckets, will often stick and tear when removed.
- Insert objects into the liquid very slowly. This avoids the splashing, which occurs when the liquid boils on insertion of 'warm' objects.
- When refilling containers pour the liquid slowly. This avoids 'blowout'. If the nitrogen boils vigorously it may shoot out the neck of the container. Don't use plastic funnels as these will crack and disintegrate.
- First Aid: Flood the affected area with large quantities of unheated water and later apply cold compresses. If the skin is blistered, or if the eyes or other delicate tissues are affected, seek medical attention immediately.

3.2.2 Preparing for Artificial Insemination

The preparatory processes or procedures for AI include: preparation of AI equipment and tools, loading the insemination gun, and thawing the semen. Prepare insemination equipment as near as possible to the inseminating area to avoid undue delays. Always prepare insemination equipment in a clear, dry and shaded area. Direct sunlight will injure sperm.

(1) Equipment and tools required for AI

Equipment and tools required for AI include liquid nitrogen tank, dipstick for measuring nitrogen level, paper towels, clean and sharp scissors, insemination gun, arm size long gloves, sheath, glove lubricant, semen straw, straw tweezers (forceps), warm water bath/flask, thermometer, rubbing alcohol, clock, record books, and personal protective equipment (PPE).

(2) Semen thawing

The process of thawing is raising the temperature from minus 196 - 38℃ in warm water. To bypass crystallization zone, thawing must be rapid and uniform. The most convenient method is to plunge the straw in warm water. The temperature should be 38 - 40℃ for 60 seconds for optimal survival of spermatozoa. During thawing entire straw must be completely submerged in water bath. Semen thawing and loading insemination gun procedure:

1) Place and prepare the kit box. Check the temperature of the thawing bath. The temperature should be 38 - 40℃.

2) Identify the canister from which the desired semen is to be taken. Ascertain the color of the straw by reading identification tag.

3) Remove the lid from the container and lift the proper canister up to the level of the frost line. Never lift the canister above the neck level.

4) With a help of forceps, grasp an individual straw and remove it, at the same time lower the canister immediately back into the container. If you are unable to take the straw within 10 seconds, lower the canister back to nitrogen, wait for some time and make the next attempt. Return the buckets to their correct positions and replace the stopper.

5) With the wrist movement give one or two jerks to the straw to expel liquid nitrogen trapped at the end of factory seal.

6) Dip the straw into a clean water bath at 37 - 40℃ for 60 seconds for thawing. Remove the straw from bath and dry the straw with a clean tissue paper or cotton. Inspect the straw carefully and discard straw with cracks or defective seals. Semen must never come in contact with water.

7) Remove the gun from its protective case and pull out the plunger to approximately the length of the straw. Place the straw into the insemination gun, manufacturer's end first (laboratory end up). A stop in the barrel prevents the straw going further than the correct distance.

8) Hold the gun and the straw vertically and tap the laboratory end gently with the

scissors. This will make the air bubble go as far up towards the laboratory plug as possible.

9) Cut the laboratory seal of the straw at a right angle through the middle of the airspace using scissor. Make sure that the clipped end of straw has a straight clean cut with no jagged edges. Straws cut at other angles will result in back flow and wastage of semen at the time of insemination.

10) At least 1 cm of the straw should protrude from the end of the insemination gun. This, together with an accurate cut, is necessary to ensure a perfect seal between the straw and the sheath which is placed over the 'barrel' . Wipe scissor blades after cutting the straw to avoid contaminating the next straw to be cut.

11) Remove the sheath from its protective container and place it over the barrel. (Handle only the split end to keep the sheath clean).

12) Holding the lock ring above the tapered section, push the sheath on over the tapered section and through the lock ring until the end of the straw and the inside edge of the sheath are flush. Twist the lock ring and push it onto the tapered section to lock the sheath into position.

13) Push the plunger in until the semen is just visible at the end of the sheath. This shortens the span of the fingers and thumb required when depressing the plunger during insemination. It also removes the chance of carrying contamination from the vagina into the uterus by eliminating the hollow space at the end of the sheath. Do not load more than two guns at any one time.

14) Insemination of the cow, as described below can then begin.

3. 2. 3 Artificial Insemination Procedure

The recto-vaginal technique is the most commonly used method to artificially inseminate cattle. Patience, practice and proper hygiene are the keys to good insemination technique. Follow the procedures to inseminate cows efficiently (Delanette and Nebel, 2015):

1) Restraining the cow: The first step in the insemination process is to restrain the animal to be inseminated for the safety of both the animal and the inseminator.

2) Use of hands and inserting into the rectum: It is recommended that you use your left hand in the rectum to manipulate the reproductive tract (Fig. 25A) and the right hand to manipulate the insemination gun. A gentle pat on the rump or a soft-spoken word as you approach for insemination will help to avoid startling or surprising the animal.

- Raise the tail with your right hand and gently massage the rectum with the lubricated glove on your left hand. Place the tail on the back side of your left forearm so it will not interfere with the insemination process.
- To relax rectal constriction rings, insert two fingers through the center of the ring and massage back and forth and insert your hand with arm size plastic glove (if no glove wash hands with detergents) in the rectum, up to the wrist.
- Keep your open hand flat against the floor of the rectum, allowing manure to pass

over the top of your hand and arm.

3) Clean the vulva: Gently wipe the vulva with a paper towel to remove excess manure and debris. Be careful not to apply excessive pressure, which may smear or push manure into the vulva and vagina. Provide a clean entry for the gun through the vulva-open the lips by pressing your arm down in the rectum or with the aid of paper towel.

4) Grasping the cervix

➢ Grasp the cervix and push it forward to straighten vaginal folds (Fig. 25B).

➢ Grasp the external opening to the cervix with the thumb on top and the forefingers underneath to close the fornix and guide the gun tip into the cervix (Fig. 25C).

5) Inserting the insemination gun

➢ Direct the gun upwards at 45° to avoid the opening to the bladder.

➢ Twist and bend the cervix until you feel the second ring slide over the gun tip. Use your index finger to check gun placement (6 mm past the end of the cervix) before depositing semen.

➢ Do not push your hand towards the cervix ahead of the gun.

➢ Work the gun through the cervix. Place the index finger at the front of the cervix to feel the gun passing through, preventing the gun progressing too deep into the uterus (Fig. 25D).

6) Injecting/releasing the semen: Push the plunger slowly so that drops of semen fall directly into the uterine body (Fig. 25E). Make sure you push in with the plunger and do not pull back on the gun. Pulling back may result in much of the semen dose being deposited in the cervix and vagina. Contractions will transport spermatozoa forward to the horns and oviducts (Fig. 25F).

7) Remove the insemination gun

➢ Remove the gun from the tract and massage the cervix and uterus for a few seconds to stimulate the release of oxytocin. Rough handling upsets the cow and causes the release of adrenaline, which counteracts the effect of oxytocin.

➢ Loosen the locking ring of the gun but do not allow it to slide onto the soiled part of the sheath. Slide the sheath off the barrel. The straw will come away with the sheath. Do not discard the sheath and straw until records have been completed. Burn the sheath and straw and disposable glove to avoid possible disease transmission.

8) Release the cow from the bail or crush.

9) After insemination

➢ Observe personal hygiene measures by washing in disinfectant and cleaning boots.

➢ Maintain equipment properly.

➢ Dismantle guns for cleaning and wash them in methylated spirits.

➢ Discard the sheath and straw and dispose of them hygienically.

Precautions while inseminating a cow

- Work under very hygienic conditions, handle semen and AI equipment correctly and do not try your own methods.
- Never use force while introducing the AI gun. This would lead to injuries in the reproductive tract and affect the conception rate. If an animal jumps or falls down, withdraw the AI gun immediately to avoid injury to the cow.

4 PREGNANCY AND PARTURITION IN DAIRY CATTLE

The period from the date of conception to the day of parturition is called "Gestation period" and the condition of the female of carrying the fetus during this period is called *Pregnancy*. The sequences of events from ovulation to conception are:

Ovulation: ovulation is the release of the egg from the ovary. Luteinizing hormone gradually softens and breaks down the wall of the follicle, so the ovum can mature and float free. When released, the ovum is caught by the infundibulum and passes into the Fallopian tube which moves it towards the uterus. Ovulation usually occurs about twelve hours after the end of heat, although its timing varies within a normal range of 2 – 26 hours.

Ovum transport: the beating action of the cilia propels the ovum by setting up currents in the fluid secreted by the oviduct. As the egg moves down the tube, excess cumulus cells may be removed by the cilia. Muscular contractions of the oviduct also contribute to the transport of the ovum.

Sperm transport: in natural mating, semen is deposited in the anterior part of the vagina near to the cervix. Sperm are carried into the uterus by the cervical mucus and some reach the oviduct within two to four minutes. The rapid movement of sperm to the site of fertilization is believed to be due to contractions of the uterus and oviducts. Oxytocin released at mating stimulates muscular contractions. Prostaglandins in sperm also assist sperm transport in the uterus and oviduct.

Capacitating: sperm must be in the female tract for one to six hours before they are capable of fertilizing the egg. During this time, they undergo a series of chemical changes which prepare them for penetration and fertilization of the ovum. This process of change is known as capacitation. It is thought to include the removal of a membrane to expose the enzymes, which facilitate penetration of the egg.

Penetration: the enzymes released during capacitation allow sperm to move through the layers of the egg to reach the nucleus. The reactions in this case result in the layers of the egg dissolving to allow the sperm to pass through. Hyaluronidase and trypsin act on cumulus cells and zona lysin acts on the zona pellucida.

Fertilization: fertilization, by which male and female gametes unite to form the zygote, takes place at a site one-third of the way down the Fallopian tube or oviduct. Sperm are not strongly attracted to the ovum, and fertilization occurs by a chance collision of the sperm and ovum. The wall of the egg becomes impervious once a spermatozoon has entered, so as to prevent polyspermy, i. e. fertilization by more than one sperm.

5 STAGES OF PREGNANCY AND EMBRYO/FOETUS DEVELOPMENT

The duration of pregnancy (the gestation period) in cows is about 283 days, with a normal range of 273 - 291 days. Zebu and larger European breeds tend to have longer gestations; small breeds such as Jerseys have shorter gestations (Perera, 1999). Gestation may be divided into two stages: 1) embryonic stage, from fertilization to 45 days of pregnancy; and 2) fetal stage, from days 45 to calving.

Within 15 to 30 hours of fertilization, the zygote divides. At four to five days the embryo has reached the uterus, has 16 to 32 cells, and is termed a morula. At day 8 the zona pellucida disintegrates and a blastocyst is formed.

On the 14th day the blastocyst attaches loosely to the uterine wall and part of it elongates to form a membrane called the chorion. Uterine 'milk' nourishes the embryo at this stage.

Implantation begins by the 35th day with the chorion and the uterus forming cotyledons or 'buttons' . The embryonic and maternal tissues are closely associated, allowing nutrients to pass from the maternal to the fetal blood supply and waste products to pass in the opposite direction.

Cells divide rapidly, and the head, heart and limb buds are all present by the 40th day. The embryo is very sensitive during this phase of rapid growth and adverse influences can produce deformities.

Organs continue to differentiate during the fetal stage, and there is a rapid increase in weight, particularly in the last 60 days, when the fetus trebles in size. It is called the last trimester. Because of the rapid growth during this period, pregnant lactating cows should be dry.

5.1 Pregnancy Staging to Determine the Age of Pregnancy

1) 30 to 45 days

- Cervix & uterus are usually in the pelvic canal.
- One horn is slightly enlarged and fluid-filled. CL is on the same side as gravid horn.
- A membrane slip technique used to detect pregnancy. Proceed with caution because damage to the membrane can cause pregnancy to be terminated.

2) 45 to 60 days

- The cervix is typically in the pelvic cavity.
- Pregnant uterine horn begins to fill with fluid with little fluid in non-pregnant horn.
- Amniotic cavity is about the size of a hen egg.

➢ Use membrane slip technique. Caution: too much pressure can damage membranes

3) 60 to 90 days

➢ Cervix moves in anterior position; torsion evident when you attempt to pick up the tract.

➢ Pregnant horn is at the anterior part of pelvic brim and may start to fall over the rim.

➢ Amniotic cavity is about the size of a grapefruit & fetus about the size of a small rat.

➢ Difficult to move hand completely around the pregnant uterus.

➢ This stage of gestation is the earliest stage that most individuals like to start to palpate pregnancy with a high degree of accuracy.

4) 90 to 120 days

➢ The cervix is at the pelvic brim.

➢ Uterine body enlarged & fluid-filled about the size of a football. Uterus usually dropped over the pelvic brim.

➢ The fetus is the size of a small cat.

➢ Start to feel the placentomes.

5) 120 to 150 days

➢ The cervix is almost completely over the pelvic brim.

➢ Uterine body and horns are not easily palpated.

6) 150 to 180 days

➢ The cervix is well over the pelvic brim.

➢ Entire uterus is stretched and fluid-filled and well down into the body cavity.

➢ Very hard to palpate fetus if you have short arms.

➢ Placentomes are about the size of a fifty-cent piece. There is a nice buzz to the uterine artery.

7) 180 to 210 days

➢ Pregnancy easy to detect. The fetal head is at or near pelvic cavity and can be palpated. Fetus about the size of a beagle dog. Placentomes are about the size of a silver dollar.

➢ There is a very strong buzz to the uterine artery.

8) 210 days to term (calving)

Pregnancy is easy to detect; fetal calf quite often located in the pelvic cavity and its head is easy to feel, very strong buzz to the uterine artery when digital pressure is applied to it. You can bounce fetus like a basketball (Fig. 26).

5.2 Pregnancy Diagnosis Techniques in Dairy Cows

Early identification of non-pregnant cows post-breeding improves reproductive efficiency and pregnancy rate in cattle by decreasing the interval between AI services and increasing AI service rate. The significances of pregnancy diagnosis are described by Mondal (2015):

➢ Helps in culling animals which become sterile.

➢ Enables a person to take curative/remedial measures in case of low fertility or infertility.

- Determination of proper care and proper feed according to pregnancy needs
- For regulating calving for uniform production throughout the year
- Keeping herd of high breeding efficiency cows
- It has a relation to the economy of the farm.

The most common methods/techniques for pregnancy diagnosis are: signs of pregnancy exhibited by female, rectal palpation and examination of the reproductive tract, ultrasound, laparoscopy, and use of rapid milk progesterone test kits.

(1) Signs or symptoms of pregnancy

Cessation of estrus cycle or non-return to estrus: If estrus signs are not observed around 3 weeks after services or insemination, the cow is generally assumed to be pregnant. However, even if estrus detection is good, not all of these cows will be pregnant. On the other hand, up to 7% of pregnant cows will show some signs of estrus during pregnancy. Insemination of these animals may result in embryonic or fetal death (Partners in reproduction, 20015). Other signs that may indicate pregnancy are:

- Sluggish temperament
- Tendency to fatten
- Gradual drop in milk yield
- Gradual increase in weight
- Drooping quarters
- Increase in size of udder
- Waxy-appearance of teats in last month of pregnancy

(2) Rectal palpation (examination of ovaries, uterus and vagina)

This clinical diagnosis of pregnancy is the most convenient method. Examination of the genitals can easily be done in bovine by rectal palpation. Advantage: immediate result enabling early treatment of non-pregnant cattle. Accuracy: depends on the experience of the practitioner and can reach 95%. In the presence of experienced personnel, rectal examination is usually done between 35 & 65 days post AI. The limitation of per rectal examination of the reproductive tract for pregnancy diagnosis is the relatively long interval from breeding until an accurate diagnosis is made.

1) Early pregnancy diagnosis (1-3 months)

It is based on a combination of the following:

- Asymmetry of the uterine horns
- Decrease in the tone of the pregnant horn.
- Fluctuant contents in the pregnant horn (later both horns).
- A palpable CL on the ovary on the same side as the pregnant horn
- Membrane slip (effective from 45 – 90 days of pregnancy)
- Appreciation of an amniotic vesicle (effective from 30 – 90 days of pregnancy)
- Diagnosis in later pregnancy (>3 months)
- The cervix is located anterior to the pelvic rim and the uterus cannot be retracted.

- The uterus is flaccid
- Placentomes, and sometimes the fetus, are palpable (effective from 75 days and onwards).

2) Procedures of rectal palpation

- Bring the cow in breeding crate/chute/crush for proper control.
- Clean the hands with soap and water and wipe off using a clean dry towel.
- Put the rectal palpation gloves on the hands.
- Lubricate the hand gloves with liquid paraffin.
- Insert the hand into the rectum and take out the dung.
- Palpate the cervix and push the hand forward to palpate the common body of the uterus and uterine horns through the rectal wall.
- Explore the uterine horns carefully from beginning of the uterus.
- Examine the ovaries for their size, position, presence of cysts, corpus luteum and record observations.

(3) Hormone measurements (progesterone assay)

The progesterone secreted by a functional CL between 18 and 24 days after service or insemination is an early indication of pregnancy. It can be assayed in milk or plasma. The optimal assay time is 24 days after service or artificial insemination; this eliminates the possibility of long estrus intervals which might result in false positives.

(4) Ultrasound examination (ultrasonography)

Real time (B-mode) ultrasound is a reliable and relatively simple method of diagnosing pregnancy as early as day 26. There are two methods of Ultrasound examination, i. e. Doppler method and Pulse echo method. In Doppler method, the equipment works on 9-volt battery with a headphone/speaker and a transducer with abdominal probe. Ultrasound wave produced by transducer when comes in contact with vibrating fetal heart, an reflected back and converted into audible sound to be heard by headphone.

(5) Laparoscopy

An endoscope/Laparoscope is used to visualize the reproductive organs of a dam to see changes related to pregnancy. The equipment consists of a cold light source, an optical cable, a trocar-and-cannula, laparoscope, and endo photographic equipment.

5.3 Parturition in Dairy Cows

Parturition is the term used for giving birth to the young calf. The calf is normally presented with the forelimbs extended and its head resting on, or between them. The normal position of the calf when expelled from the uterus is the front feet first, followed by the nose, then the head, shoulders, middle, hips, rear legs and feet. In some instances, the hind limbs may be presented first. However, about 80% of all calves lost at birth are anatomically normal. Most of them die because of injuries or suffocation resulting from difficult or delayed parturition called *dystocia*. About 5% of the calves at birth are in abnormal positions, such as forelegs or head turned back, breech, rear end position; sideways or rotated. This requires the assistance of a veterinarian or an experienced herdsman to position the fetus correctly prior to delivery (Hudson, 2002).

5.3.1 Signs of Approaching Parturition

Some signs are observable in cows and heifers, others in one or other (Dunn et al., 2009), these signs indicating approaching parturition are:

- In cows, the udder enlarges one to four weeks before calving. This is not a reliable sign in heifers as the udder begins to develop halfway through pregnancy.
- Production of colostrum, a creamy or pink secretion, begins from the udder.
- The white stringy vaginal mucus becomes more profuse.
- The mucus plug in the cervix liquefies.
- The animal usually moves to a quiet spot away from the rest of the herd.
- Heifers may become restless and lose their appetite.
- Pelvic ligaments relax under the influence of relaxin, 24 - 48 hours before calving, making the tail appear to be set higher and causing a looser walking action, i.e. 'springing'.
- The vulva swells to six times its normal size.

5.3.2 Stages of Parturition

Normal parturition/calving can be divided into three general stages: preparatory, fetal expulsion and expulsion of the placenta or afterbirth. The time interval of each stage varies among types and breeds of cattle and among individuals of the same breed. Although the exact stimulus that initiates parturition is unknown, it does involve hormonal changes in both the cow and fetus as well as mechanical and neural stimulation in the uterus. A general understanding of the birth process is important to proper calving assistance (Whittier and Thorne, 2007).

Stage 1: Preparatory (2-6 hours)

During pregnancy, the fetal calf is normally on its back. Just prior to labor, it rotates to an upright position with its forelegs and head pointed toward the birth canal. This position provides the least resistance during birth. Toward the end of gestation, the muscular lining of the dam's uterus increases in size, this aids in delivery.

- In the preparatory stage, the cervix dilates, and rhythmic contractions of the uterus begin. Initially, contractions occur at about 15 - minute intervals.
- As labor progresses, they become more frequent until they occur every few minutes. These contractions begin at the back of the uterine horn and continue toward the cervix, forcing the fetus outward. Any unusual disturbance or stress during this period, such as excitement, may inhibit the contractions and delay calving.
- At the end of the preparatory stage the cervix expands, allowing the uterus and vagina to become a continuous canal.
- A portion of the placenta (water sac) is forced into the pelvis and aids in the dilation of the cervix. This water sac usually ruptures, and the membranes hang from the vulva until Stage 2

Stage 2: Delivery (1-2 hours, may be longer in heifer)

- This stage begins when the fetus enters the birth canal and usually occurs while the cow is lying down.
- Uterine contractions are now about every 2 minutes and are accompanied by voluntary contractions of the diaphragm and abdominal muscles.
- Surrounded by membranes, the calf's forelegs and nose now protrude from the vulva. After the nose is exposed, the dam exerts maximum straining to push the shoulders and chest through the pelvic girdle.
- Once the shoulders have passed, the abdominal muscles of the calf relax, and its hips and hind legs extend back to permit easier passage of the hip region.
- The calf is normally born free of fetal membranes (placenta), because they remain attached to the cotyledons or "buttons" of the uterus. This ensures an oxygen supply for the calf during birth. Upon passage through the vulva, the umbilical cord generally breaks, respiration begins, filling the lungs with air and causing the lungs to become functional.
- Delivery is normally completed in one hour or less in mature cows. Special assistance is warranted if this stage goes beyond 2 - 3 hours. First-calf heifers can take 1 - 2 hours, or longer. Proper judgment should be used so that assistance is neither too hasty, nor too slow.

Stage 3: Cleaning (2-8 hours)

- Thecaruncle-cotyledon attachment between the uterus and placenta relaxes and separates after parturition.
- The placenta is then expelled by continued uterine contractions. Cows normally expel the placenta within 2 - 8 hours.

5.3.3 Preparing for Parturition

The heifer/cow requires a comfortable, clean place where she can lie down easily. She should be cleaned before entering this area to avoid introducing unnecessary contamination. Farmers should also practice good hygiene by washing their hands and the rectal/vaginal area of the cows. The preparation for parturition is described as follows (Moran, 2012). For its detailed information, refer module 4.

6 EVALUATING BREEDING EFFICIENCY OF DAIRY COWS

The ultimate goal of the dairy industry is to operate an economically efficient production system, and this depended upon high reproductive/breeding efficiency of the cows. Reproductive/breeding efficiency is defined as a measure of the ability of a cow to conceive and maintain pregnancy when it is served at the appropriate time in relation to ovulation. Poor fertility decreased the profit margin due to loss in milk yield, cost of replacing culled cows and decreased calf sales per cow. Thus, cows must be managed in such a way that the reproductive performance measures should not exceed the optimum recommended values for profitable dairy herd (Evelyn, 2001).

The efficiency of reproduction can be assessed from several measures or parameters, which are termed reproductive indices. The most common indices of reproductive performance are age at first service and age at first calving, the interval from calving to first observed estrus (service), calving to conception (days open), calving interval, number of services per conception, heat detection efficiency, pregnancy/conception rate compared to total herd and number inseminated or mate, calving rate and breeding efficiency (House, 2011).

Age at puberty and age at first service: In the case of heifers, the important indices are the age of attainment of puberty, which depends on the time of onset of ovarian activity. Age at first service (AFS) is defined as the number of days from birth to the date of first service. Age at first puberty is an important determinant of reproductive efficiency. A substantial delay in the attainment of sexual maturity may mean a serious economic loss, due to an additional, non-lactating, unproductive period of the cow over several months (Mugerwa, 1989). The recommended optimum targets of 14 – 15 months for AFS.

Age at first calving: Age at first calving (AFC) marks the beginning of a cow's productive life. Age at first calving is closely related to generation interval and, therefore, influences response to selection. Under controlled breeding, heifers are usually mated when they are mature enough to withstand the stress of parturition and lactation. This increases the likelihood of an early conception after parturition. In traditional production systems, however, breeding is often uncontrolled, and heifers are bred at the first opportunity. This frequently results in longer subsequent calving intervals. The average age at first calving in *B. indicus* cattle is about 44 months, compared with about 34 months in *B. taurus* and *B. indicus* × *Bos taurus* crosses in the tropics. It is recommended that heifers calve between 23 and 25 months of age, which is considered as optimum that increase profitability of the dairy business. A reduction in AFC will minimize the raising costs and shorten the generation-interval and subsequently maximize the number of lactations per head. In general, earlier first calving increases lifetime productivity of cows. Generally, since the heritability of age at puberty, at first conception and at first calving are generally low, indicating that these traits are highly influenced by environmental factors. Thus, prolonged AFC of cows could be attributed to factors such as poor nutrition and management practices, including poor heat detection at the time of mating the heifers. With good nutrition, it is expected that heifers would exhibit fast growth and attain higher weights at relatively younger ages.

Calving interval: A calving interval (CI) is the period of time between successive parturitions. The calving-interval affects both the total milk production of the dairy herd and the number of calves born. In most modern dairies, the general practice is to breed cows early, with the aim of establishing optimum calving interval of 12 – 13 months. According to Tadesse et al. (2010), CI values were interpreted as short CI<11.7 months, 11.7 – 13 months optimum standard CI set for commercial dairy farms, 13 – 13.5 months slightly high, 13.6 – 14 months moderately high, and high CI which is more than 14 months. Long interval means long dry periods and great loss of feed and carrying of unproductive females. Attaining short calving interval enables to exploit the early peak milk yield in each lactation. Calving interval can be divided into three periods: calving to first estrus-period, calving to conception period (days open), and gestation period.

Voluntary Waiting Period (VWP): It is the time period between when the cow calves and when she becomes eligible for insemination. Any farm has a VWP where the breeder retains a cow seen in heat un-bred to allow sufficient time to recover from parturition stresses. The optimum recommended VWP for efficient dairy herd ranged 45 – 60 days postpartum.

Calving to first service interval: It is defined as the average number of days from calving to the day of first insemination served. Unless heat detection is poor, the average days to first service interval (CFSI) should be within 30 days after the VWP. Hence, the optimum recommended CFSI had to be within 65 days. This helps determine if cows are cycling after calving and provides an indication of heat detection.

Calving to conception interval also called days open: It is defined as the interval from calving to conception, i. e. the number of days between parturition and the insemination that resulted in a pregnancy. According to Evelyn (2001), a herd average of less than 85 open days indicates cows are being bred early, between 85 – 115 open days considered as optimum for dairy herd, 116 – 130 open days indicate slight problem, 131 – 145 open days shows moderate problem, but more than 145 open days considered as sever reproductive problem in the dairy herd. The days open was reported to be influenced by cow and management/environment-related factors, such as method and efficiency of heat detection, type and efficiency of breeding service and the ability of the cow to resume regular ovarian cyclicity after calving, display an overt heat signs, and conceive with the given service.

Conception rate to first service: The conception rate to first service (CRFS) is the percentage of females actually pregnant after first breeding or the ratio of animals confirmed pregnant at the first service to the number of cows bred (Mekonnen et al., 2006). It is a result of combining estrus detection on accuracy and breeding technique. The commonly recommended goal for conception rate to first service is 45 – 60%. Conception rate to first service provides a useful estimate of the conception rate for a herd. However, it is a measurement that combines the effect of semen quality, fertility of the cow, timing of insemination, semen handling and insemination techniques as well as factors such as high environmental temperature and stress.

Conception/pregnancy rate: It is the proportion of services/mating that resulted in a pregnancy. During monthly visits, the conception rate is normally calculated for the cows that were bred 6 – 10 weeks before the day of examination. Thus, the optimum overall conception/pregnancy rate should be 80%.

Number of services per conception: Number of services per conception (NSC) is a very important economic factor both in natural breeding and AI, for any AI center number of services per conception exceeding 2.0 means a disaster. It is usually higher under uncontrolled natural breeding and low where hand-mating or AI is used. The optimum recommended NSC for profitable dairy cows ranged 1–1.7 (Evelyn, 2001).

Higher NSC results from either failure to conceive at a given service and/or failure to maintain pregnancy, thus requiring repeated services (Hammoud et al., 2010). Management factors such as accuracy of detection of heat, timing of insemination, proper insemination techniques, semen quality, proper semen handling and skills in pregnancy diagnosis have been reported to decrease the NSC. Furthermore, proper heat detection, feeding and postpartum reproduction management may reduce NSC, and hence there is a possibility of reduction in days open and calving interval.

Heat/estrus detection efficiency: It is defined as the percentage of cows displaying estrus that are identified as being in heat. It is the proportion of eligible cows bred every 21 days. For example, if 100 cows are cycling normally and only 50% of these cows are detected in heat, the estrous detection efficiency is 50%. The recommended optimum is 70%. Heat detection efficiency is largely influenced by management's ability to identify cows in heat. It provides immediate feedback as to what has been done during the preceding three-week period. Research result indicated that a 1% increase in estrous detection efficiency resulted in a 0.5 day decrease in days open. Therefore, a 20% improvement in estrous detection efficiency would reduce days open by 10 days (Senger, 2002).

Estrous detection errors: It is defined as the proportion of cows that are inseminated that are not in estrus. Estrous detection errors are brought about by identifying cows based on secondary signs of estrus rather than identification of cows standing to be mounted. Also, poor cow identification results in confusion on the part of the person identifying animals in estrus; thus, the "wrong cow" may be presented for insemination. Estrous detection errors may result in subsequent uterine infection because cows are often inseminated during the luteal phase of the cycle when the uterus is most susceptible. Errors also result in wasted semen and labor. Furthermore, they create recordkeeping errors via incorrect estrous dates that further confuse the management team when trying to predict future estrus or breeding dates or the timing of pregnancy diagnosis.

Calving rate: It is defined as the number of calves' born/100 services. Furthermore, calving rate also could be defined as the number of calves born per cow per year. From a biological point of view, calving rate was the most appropriate measure of fertility. However, even under ideal conditions, with 100% 'normal' cows and 100% efficiency of estrus detection, a 100% calving rate was not achievable. This was mainly due to embryonic or fetal deaths that occurred in the cows. The recommended the optimum calving rate is 75–80% (Evelyn, 2001).

Percentage of pregnancies lost: This figure is affected by when pregnancies are diagnosed. If pregnancies are diagnosed after 150 days of pregnancy, the percentage tends to be quite low (<2%) as most pregnancies are lost during early pregnancy and are often not noticed. When reproductive examinations are conducted around 42 days of pregnancy, the presence of the pregnancy is documented, and the loss will be accounted for, the recommended is target <5%.

Breeding efficiency: Breeding efficiency (BE) is a measure based on the regularity of calving and the age at which cows first calve. If an animal calves late for the first time its maintenance costs as a fraction of total costs tend to increase and its lifetime production tends to decrease. Thus, profitability of dairy farm depends on herd life which in turn is affected by breeding efficiency of cow at the farm. The following method was used for evaluation of BE= [(*N*-1) 390 + 960] / (age at each calving). Where, "*N*-1" is the number of calving intervals with *N* calving; 390 is the upper limit of desirable calving intervals (days); 960 is the upper limit of age at first calving (days). The estimated coefficients were expressed as a percentage. The optimum recommended BE is more than 80% (Million et al. , 2006).

Lower BE values indicate lower herd management practice for genetic expression of animals. Furthermore, high abortion rate and longer calving interval reduce BE value. BE value is highly influenced by nutrition and climatic stress. Some other factors affecting breeding efficiency are: fertilization rate, number of services per conception, conception rate, calving rate, service period and calving interval. Improved management in all these factors would be helpful (Table 5. 4).

Table 5. 4 Summery of reproductive indices for dairy cattle the recommended standard and suggested acceptable smallholder under improved smallholder systems in the tropics

Reproductive indices	Recommended standard	Acceptable
Age at puberty (month)	12 - 14	< 22
Age at first service (months)	14 - 16	< 26
Age at first calving (month)	23 - 25	<36
Voluntary waiting period (days)	45 - 60	<60
Calving to first service (days)	65	< 90
Calving to conception (days)	85 - 115	< 115
Calving interval (month)	12 - 13	13 - 14
First service conception rate (%)	45 - 60	> 55
Overall conception rate (%)	80 - 85	> 75
Calving rate (%)	75 - 80	> 70
Number services per conception	<1. 7	< 1. 8
Heat detection efficiency (%)	>70	>50
Heat detection error (%)	<30	<40
Pregnancy loss (%)	<5	<10
Breeding efficiency (%)	>80	>80

Source: Perera, 1999.

Modification of the environment (cooling/provision of shade/air conditioning) would alleviate the adverse effects. Feeding management and breeding of cows in the cooler months of the year will help improve this trait (Gebeyehu, 2005).

7 IDENTIFYING MAJOR REPRODUCTIVE PROBLEMS IN DAIRY CATTLE

Infertility means a reduced ability or temporary inability to reproduce, while sterility is a complete and permanent inability to reproduce. Infertility of the lactating dairy cow continues to be a critical problem limiting profitability and sustainability of dairy farms. When infertility or sterility occurs in an individual animal or a herd, it is essential that the underlying causes are accurately identified. It is only then that a rational decision can be made to overcome the problem and prevent any future occurrence (Senger, 2002).

Reproductive performance of lactating dairy cows is influenced by several factors that could be broadly classified into: 1) hereditary and anatomical factors, 2) cow related factors, 3) infectious disease, 4) hormonal disorder, 5) environmental factors such as nutritional factors, injuries, stress, and managerial factors.

7.1 Hereditary and Anatomical Factors

About the hereditary disease, only heifers will be the objectives of attention. Because cows already had given birth, which means at least they don't have any infertile hereditary disease. Therefore, our main concern is if heifers have any hereditary diseases and if they are infertile or not. Most of hereditary disease, we don't have any treatment methods. And in case that these diseases are heritable, early diagnosis and early culling are recommended.

1) Hypoplasia, aplasia or malformation of reproductive organs

Sometimes "double external orifices of cervix" can be seen. It is a kind of malformation, though if the one of cervix is open to the uterus, there is a possibility to be pregnant. There are many types of this malformation, sometimes completely cervix, uterine body and uterine horn are separated, or in another case only the external orifices of cervix are separated. In this case we have to check if the cervix is blind-end or not.

2) Freemartin

In case of twin pregnancy if the gender of the fetuses were mixed (female and male) most of the female calves will lose their fertility. Infertile freemartin female's reproductive tract will be poorly developed, a kind of hypoplasia. This is because of the blood vessels of the female and male tied together (anastomosis) inside the uterus. More than 90% of such female become infertile, so early diagnosis is recommended. Female calf can be checked of its fertility by checking the vagina depth. Inserts something like test tube (the tip should be round-shape) into vagina. If the vagina depth is less than 10 cm, they can diagnose as infertile. Normal female calf has deeper vagina depth.

3) Infantile or absence of ovaries

Sometimes during development, because of a hormone imbalance, the ovaries fail to develop enough or will not function. In some cases, they may be absent.

4) Cryptorchidism

This is the failure of one or both of the testes to descend into the scrotum. The testes are retained in the abdominal cavity resulting in complete sterility. Sometimes only one testis is retained, and the defect is referred to as unilateral cryptorchidism. This does not cause complete sterility but results in a reduction in the number of viable sperm produced. Both of these conditions are inherited and should be avoided in a breeding program.

5) Scrotal hernia

This condition may not cause sterility but can cause an animal not to breed. If the hernia is large enough to allow part of an intestine to drop through, it can be very dangerous.

6) Malformed penis

Malformed penis results from an injury or birth defect, the penis can be malformed to the extent that copulation cannot be performed.

7) Genetic factors

Some bloodlines are known to have a high genetic factor or weakness for sterility or low productivity. Inbreeding also may result in lower fertility.

7.2 Infectious Disease

There are several infections that affect the reproductive organs of the animal. Some may prevent pregnancy and others may cause abortion. Many infectious diseases cause abortion, such as: *Brucella abortus*, endometritis, pyometra, and others (see Module 9 for detailed discussion on these diseases).

7.3 Hormonal Disturbances

The sexual behavior of animals is affected by the secretion of hormones. When these hormones are not properly secreted, the animal may not be able to reproduce. Thus, some female reproductive disturbances are hormonal or glandular in nature. These include: cystic ovaries

7.3.1 Cystic Ovaries

A cyst is a swelling containing a fluid or semisolid substance. Ovarian cysts are structures, usually greater than 2.5 cm in diameter, which persist on one or both ovaries for 10 days or more. Fertility in cystic cows is reduced due to hormonal changes, changes in uterine tone and, in many cases, failure to release an egg (Gillespie and Flanders, 2010).

(1) Possible factors involved

- Excessive calcium intake or wide calcium-phosphorus ratio. Total dietary intake of greater than two parts calcium to one-part phosphorus may lead to increased incidence of cysts.
- High estrogen intake, whether given by injection, through fresh legume forage or from some mould toxins, may increase the incidence of cystic ovaries.
- Genetic predisposition

- Stressful conditions or health problems at calving or early postpartum

(2) Prevention and control measures

Have forages analyzed, including calcium and phosphorus analysis. Check the feed program to ensure that the calcium-phosphorus ratio is between 1.5 : 1 and 2 : 1 in the total diet. Include all forages, grains and free-choice minerals in estimating mineral intake. Do not use feeds containing mould.

7.3.2 Anestrus

Anestrus cows do not show heat signs or silent heat. There are three types: undetected estrus, true anestrus, and silent estrus.

(1) Possible factors

1) Undetected estrous

The signs in cows with normal ovarian activity resulting from:

- Inadequate estrus detection since 66% of estrus signs are shown between 6 p.m. and 6 a.m. Cows with short estrus (less than 12 hours in length) may be missed even with twice-a-day estrus detection. This is particularly true when cows are observed in estrus, during a time when they are unlikely to exhibit standing behavior;
- Inadequate animal identification and/or inadequate records;
- Lack of opportunity for cows to express estrus, i.e. cows not turned out; slippery footing; lameness or stiffness; groups too small for adequate interaction.

2) True anestrus

It is lack of ovarian activity caused by:

- Anemia: due to anaplasmosis, internal or external parasites, deficiency of protein, Iron, Copper, Cobalt or Selenium
- Phosphorus deficiency
- Energy deficiency, cows losing flesh due to high production and/or underfeeding
- Low hormone levels associated with prolonged feeding exclusively on stored feeds
- Cystic ovaries (70% of cystic cows are anestrus)
- Pyometra or pus in the uterus

3) Quiet estrus or silent estrus

There is a normal ovarian activity with little or no signs of estrus. Most of the factors associated with true anestrus (anemia, phosphorus deficiency, energy deficiency and low endocrine levels) can also be associated with quiet estrus.

(2) Prevention and control measures

- Maintain adequate reproductive records. Record all estrus dates, examination dates and findings, unusual events such as difficult calving or retained placenta, and treatments.
- Closely observe cows for estrus.
- Observe cows at least twice and preferably 3 times a day for estrus. Make sure an estrus detection time is in the evening.

- Observe cows for at least 20 minutes at each detection time.
- Don't let any other farm chores distract from estrus detection. Observe cows when they are not occupied with eating, being milked or other activities.
- Provide anon-slippery surface for estrus detection, so cows are more likely to exhibit signs of estrus.
- Check problem cows for anemia. If anemia is present, check for internal and external parasites. Treat for anemia if detected.
- Submit forage samples for standard and mineral tests. Seek assistance to develop a feed program that meets herd nutritional requirements.
- Have cows examined for uterine infection, cystic ovaries and for evidence of ovarian activity.
- Whenever possible, have cows in a weight-gaining condition as desired breeding time approaches.
- Control conditions around calving time and early lactation that may contribute to anestrus problems (retained placenta, metritis, and ketosis).
- Provide access to fresh forage for at least 4 - 6 weeks each year.
- Have pregnancy exams done 40 - 60 days after breeding.

7.4 Cow Related Factors

Some of reproductive problems are as a result of cow related factors, such as dystocia, retained placenta, uterine prolapsed and repeat breeders.

7.4.1 Dystocia

It means "difficult birth." The major cause of this is fetopelvic disproportion (calf too large for the birth canal).

Possible factors involved

Birth weight of the calf and pelvic area of the dam are two of the most important factors that contribute to dystocia. Improper positioning of the fetus is also a major contributor.

Prevention and control measures

Other preventive measures besides assisting parturition include:

- Reduction in incidence of dystocia can always occur when bulls are selected for a high degree of calving ease, especially in heifers.
- Growing heifers to achieve 540 - 570 kg immediately after calving (24 months) coupled with the use of calving-ease bulls can reduce the incidence of dystocia in first-calf heifers.

Thus, management can exert a strong preventive influence by selecting calving-ease bulls for use in heifers and by employing proper heifer management and maternity pen care. A certain amount of caution should be exercised because continued use of calving-ease bulls can predispose a herd to smaller cows over time. Such a practice would amplify the dystocia problem during the same time frame.

Regardless, there will always be a certain proportion of cows afflicted by difficult birth. Almost without exception, cows that have difficult births have "downstream" reproductive problems, including retained placenta, delayed uterine involution, and poor cyclicity (Senger, 2002).

7.4.2 Retained placenta

When a freshening cow fails to expel her placenta (afterbirth) within 12 hours after calving, the condition is known as retained placenta. Incidence of retained placenta in dairy herds should not normally exceed 8% (Whittier and Thorne, 2007).

Possible factors involved

- It commonly accompanies difficult births, multiple births and short gestations.
- Specific infections may cause abortion but can also cause retained placenta following delivery at term.
- Nonspecific infections by a wide range of bacteria and viruses that occur during pregnancy or at calving can be associated with retained placenta.
- Twin births and abnormal deliveries, including prolonged or difficult deliveries or caesarean sections, are often followed by placental retention.
- Deficiencies of selenium, vitamin A or vitamin E may cause higher than normal incidence of retained placenta.
- Over-conditioning of dry cows due to excess energy intake and/or prolonged dry period is often associated with retained placenta.

Prevention and control measures

- Test for specific infections. If an infection is identified, treat, vaccinate or cull infected cows as indicated.
- Keep calving areas clean and well bedded. Calve on grass if practical. Don't use maternity pens for any other purposes.
- Breed heifers to bulls with a record of calving ease. Observe freshening cows and heifers closely. Provide assistance if hard labor continues for over 30 minutes without progress. If assistance is required, do so in a clean, gentle manner.
- Provide cows with fresh forage as pasture or green chop for at least 4 - 5 weeks each year.
- Provide about 160,000 units of vitamin A (1 mg of carotene is equivalent to 400 units of vitamin A) from all sources (natural and supplemental).
- Avoid over-conditioning. At drying off time have cows in about the body condition in which they should freshen. Limit access to high-energy feeds such as corn silage or grain during the dry period.

7.4.3 Uterine Prolapse

This is an inversion of the uterus that can occur when a partial vacuum is formed in the uterus. It is sometimes caused by pulling the calf too rapidly and may result in death of the cow if not treated promptly and correctly. Encouraging the cow to stand soon after delivery will reduce the chances of a prolapse. Always contact a veterinarian for treatment and necessary drugs (Whittier and Thorne, 2007).

7.4.4 Repeat Breeders

A reasonable goal for first-service conception rate is 50%. After two services the cumulative conception rate should be around 75%. Cows requiring three or more services before conception or culling are generally designated as repeat breeders.

Possible factors involved

Embryonic or early fetal mortality caused by gross overfeeding, excessive manipulation of the reproductive tract by rectal examination, trichomoniasis or vibriosis. Either of these venereal diseases can cause early embryonic death and repeat breeding.

- Breeding too early or too late in relation to the time of ovulation
- Use of low fertility sires
- Use of semen damaged in storage or handling
- Poor insemination technique
- Serious imbalances or deficiencies of vitamins or minerals

Prevention and control measures

- If natural service is used, test for vibriosis and trichomoniasis. If present, discontinue natural service and cull the bull.
- Follow vaccination programs for reproductive diseases.
- Avoid gross overfeeding of grain, especially after breeding.
- Provide cows with access to fresh forage such as pasture or green chop for at least 4-6 weeks a year.
- Carefully observe estrous onset and duration. Breed 12 hours after initial observation. On long-estrous cows (more than 24 hours), inseminate second time 12-18 hours after the first insemination.
- Have a veterinarian examine repeat breeder cows for presence of endometritis (low-grade uterine infections), delayed ovulation or other abnormalities.
- Use high fertility bulls. Have semen handling and insemination technique checked by competent AI personnel. The semen viability of representative straws can be analyzed.

7.5 Environmental Factors

Environmental factors such as nutritional problems, housing problems, mechanical injuries and stress have significant impact on dairy cattle reproductive performance described as:

Mechanical injury: physical damage to reproductive organs. Injuries usually occur because of improper handling, unsafe facilities, fighting among animals, or complications during parturition or copulation (Senger, 2002).

Handling stress: severe climatic conditions (primarily extreme heat), high population density, rough handling, and other stressful environmental factors can cause reproductive distress.

Heat stress: the single most important environmental factor impacting reproductive performance is heat stress. Heat stress could cause: increased embryonic death, decreased length of estrus, decreased number of mounts per estrous period, and decreased conception rate.

Cooling cows during periods of heat stress, improves conception rates. Methods of managing heat stress are almost totally under the control of the herd management team. For example, decisions regarding improved ventilation, misting, shade, etc. are made exclusively by the management team (Britt et al., 1986).

Footing surface: Estrous behavior is dramatically affected by the composition of the footing surface on which cows interact. Cows that interacted on a dirt surface had a more sustained estrous period (13.8 h) than cows that interacted on a grooved concrete surface (9.4 h). It should be emphasized that slippery concrete is a major factor that influences poor estrous behavior (standing and mounting), cow safety, and comfort. Scoring slippery concrete can have profound positive effects on estrous detection.

Nutritional deficiencies: nutrition has a profound effect on fertility. Nutritional requirements are dependent upon the physiological state of the individual animal. Fertility, maintenance of pregnancy, lactation, and other events involved in reproduction exert high demands upon an animal's metabolism.

When the level of feed intake and quality of nutrients are insufficient to meet the demands of the body, an animal's reproductive abilities are often impaired. An animal's level of body condition is dependent upon the animal's plane of nutrition. Lack of condition or obese condition usually reduces reproductive efficiency. During periods of low nutrition, the body lacks the energy stores necessary for reproductive activities. During periods of obesity, fatty deposits collect in and around the reproductive organs, impairing function and productivity.

Certain quantities of vitamins and minerals are essential for efficient reproduction. The following vitamins and minerals, if not balanced in the animal's diet, are known to affect reproduction in the following ways:

- Vitamin A: shortened periods of gestation, higher incidence of retained placentas, stillbirths, abortions, mastitis, calves born blind and uncoordinated
- Vitamin E: poor conception rates, higher incidence of stillbirths and newborn mortality

- Phosphorus: poor conception rates, delayed puberty, lower weaning rates, erratic heat
- Calcium: increased calving difficulty, uterine prolapse, retained placenta
- Cobalt: poor conception rates, general reproductive failure
- Iodine: retained placentas, delayed puberty, arrested fetal development, irregular heat, abortion, stillbirths; calves that are blind, hairless, and have enlarged thyroid glands.
- Copper: delayed puberty, abortion, retained placentas
- Iron: general reproductive failure, anemic young
- Manganese: irregular or suppressed heat
- Ingestion of toxic plants: poisonous plants can also cause reproductive stress or abortion.

8 CULLING DECISIONS

It is normal for a proportion of cows to be difficult to get back in calf; the size of this group is largely influenced by the nutritional management of the herd, heat detection efficiency, and the expertise of the inseminators. From an economic perspective, farms need to establish guidelines as to when a cow is to be culled from the herd.

It is common for 20 - 30 % of a dairy herd to be replaced each year. In herds with good nutrition and reproductive management, most cows are culled because they are less productive than other cows or they have problems with recurrent mastitis or lameness. Voluntary culling of less productive cows leads to improvement in herd productivity.

Poor reproductive performance leads to forced culling of non-pregnant cows that are in late lactation but producing insufficient milk to pay the costs of maintaining them. In this case, cows with problems of mastitis or lameness are often retained to maintain herd size. Forced culling decisions are generally associated with static or declining herd productivity.

Herd-level factors that influence culling decisions include: herd expansion plans, milk price feed costs, cost of buying replacement stock, historical reproductive performance and availability of home-bred replacement stock, and value of cull cows.

Decisions on culling individual cows are influenced by: current milk production, age, previous milk production, prior health history (mastitis, lameness), and days since calving, pregnancy status, and genetic merit. Generally, older, over-condition (too fat) cows in late lactation that have been bred numerous times are notoriously difficult to get pregnant.

>>> SELF-CHECK QUESTIONS

Part 1: Multiple choices.

1. The start of a narrow canal that passes through the center of the cervix is?
   A. Vestibule B. Fornix C. Serosa D. Oss
2. The innermost layer of the uterus that aids in the transportation of sperm to the oviduct is
   A. Myometrium B. Endometrium C. Serosa D. All
3. The muscle layer of the uterus, which aids in the expulsion of the calf at the time of birth is
   A. Myometrium B. Endometrium C. Serosa D. All
4. The enlarged upper end of the oviduct/Fallopian tube is
   A. Uterine horn B. uterus body C. Ampulla D. All
5. The gland near urethra & bladder that produces a fluid mixes with the seminal fluid is
   A. Prostrate B. Vas Deferens
   C. Penis D. Scrotum
6. What is the period of time when a female will accept the male for breeding?
   A. Gestation B. Estrus
   C. Conception D. Parturition
7. A hormone that causes the development of secondary male characteristics and sex behavior (*libido*) is
   A. Estrogen B. Testosterone C. Progesterone D. All
8. The testes are suspended from the body by
   A. Scrotum B. Vas deference C. Epididymis D. Spermatic cord
9. The process of spermatogenesis takes place within the
   A. Seminiferous tubules B. Epididymis
   C. Vas deference D. Spermatic cord
10. Leydig cells function in the production a male hormone called
    A. Estrogen B. Testosterone C. Progesterone D. All
11. A coiled tube connected to each testis and is responsible for the maturation, storage, and transportation of sperm cells.
    A. Epididymis B. Vas deferens C. Penis D. Scrotum
12. A gland that secretes a thick, milky fluid that mixes with the seminal fluid and also provides nutrition and substance to the ejaculate.
    A. Prostate gland B. Seminal vesicles
    C. Cowper's glands D. All
13. Paired-glands that secrete and adds fructose and citric acid to nourish the sperm and

functions as a protection and transportation medium for sperm upon ejaculation.

A. Prostate gland B. Seminal vesicles
C. Cowper's glands D. All

14. A gland that secretes a fluid which cleanses and neutralizes the urethra from urine residue that can kill sperm cells.
A. Prostate gland B. Seminal vesicles
C. Cowper's glands D. All
15. The hypothalamus produces the hormone
A. Prostaglandins ($PGF_{2\alpha}$) B. Progesterone
C. Gonadotropin Releasing Hormone D. All
16. The Follicle Stimulating Hormone and Luteinizing Hormone are produced by
A. Posterior pituitary gland B. Anterior pituitary gland
C. Hypothalamus D. Ovary
17. A hormone that is released and functions to destroy the corpus-luteum is
A. Prostaglandins
B. Progesterone
C. Gonadotropin Releasing Hormone
D. All
18. A synthetic form of the naturally occurring hormone, progesterone is
A. Melengestrol Acetate B. Estrumate
C. Fertagyl D. Receptal
19. A synthetic form of the naturally occurring hormone, Prostaglandins ($PGF_{2\alpha}$) is
A. Melengestrol Acetate B. Estrumate
C. Fertagyl D. Receptal
20. A synthetic form of the naturally occurring Gonadotropin Releasing Hormone (GnRH) is
A. Melengestrol Acetate B. Estrumate
C. Fertagyl D. Lutalyse
21. The onset of the first behavioral estrus and ovulation in heifer and start of sperm production in bull is called
A. Estrus B. Mating C. Puberty D. Poverty
22. The regular sequence of stages which the cow undergoes from one heat to the next is
A. Estrus B. Heat C. Estrus cycle D. A & B
23. Which of the following structure is found on ovary at a different time is?
A. Follicle B. Corus Luteum C. Ova D. All
24. Which of the following hormone is produced by the uterus?
A. FSH B. LH C. $PGF_{2\alpha}$ D. Testosterone
25. The non-breeding period in the estrus cycle is
A. Estrus B. Met-estrus C. Di-estrus D. Anestrus

26. Stage of estrus cycle when estrogen level highest is
    A. Estrus　　B. Met-estrus　　C. Di-estrus　　D. Anestrus
27. Stage of estrus cycle when ovulation occurs is
    A. Estrus　　B. Met-estrus　　C. Di-estrus　　D. Anestrus
28. The mating/breeding method in which heat detection of cows is not required is
    A. Pasture mating　　B. Hand mating
    C. Artificial insemination　　D. All
29. The semen deposition site during AI is
    A. Oviduct　　B. Cervix　　C. Uterine body　　D. Uterine horn
30. Which of the following vitamin deficiency may cause retained placenta?
    A. Vitamin A　　B. Vitamin C　　C. Vitamin D　　D. Vitamin E

Part 2: True or false.

1. The cervix relatively relaxed during estrus to allow the passage of sperm into the uterus.
2. During pregnancy the cervix is tightly closed.
3. Clitoris is the sensory and erectile organ of the female.
4. The cervix is larger in *B. taurus* than *B. indicus* cattle.
5. Infundibulum is close and attached to the ovaries.
6. Accessory glands produce secretions that contribute to the liquid non-cellular portion of semen.
7. The follicle remains relatively soft throughout the development of the ovum but becomes hard just before it ruptures at the time of ovulation.
8. All oocytes within a female at the time of birth will develop into ova and mature to ovulation.
9. The function of the follicle is to store the growing ovum and to produce and store the hormone testosterone.
10. The oviduct is the site where fertilization takes place in mammals.
11. A male that is castrated at an early age, their reproductive organs continue to grow.
12. When an adult male is castrated, the reproductive organs diminish in size and lose some of their functions.
13. Removing calves for 48 hours improves pregnancy rates in a CO-Synch system.
14. A fixed-time AI results in better pregnancy rates that an estrous detection system.
15. The CIDR is a better progestin for synchronizing heifers.
16. Body condition score and days postpartum do not play a role in response to estrous synchronization.
17. In pasture mating method the bull runs with the cows and heifers.
18. Heifers attain puberty when they reach 60% of their mature cow body weight.
19. Growth rate and body weight of heifer are important determinants of puberty than age of heifer.

20. Zebu cattle attain puberty earlier than *Bos taurus* and their crosses.

21. The optimum time for mating cow/heifer is during the latter part of heat or immediately after the end of heat.

22. After normal calving, the uterus returns to its non-pregnant size after 120 days.

23. The requirement for estrus synchronization of cows using $PGF_{2\alpha}$ is that, the cow should be cycling and has active corpus-luteum.

24. Administration of progestin for estrus synchronization is only effective for cycling cows.

Part 3. Matching.

Match the term in the left column with its definition in the right column.

A	B
1. Semen thawing	A. Period from fertilization to 45 days of pregnancy
2. Artificial vagina	B. Period from 45 days to parturition
3. Retained placenta	C. Semen collection
4. Follicles	D. At temperature of 40℃ for 30 seconds
5. Dystocia	E. Extend the penis from the sheath during erection
6. Estrus synchronization	F. Tiny cavities found in the ovaries.
7. Cryptorchid	G. Sterile female calf born twin to a male calf
8. Caruncles	H. part of an intestine is dropped into scrotum
9. Cotyledon	I. Difficulty in calving
10. Uterus involution	J. Failure to expel the afterbirth
11. Fetal stage	K. Uterus side attachment of placenta
12. Embryonic stage	L. One or both testicles remain in the body cavity
13. Freemartin	M. Placental side attachment to the uterus
14. Sigmoid flexure	N. All cows/heifers show heat sign at the same time
15. Scrotal hernia	O. Reproductive tract recovery period after calving

Part 4. Discuss the following questions.

1. What hormone do the testes produce? What is its function?
2. What is the purpose of the infundibulum?
3. What two hormones do the ovaries of mammals produce? What are their functions?
4. How does the cervix function during pregnancy?
5. Name three functions of the uterus.
6. Name three functions of the epididymis.
7. Why are the testicles in males raised and lowered as the weather changes?
8. How long can sperm live in the female reproductive tract in cattle?

9. Discuss the reproductive axis that control estrus cycle
10. What would happen if the cow is not pregnant by day 19 of the estrus cycle?
11. Describe the primary and secondary signs of heat in cattle.
12. Discuss the advantages and dis-advantages of teaser bull for heat detection
13. Describe the importance of estrus synchronization
14. Describe ovulation in the mammal.
15. List equipment and tools required for AI.
16. Describe AI procedures.
17. Explain how the egg cell is fertilized in cattle.
18. Describe what happens during gestation in cattle.
19. Describe rectal palpation procedures for pregnancy diagnosis in cows.
20. Describe the steps of parturition in cattle.
21. For estrous synchronization to be effective, what four areas of management should a producer focus on prior to synchronizing their cows?
22. What are an advantage and a disadvantage of feeding MGA?
23. What are an advantage and a disadvantage of using a CIDR?
24. Which is the best estrous synchronization system for synchronizing heifers?
25. List the signs of approaching parturition in cows.
26. What are the conditions for assisting parturition?
27. Discuss the stages of parturition in cows.
28. What are the factors for long calving interval?
29. How long calving interval affects profitability of dairy farm?

>>> REFERENCES

Ball P J H, Peters A R, 2004. Reproduction in Cattle [M] . 3rd ed. Blackwell Publishing Ltd, Oxford, UK.

Barszcz K, Wiesetek D, 2012. Bull Semen Collection and Analysis for Artificial Insemination [J] . Journal of Agricultural Science, 4 (3).

Belihu K, 2002. Analysis of Dairy Cattle Breeding Practices in Selected Areas of Ethiopia [D] . Humboldt University, Berlin.

Britt J H, Scott R G, Armstrong J D, et al., 1986. Determinates of Oestrous Behaviour in Lactating Holstein Cows [J] . J Dairy Sci, 69: 2195.

Dunn B, Fawcett G, Fahey G, et al., 2009. Artificial Breeding of Beef Cattle [E] . https://futurebeef.com.au/wp-content/uploads/Artificial_breeding_of_beef_cattle.pdf.

Ewer T K, 1982. Practical Animal Husbandry [M] . John Wright & Sons Ltd: 42-44.

Garwe E C, 2001. Reproductive Performance of Crossbred Cattle Developed for Milk Production in the Semi-arid Tropics and the Effect of Feed Supplementation [E] . https://assets.publishing.service.gov.uk/media/57a08d6e40f0b6497400186a/R6955a.pdf.

Gebeyehu G, Asmare A, Asseged B, 2005. Reproductive Performances of Fogera Cattle and their Holstien

Friesian Crosses in Andassa Ranch, Northwestern Ethiopia [J] . Livestock research for Rural Development, 17: 131.

Gillespie J R, Flanders F B, 2010. Modern Livestock and Poultry Production [M] . 8th ed. Delmar Cengage Learning.

Graves W M, 2009. Dairy Herd Synchronization Programs [E] . http: //extension. uga. edu/publications/files/pdf/B 1227 _ 3. pdf.

Hafez E S E, 1993. Reproduction in Farm Animals [M] . 6th ed. Philadelphia, PA: Lea and Febiger.

House J, 2011. A Guide to Dairy Herd Management [M] . Meat & Livestock Australia Limited.

Hutchinson L J, 2009. Trouble-shooting Infertility Problems in Dairy Cattle [E] . http: //extension. psu. edu/animals/dairy/health/reproduction/infertility/trouble-shooting-infertility-problems-in-cattle.

Michael L, 2007. Dairy Reproductive Management Using Artificial Insemination [E] . https: //www. uaex. edu/publications/pdf/FSA-4007. pdf.

Moran J, 2012. Rearing Young Stock on Tropical Dairy Farms in Asia [M] . CSIRO Publishing.

O'Connor M L, 1993. Heat Detection and Timing of Insemination for Cattle [E] . http: //extension. psu. edu/animals/dairy/health/reproduction/insemination/ec402.

Rasby R, Vinton R, Steele J, 2015. Estrous Cycle Learning Module [E] . http: //beef. unl. edu/learning/estrous. shtml.

Tadesse M, Dessie T, Tessema G, et al. , 2006. Study on Age at First Calving, Calving Interval and Breeding Efficiency of *Bos taurus*, *Bos indicus* and their Crosses in the Highlands of Ethiopia [J] . Eth. J. Anim. Prod, 6 (2): 1-16.

MODULE 6:

IMPLEMENTING CALF AND HEIFER REARING

>>> INTRODUCTION

Calf is the future foundation stock for any dairy enterprise. Healthy, strong, genetically superior, and properly managed replacement heifers are significantly important for the culled cows are properly replaced and the dairy herds to be improved effectively and efficiently. Selection of replacements for culled cows can only be possible, if good replacement heifers are available adequately to allow for a more rigid selection. Accordingly, well managing and nurturing of dairy calves is crucial to be viable to sustain productivity and profitability of the dairy operation. Besides, in order for the dairy heifer to express her full genetic potential, she must be kept as contented and comfortable as possible (Godden, 2014; James, 2001; Roussel and Constable, 1999).

Therefore, this module covers all information required for calf and heifers rearing and management activities including calving-management, calf-feeding, and calf-management; calf-rearing facilities and housing; and heifers rearing and bull-management.

1 CARE OF THE COW BEFORE CALVING DURING CALVING

Calf is the mainstay of the future dairy herd. Thus, the advantage of calf rearing is to select calves from the cows-best for faster-genetic improvement, and to make replacements rearing on the farm are usually more economical than buying heifers. It also aids to prevent disease introduction onto the farm, and replacements may be difficult to buy, not calve when desired and their breed quality is uncertain. Successful calf-rearing is dependent upon calving, feeding and other management routines, particularly from the last 6 – 8 weeks of pregnancy (Godden, 2014; Fox et al. , 1999; NRC. 2001).

1.1 Care of Pregnant Cow

The first step in producing dairy calves is to have strong calves at birth. Weak calves are more difficult to grow than strong and healthy calves. The cows to give strong calves, needs to have at least six-eight weeks of rest (drying-off) between lactation, and must be provided

with a well-balanced ration and proper care.

The calf is the foundation stock of the future dairy farm. The care that is given to the calf in the womb is also considered. During the last 6 – 8 weeks of pregnancy [(281±10) days] on average a dairy cow should be fed liberally with balanced concentrate, in particular the concentrate mixture should be fed at the rate of 1 kg concentrate per 100 kg of body weight (James, 2001; Banerjee, 1998).

The method of feeding supplement to pregnant cows during the last 6 – 8 weeks of pregnancy with balanced concentrate feed is known as*steaming up*. If the cow is in milk during the last 6 – 8 weeks, it should be dried off (stop milking). Steaming up is useful for the following purposes:

- It provides sufficient nutrients for the rapidly growing (developing) fetus. It is known that 3/4 of the fetus development takes place during the last two months of pregnancy. During this time the nutrient demand of the fetus is very high. If the cow is not supplemented with supplementary feed, she draws on stored nutrient reserves (fat deposits) from her body and she may be harmed, and the calf may be weak and diseased.
- It prepares the cow for maximum milk production during the following lactation. The maximum milk yield is obtained during the second and third months after calving. This is possible if the cow is fed during steaming up.

Besides, the cow needs to be provided with laxative forage and green-fodder diet as a source of vitamins, minerals such as calcium (0.2% of the diet), phosphorus content (0.1% of the diet); and any trace-elements known to be deficient in an animal's body (John, 2011).

1.1.1 Maternity Management

During the last 4 – 6 days of pregnancy, the cow's udder starts to develop and fill with colostrum and the vulva swells even more and the pelvic ligament relax, causing the area between the tail-head and pin-bones to become loose and sunken (Zollinger and Hanson, 2003; Banerjee, 1998). As the calving process is approaching, these signs are followed by restlessness, relaxation of muscle around the tail head, swelling of the vulva, extremely extended udder, secretion of clear, glassy, stringy mucus from the vulva.

As soon as these signs are shown we have to expect the cow is giving birth within limited period and the cow should be closely watched and separated from the rest of the milking cows and put in a "maternity pen", cleaned, disinfected and well bedded with hay, saw dust etc. The maternity pen should be equipped with hayrack and concentrate feeding box.

According to Zollinger and Hanson (2003), John (2011) and Banerjee (1998), normally in 95% of the cases dairy cows give birth without any assistance; however, sometimes some cows may need external assistance. If there is calving problem or difficult birth or dystocia, we have to offer the necessary assistance, that is, if calving is not over within a short period after the onset of labor. Generally, a malformation or malposition is responsible for dystocia.

Thus, the cow must be watched closely, and moved to a closed clean-dry, well-lighted, ventilated and drafts-free maternity pen (box-stall of at least 12 ft ×12 ft in size, calf pen and equipped with a hayrack, concentrate feeding box) to protect the cows and new born calves from colds and drafts, and from the other cows (Zollinger and Hanson, 2003) (Fig. 27).

1. 1. 2 Calving Environment

Scours is a common disease problem, and the young calf may be infected by the cow and the environment into which it is born. As the cow's immunity is depressed around the time of calving, they are more likely to shed pathogens in their manure. Some of the pathogens may also proliferate in a moist environment.

It is important to clean the pens after each use and apply lime or other granular material to the floor before covering it with adequate amounts of dry-bedding such as clean straw. A good nonslip base and adequate amounts of bedding can prevent injuries and udder trauma during calving. Wet sawdust, mouldy/spoiled silage/hay is not used for bedding (FAO and IDF, 2011; Zolinger and Hanson, 2003). Many cases of infectious mastitis can be traced to contaminated bedding, especially green wood shavings and sawdust. Mouldy hay and silage, and manure contaminated bedding contain organisms that can infect the uterus, udder and the calves as well.

1. 2 Care of Calf at Calving and after Calving

1. 2. 1 Care during Caving Process

The fetus triggers the calving process by initiating a cascade of hormones that result in several biologic events. Briefly, when the fetus grows to a stage when uterine space becomes limited, the fetus becomes stressed and produces a hormone called cortisol ("stress hormone") that leads to several hormonal changes in the cow's placenta, stimulating stretching of pelvic ligaments, uterine contraction, cervix dilatation and consequent delivery. Thus, the fetus actually determines when it will be born (Zolinger and Hanson, 2003). The actual calving process can be divided into 3 stages that last up to 20 hours.

Stage 1: Preparatory stage

This is the opening period of the cervix during which the uterine wall leading to discomfort. A thick, cloudy, slimy discharge "the plug originally blocking the cervix" may occur. The calf alters from the position bringing its front feet up, so that they are extended forward, ready to lead the way through the cervix, and its nose also comes upwards (Fig. 6. 1).

During stage 1, cows typically show signs of discomfort due to the contractions. You may notice restlessness, arching the back, straining slightly and kicking at the belly. Cows may separate themselves from the rest of the herd (if not separated from the herd), and also urinate

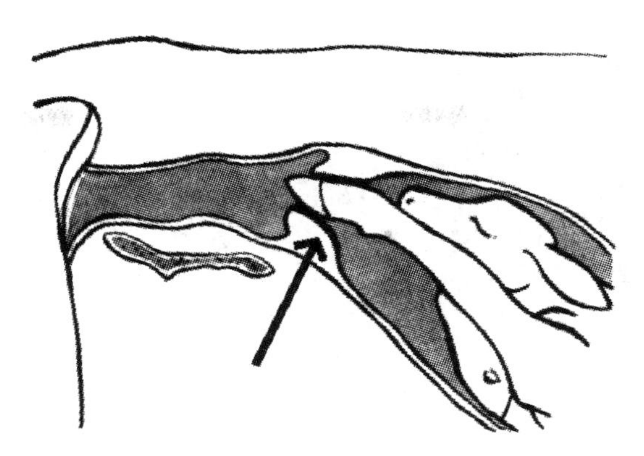

Fig. 6. 1 The cervix begins to dilate, allowing the foetus to enter the birth-canal
Source: Calving School Handbook, Oregon State University.

frequently. However, cows are still alert and fully aware of their surroundings, and may eat, drink, and behave normally. The end of stage 1 is typically marked by expulsion of the water-bag, which is the most external of the fetal membranes (John, 2011).

Stage 2: Fetal expulsion (60-120 minutes of duration)

Maternal estradiol stimulates mucus production by the cervix and vagina, which, together with placental fluids, thoroughly lubricates the birth canal to facilitate the delivery process. As the fetus comes into the birth canal, it puts pressure on the cervix and induces a natural reflex in the cow to push, resulting in visible abdominal contractions that further aid in fetus expulsion. The combined contractions of the uterus and abdomen stimulate the feet and its head to progress through the birth canal and put pressure on the placenta, reaching a certain level where it ruptures. Placental-fluids are then released and further help in lubricating the birth canal. The contractions continue to strengthen, and cows may lie-down to cope with the pain. Cow behavior may also change during this stage, as she may become oblivious of her surroundings, and focused on her contractions. After rupture of the placenta the birth is imminent, with the cow continuing to push and, hopefully, progressing normally through delivery (Fig. 6. 2).

Stage 3: Expulsion of the placenta (6-12 hours of duration after birth)

The placenta should detach from the uterus almost immediately after the calf is delivered. More specifically, the cotyledons on the placenta separate from the caruncles on the uterus and contractions expel the placenta from the cow. Sometimes the placenta expulsion is delayed because the cow is fatigued. However, if the placenta is retained for > 12 hours, special precautions may have to be taken (Zolinger and Hansen, 2003; James, 2001).

1. 2. 2 Abnormalities Requiring Correction in Birth

➢ *Leg-backwards*: The bend may be at the calf's knee. In this case, cup the calf's hoof in the palm of your hand and draw it forwards. If the abnormality has been detected at

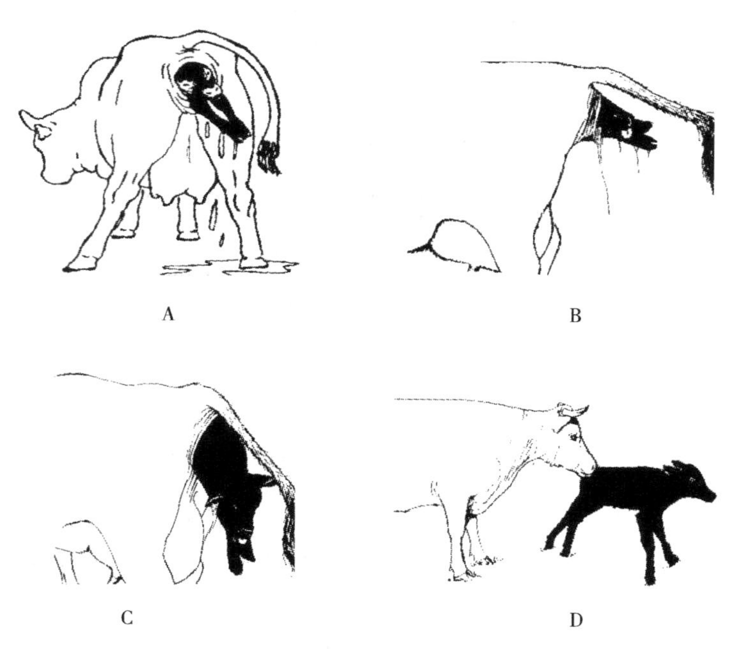

Fig. 6. 2 Normal calf delivery, leading to the end of the stage 2

Source: Calving Handbook, Oregon State University.

an early stage and the calf's head with leg are not already tightly locked in the pelvis.

➢ *Head-back*: Once the farmer is sure that the front legs are coming out, but the head not seen, the top of the calf's neck should be located. If possible, gently cup the calf's nose in the palm of your hand and draw it forward.

➢ *Backwards-delivery*: The presence of the feet in doesn't dilate the vagina because of reduced releases of oxytocin leads to reduced abdominal contractions. Secondly, when the calf's hips are passing through the vulva, its chest is entering to the cow's pelvis and so the umbilical-cord is constricted before the calf can breathe. Also, the umbilical cord is twisted around the leg, causing even more problems (Severidt et al., 2009; Senger, 2003).

➢ *Breech-presentation*: It is the most difficult abnormality to correct. In this posture, the calf is coming backwards, but with its hind legs pointing forwards, so that only the tail enters the birth canal. The absence of any object dilating the vagina means that the cow does not strain (USAID, 2012; Severidt et al., 2009; Senger, 2003) (Fig. 6. 3 – Fig. 6. 6).

1. 2. 3 Assistance in Delivery

Researchers have confirmed the key-points regarding calving assistance include:

➢ *Timing of intervention*: The cervix must be dilated before attempts are made to deliver the calf. Most mature cows will either have delivered the calf or be making good progress in delivering it within 30 minutes of starting vigorous pushing. If after

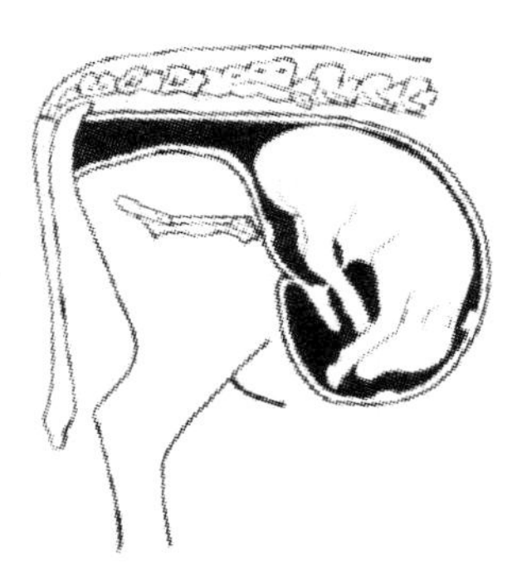

Fig. 6. 3 Posterior-presentation with rear legs extended under the calf's body (breech presentation)

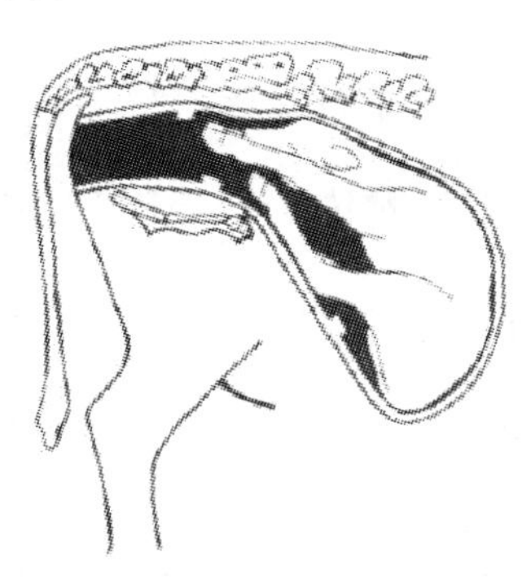

Fig. 6. 4 Anterior-presentation with the rear legs extended beneath the body

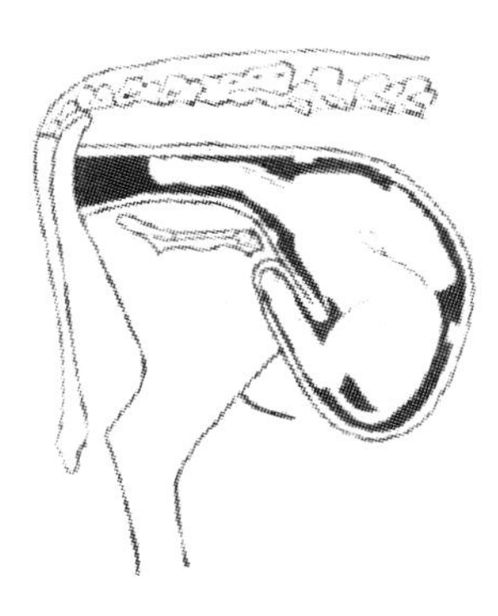

Fig. 6. 5 Anterior-presentation with the head & neck over the body

Fig. 6. 6 Posterior-presentation of fetus in an upside position

30 minutes of pushing the cow shows no signs of progress, she should be checked for complications. As heifers tend to take longer, wait for 60 minutes.

- *Sanitation*: Washing and cleanliness are important when calving is assisted to prevent dirt and manure being introduced into the uterus (John, 2011). The handler should be as clean as possible and wash hands and arms using a couple of buckets of warm soapy water. They should clean the vulva before introducing their hand to check the position of the calf.
- *Lubrication and patience*: The skill of a calving assistant is not measured by the speed of the delivery, but rather by the health status of the cow and calf following the delivery. Lubrication and dilation help to avoid tearing of the cow's reproductive tract. A cycle of lubrication, traction and relaxation is recommended (Severidt et al., 2009; John, 2001).
- *Positioning and angles*: When the cow is standing, the calf should be pulled in a downward direction (Fig. 28). If a cow lies down on her side during delivery, the

sides of the crush or race should able to be opened so that the calf can be pulled in the same direction. General assistance at calving is summarized at the end of this section.

2 CALF REARING AND MANAGEMENT ACTIVITIES

2.1 Post-partum Care of the Calf

It is important to help the calf after birth, especially if the cow does not get up to clean the calf. There are five-major things considered to take care of the calf:

Make sure that the calf can breathe: Clear the mucus of nostrils and stimulate breathing by tickling the inside of nostril with a piece of straw: the calf will sneeze, which helps to clear out the mucus. It is not recommended to hang the calf upside-down to drain fluids from the lungs, as most of the fluids are in-fact from the stomach and the position puts too much pressure on the lungs and makes breathing difficult. If the calf doesn't breathe on its own, an inexpensive option is to place a hand around the mouth and nose clamping it together, closing off one nostril, and blowing into the other nostril at 6 - 7 seconds intervals (USAID, 2012; Severidt et al., 2009).

The navel dipped with disinfectant: Dipping of the navel with 2% iodine-tincture or other suitable disinfectants such as chlorhexidine solution shortly after birth and before the navel is dry to prevent infection. Since these umbilical vessels connect directly to various organs in the body, infection at this site very harmful to the newborn. Note: Poor sanitation and mismanagement of the calving area cannot be overcome by navel-dipping (John, 2001).

Make sure the calf gets colostrum within the first 3 hours after birth: Colostrum is the calf's only source of protection from many infectious agents. Because the newborn calves are only able to absorb the immunoglobulin in colostrum within the first 24 hours of life, it is estimated that a calf should receive colostrum 10% of its body weight within that period. Colostrum may be frozen and stored when available excess, so it can be thawed and used when unavailable for another calf. Consider using a colostrometer to test the quality of colostrum obtained (Zolinger and Hansen, 2003).

Washing and sanitizing the teats and udder of the cow: Before milking of colostrum from the cow and then fed to the calf, or the calf is allowed to nurse the mother, wash the cow's teats and udder and sanitized using chlorine, or other sanitizing-solution, to decrease the amount of bacteria transferred to the calf through the digestive tract.

Warming the calf: In extreme cold-weather and cold-housing conditions, as the case may be, it is necessary to use blankets or a heat lamp. This will keep the calf warm until it is dry. As a precaution against infections, any cow that needs assistance should be given antibiotics, especially when the calving was difficult, and hands had to reach inside the uterus.

2.2 Feeding of the Young Dairy Calves

The plan of nutrition is crucial in the calf's weight gain rate and its future over-all productivity. The higher the plan of nutrition, the earlier will be the onset of puberty and entering into production. It is (the calf's feeding program is), therefore, designed with the goals: to reduce mortality rate (<5%) and maintaining good growth rate (400 - 500 g/day). The growth rate will vary as breeds, for the bigger breeds the aim should be to wean calves at 3 months at approximately 80 kg body weight (Zolinger and Hansen, 2003; James, 2001; Wilson, 1999).

2.2.1 Feeding Program of the Calves

Feeding of calves requires feeding program and it generally needs to understand the following points:

- On a commercial scale, necessary to separate newborn calves form their dams as soon as possible. Essentially, no space for calves in milking system/facilities!
- Fresh cows need special nutrition and feeding facilities to maximize their milking ability, thus calves can be housed more efficiently in separate facilities.
- Health and vigor of calves at birth depend on the nutrition of the cow during the last 60 days or so of gestation; Developing about 70% of birth weight of the calf during that time.
- Colostrum: 1) Not only provide antibodies that a new born calf lacks, but also "laxative" to help starting digestive functions; and 2) Under commercial conditions, calves rarely receive colostrum from their own dams, but no apparent difference in the effectiveness among "fresh, frozen/thawed, and fermented" colostrum.

Factors affecting feeding program

Researchers recommended that while developing the calf's feeding-program factors to be considered include:

- *The low immunity of the calf at birth*: The newborn calf is highly dependent on maternal antibodies for about two weeks. Colostrum contains maternal antibodies with short duration of absorption in the calf's body after birth. Thus, the calf mu0st suckle colostrum immediately after birth. If new animals are introduced into the herd just before calving, it may be necessary to vaccinate them against the common diseases so that they can develop antibodies and pass them on to their newborn.
- *The newborn calf is dependent on milk for nutrition in its early life as the rumen is not functional.* The suckling reflex forms a fold (groove) which serves as a pipe for delivering milk straight from the esophagus to the abomasum in young calves (bypassing fore-stomachs). Thus, young calves should only be fed on liquid diets as the groove will not allow solids to pass.
- *Calves are in short of enzymes for digestion except they secrete high amounts of*

lactase for breaking down lactose into glucose and galactose to supply energy. Thus, calves are fed milk having a high lactose level. During formulation of milk replacers, the sole energy source is milk sugar (lactose). Calves could not digest ordinary sugar "sucrose" since they haven't sucrase.

- *The calves' rumen is not functional; hence, calf cannot synthesize the B vitamins and they must be supplied in the diet*. The diet of the newborn calf should contain milk proteins since enzymes to break down complex proteins do not develop until 7 - 10 days after birth.
- *Introduce calf to solid feed*: As calf is introduced to solid feed, the rumen starts developing and the calf can be weaned as soon as it can consume enough dry feed (1.5% of body weight).

2.2.2 Digestion in Young Calf

Feeding of young calves is critical in raising replacements. During the first two months of life, the calves function primarily as a mono-gastric (simple-stomached animals). After about 2 months of age, they begin to function more like a full-fledged ruminant. In the first few weeks of life the rumen, reticulum, and omasum of the calf are relatively small in size and are quite inactive compared to the abomasum or "true stomach" . For this reason, young dairy calves have special requirements for protein, energy, and vitamins.

The liquid diet that the calf consumes bypasses the rumen, because of an esophageal groove closes when the calf consumes, and enters into the abomasum. Here it forms a curd, with the whey passing into the duodenum. The protein digestion by the neonatal calf is restricted to casein, for which an enzyme rennin is produced in the abomasum. If the milk protein is denatured, such as in pasteurized (UHT) milk, it takes a long time for the clot to form and the calf may get diarrhoea or scours, often with an accompanying *E. coli* infection.

Newborn calves cannot utilize vegetable protein before their rumens are functional as a result of having limited digestive enzymes. It is, therefore, following feeding colostrum, whole milk, fermented colostrum, or milk replacers containing milk protein or specially processed soy concentrates are to be used. Young calves limited to digest starch or some sugars such as sucrose, and/or unsaturated fats like corn and soybean oil because of the lack of certain digestive enzymes.

2.2.3 Feeding of the Calf from Birth to Four Days

(1) Colostrum feeding method

It is absolutely vital to provide the calves "maternal-antibodies" at least within the first four hours of life. *Colostrum*, the first feed of the newborn-calf, is secreted from the mammary gland (udder) of the cow in the first 4 - 5 days after parturition. It contains a mixture of blood-plasma, nutrients, and antibodies that are essential for the calves to be viable.

Within hours of suckling colostrum from its dam, the calf absorbs protective antibodies (passive immunity) into the blood stream and other immune cells into its lymph nodes that help to fight-off infection. If the calf fails to suckle or, for some reason, does not receive an adequate amount of colostrum, it must rely on its naïve-immune-system to develop protection soon enough to avoid clinical-disease (John, 2011; Severidt et al., 2009).

For most-infections it takes the immune-system 6 – 10 days to respond adequately. If management and environmental factors depresses the calf's resistance and, the infectious agents present in large number or particularly strong, the calf's immune-system is overwhelmed, and the calf succumbs to disease (Fig. 29).

(2) Natural protection of calf

A newborn calf does have some natural protection against infectious disease. For example, its skin, tears, saliva, and digestive juices are natural barriers for some harmful microbes. However, colostrum provides an additional and immediate source of natural protection. Ingestion of this antibody (immunoglobulin) and immune-cell rich milk is critical for newborn calf survival. The dam's serum antibodies (IgGs) and some important immune stimulating cells are concentrated in the udder as colostrum during the last month of pregnancy. For maximum protection, a calf must receive an adequate amount of colostrum within 4 – 12 hours of birth (Zollinger and Hansen, 2003). Consistency is lacking at the time that calf should get colostrum.

Colostrum is much valuable that it differs from ordinary milk in that it is extremely valuable to give the calf immediate resistance to bacterial diseases in its higher content of lacto-globulins (immunoglobulin's namely IgG, IgA, IgM, IgD and IgE).

- It has nutritive value, approximately 40% greater than ordinary milk, highly digestible and giving a soft curd during the process of digestion.
- It contains higher protein mostly in globulin forms with protective antibodies to protect calf against bacterial disease.
- Permeability of immunoglobulin in the calf's small intestine declines rapidly after 24 hours, mainly after the meconium is passed away (Roy, 1990).
- Its laxative effect aids in an expulsion of the first fetal dung meconium, which is a black-green sticky substance that has accumulated in the gastrointestinal tract during the calf's life in-uterus and is usually passed at about 24 hours after birth as waste material. Licking of the calf's anus by the cow during suckling probably stimulates its expulsion.

Colostrum quality is influenced by age and health status of the dam, the timing of colostrum harvest relative to birth, hygiene of the equipment used to harvest and stores the colostrum. As bacteria can multiply in colostrum, it must either be fed directly to the calf after being harvested or cooled rapidly at 4℃ for short-term storage. Inadequate colostrum consumption can arise either from lethargy in a cow/calf because of difficult-calving, or from the calf having difficulty in locating the teats, in this case the calf needs to be assisted and guided the teat whilst expressing colostrum from the teat to encourage the calf (Table 6. 1).

Table 6.1 Average composition of colostrum and milk

Nutrients	Colostrum	Milk
Water (%)	74.5	90
Ash (%)	1.6	0.7
Fat (g/kg)	36	35
Protein (g/kg)	140	33
Immunoglobulin (g/kg)	60	1
Casein (g/kg)	52	26
Albumin (g/kg)	15	5
Lactose (g/kg)	30	46
Vitamin A (μg/g fat)	42 - 48	8
Vitamin B_{12} (μg/kg)	10 - 50	5
Vitamin D (ng/g fat)	23 - 45	15
Vitamin E (μg/g fat)	100 - 150	20

Source: Emily Barrick.

(3) Colostrum substitutes

The colostrum substitute can be formulated and prepared as (Wilson, 1984):

➢ Whip-up a fresh egg in 0.85 liters of milk

➢ Add 0.28 liters of warm-water

➢ One teaspoonful of cod-liver oil

➢ One desert-spoonful caster-oil

The quantity given will make one-feed; give 3 feeds/day for at least 4 days.

2.2.4 Feeding Calves from Four Days onward

1) Whole-milk feeding

Researchers concluded that whole milk is fed by bucket until the calf is 12 weeks of age at the rate of 5 - 10% of their body weight. Some simple rules to avoid digestive upsets and promote good growth include washing and sterilizing the calves feeding utensils to prevent cross-contamination of pathogens. Small portions of milk left in buckets and other utensils cause extensive bacterial development, since milk is the best medium. Many of the bacteria, developed when taken into the calf's digestive tract, will produce putrefactive changes, in-turn scours, and most digestive disturbances can frequently be attributed directly to failure. Thus, clean the milk buckets used for feeding. Avoid under/over feeding calves with milk.

➢ Overfeeding can lead to scouring, particularly in the first month of life.

- Feeding milk powders in only a small amount of water helps to prevent digestive upsets in young calves.
- The smaller amount of liquid (clean water) also encourages them to eat dry feeds and commence rumen development.
- The daily allowance of milk can be given with one or more feeds.
- Frequent feeding of small amounts helps to prevent scouring.
- Avoid sudden changes of any feed, but gradually increase the proportions of the new feed.

Milk temperature must be the same at each feeding. Feed cold milk in warm weather, but it must be warm (35 - 38℃) to very young calves in wet or cold weather.

The calves can also be fed from buckets with rubber nipples and tubes. It takes longer to teach calves to feed from a bucket, but there is less cleaning up, and the chance of leaving milk residues and contamination is less with buckets than with nipples and tubes. Group feeding may be very useful, but calves should be similar in age and size. This system permits a chance of contamination if the equipment is poorly cleaned (Severidt et al., 2009).

Limited whole milk (Dry calf starter method)

The importance of this method is to reduce the amount of milk given to the calf. The calf is fed whole milk for 6 - 8 weeks from the 2nd week of age with total amount of 114 - 136 kg of milk. Together with a dry calf starter, a sort of balanced feed for calf with high quality protein. Usually, high quality protein is given in the form of animal protein (meat meal, fish meal and blood meal). A dry calf starter is allowed freely until the calf consumes 1.4 - 1.8 kg starter/day until the calf is 3 - 5 months (John, 2011; FAO, 2012).

Calf-starter concentrate mixture consists of:

Milk replacers: Where production costs of feeding whole milk are high and the product is sold milk, many dairymen are turning to milk replacers as a means of lowering the cost of raising young stock. Milk replacers consist of a dry feed mixture that is reconstituted with warm water according to the manufacturer's instructions and fed as a replacement for milk. Whole milk is replaced by the milk replacer gradually at the end of the first week. This milk replacer is equivalent to whole milk. Calves fed with milk replacer twice a day, then after 50 days old, it may be reduced gradually and be discontinued when the calves are two months old (Bong, 2001; FAO, 2012). Milk replacer is cheaper and more digestible than milk, which contain nutrients listed in Table 6.2.

Table 6.2 Components of milk replacer for calves

Nutrient	Recommendation
Crude protein, %	22
Ether extract, %	10
Calcium, %	0.7
Phosphorus, %	0.6
Magnesium, %	0.07
Potassium, %	0.65
Sodium, %	0.1
Sulfur, %	0.29
Iron, mg/L	100
Cobalt, mg/L	0.1
Copper, mg/L	10
Manganese, mg/L	40
Zinc, mg/L	40
Iodine, mg/L	0.25
Selenium, mg/L	0.3
Vitamin A, IU/lb.	1,730
Vitamin D, IU/lb.	273
Vitamin E, IU/lb.	18

* Should be considered as minimums. Many commercial products exceed the NRC on certain nutrients.
Source: Aseltine, 1998.

Artificial milk replacers: The most common method for rearing calves artificially is to feed them with milk or a milk substitute from a bucket, so that most of cow's milk is kept for sale. There are few problems if the correct amount of feed is given and hygiene is strict. The behavior of the calves of *Bos indicus* is often differs from that of *Bos taurus*, which making it difficult to train them to drink from a bucket. These can be bottle-reared, but even then, they do not grow as well as they do if suckled. If the calf is removed immediately after birth, the problems with training the cow to milk and the calf to drink are reduced. Artificial rearing means that calves can be weaned from milk to other feeds earlier, without the problem of the cows drying up. From approximately 4 days of age, many calves are fed 'artificial' milk, milk replacer, for 5 - 6 weeks. Milk replacers are based on dried milk powder, produced from surplus cows' milk (Matthewman, 1993; Hamberlain, 1989).

2) Skim-milk feeding

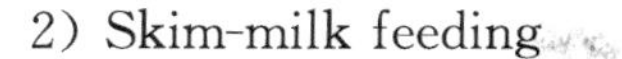

Skim milk is milk from which the cream (butterfat) has been separated to prepare butter; hence it is the by-products used to feed the calf. There are two types of skim milk: fresh skim milk and skim milk powder (Severidt et al., 2009).

- *Feeding fresh skim-milk*: In this method fresh re-constituted skim milk is used to replace whole milk or the milk replacer in calves' feeding. While using skim milk, it needs to supplement the ration with vitamin A and D, since these are removed together with cream. On a dairy farm where sufficient amount of skimmed milk is available, it can be fed to the calf at 8 - 10% of their body weight daily from 6 months to 1 year old with simple grain mixture.
- *Feeding milk powders*: The amount of milk powder fed to a calf will determine its growth rate, and one must decide whether a faster growth rate at an earlier age justifies the high cost of the additional milk replacer.

A conventional milk powder for calves contain at least 60% dried skimmed milk, 15 - 20% added fat, and minerals and vitamins. Skimmed milk is composed mainly of lactose and milk proteins, principally casein. Whey, which is a by-product of cheese making, is composed of lactose and the whey proteins, principally albumin and globulin. Conventional skimmed milk powder provides all the nutrient requirements of the newborn calf. Only some antibacterial agents in milk, such as the lacto-peroxidase complex, are denatured by the processing, but these can be added artificially (Table 6.3).

Table 6.3 Feeding schedule for a calf with a birth weight of 35-40 kg

Age	Milk (colostrum)			Concentrates +water	Roughage
	Morning	Afternoon	Evening		
2nd day	1	1	1	—	—
3rd day	1.25	1	1.25	—	—
4th day	1.25	1	1.25	—	—
5th day	1.5	1	1.5	—	—
6th day	1.5	1	1.5	—	—
2nd week	2	—	2	*ad libitum*	hay, grass or silage
3rd week	2.5	—	2.5	*ad libitum*	hay, grass or silage
4th week	2.5	—	2.5	*ad libitum*	hay, grass or silage
5th week	2.5	—	2.5	*ad libitum*	hay, grass or silage
6th week	2.5	—	2.5	*ad libitum*	hay, grass or silage
7th week	2	—	2	*ad libitum*	hay, grass or silage
8th week	1	—	1	*ad libitum*	hay, grass or silage

Source: Emily Barrack.

Calves require about 360 g of milk replacer daily in two feeds of 1.5 liters, after finishing their colostrum feeding at 4 days of age, for them to grow at a rate that will allow them to calve at 2 years of age. Later this can be increased to up to 1 kg milk replacer daily, diluted to 7 liters with water, which should be sufficient for 50 kg calves to grow at about 0.7 kg/day (Phillips et al., 1999).

2.2.5 Feeding Methods

Bucket feeding

Milk replacer is reconstituted once or twice daily in bucket feeding systems. It is normally mixed with one half of the water at 46℃ and then the remaining water added is either hot or cold, as required to achieve a final temperature of 42℃ to achieve optimum digestibility (FAO and IDF, 2011).

Bucket feeding (Fig. 30) is not a natural process for a young calf, whose instincts direct it to suckle from birth. However, calves can be successfully trained to drink from buckets using two upturned fingers immersed in the milk, to simulate the mother's teat. When the calf has learnt to suck milk around the fingers, they can be withdrawn gradually, and the calf will drink unaided.

The stress caused by the absence of a suckling stimulus is exacerbated by the calves' individual penning. Individually penned calves may also spend a lot of time licking objects in the pen, which may be caused by inadequate mineral supply, principally sodium. Sodium requirement is increased by stress in individual penning and additional salt can be added to the concentrate, but this will increase water consumption and urination, so that it is important to regularly ensure that the calf has enough clean bedding (Rutgers and Grommers, 1988).

Group-feeding systems

Milk may also be fed to calves housed in groups by a length of pipe to a large container, provided that the milk is acidified to keep it from spoiling. Usually weak organic acids are added to reduce the milk-pH up to 5.7, since casein will clot if the pH is reduced any further. Stronger acids can only be added if the powder is whey-based, to give a pH of about 4.2 (Webster, 1984). Acidified milk replacers are usually made available *ad libitum*, and intakes may be 20–30% more than if the calves are fed from buckets. The intake depends on the temperature of milk, and whereas the milk is warmed to 42℃ for bucket feeding, it is convenient for group feeding of acidified milk replacer *ad libitum*. If the milk is at ≤10–15℃, it chills decline the calves' intakes and so does the calf growth rate. Not all calves respond equally, and it needs to keep a careful watching for calves are not thriving on the system. The advantage of group-rearing and offering calves acidified milk-replacer is saving labor; 10 calves/group and regular inspection of teats for damage is recommended (Rutgers and Grommers, 1988; Dellmeier et al., 1985).

Introducing calves into solid feed

For the first few weeks of a calf's life the milk replacer feeding systems described above will provide sufficient nutrients for near maximum growth rates in the calves. However, because of the high-cost of milk-replacer and the requirement by the calf for some solid-food, it is usual to offer concentrate feed from about 1 week old. This allows the rumen papillae and rumen motility to develop properly. Many calves that do not have access to solid-food develop hairballs in their rumen, caused by the consumption of hair during their licking and grooming activities and the lack of rumen motility to transfer them to the abomasum (Fig. 31).

Concentrate diet offered can be either small pelleted (3 – 5 mm) compound-food or a 'coarse mix', which is a blended mixture of concentrated ingredients without any attempt to bind them together. Both foods produce similar growth rates in the calf, however, the pelleted products advantage in the selection by the calf is not possible, and little feed wastage. Despite this fact, it is too hard to chew as the calves eat less, and coarse mixes tend to be dusty, leaving a mixture of dust and saliva on the bottom of the bucket (Randall, 2009).

Careful management is necessary to feed calves properly. If milk is offered *ad libitum* only a small amount of concentrate feed will be consumed during the first few-weeks, and a greater check to growth rate occurs at weaning. It is also important for the calves having concentrate available to them. Intake increases over the first few-weeks of life until it is normally 1 – 1.5 kg at 5 – 6 weeks old (FAO and IDF, 2011).

2.3 Feeding Heifers, Bulls, and Dairy Beef

2.3.1 Four to Twelve Months of Age

If heifers are properly introduced to solid feeds before weaning, a growing ration can be changed gradually so that they reach puberty at 15 months of age. Not sufficient for the animal to satisfy the energy need from forages alone, thus feeding some grain is necessary untill year of age:

- Summer: Pasture, hay, and grain mix (3 – 7 lb/d depending on body size and forage quality)
- Winter: Hay, silage, and grain mix (3 – 7 lb/d depending on body size and forage quality)

The same forage and grain mix used for the milking herd can be used for heifers:

- Should vary "inversely" the protein content between the grain mix & forage.
- A free-choice mineral mix is recommended. Should include Ca, P, salt, and trace minerals with a poor forage.
- Suggested grain mixes for the growing calf: Should be limited to no more than 5 – 7 lb daily along with free-choice forage consumption (Table 6.4).

If necessary, limit grain to keep calves from becoming too fat:

- Excess fat can develop breeding problems.
- Also, produce less in later life vs. those reared on a more moderate nutrition possibly

because of excess fatty tissues in the udder.

Table 6.4 Suggested grower rations for 200 kg dairy calves (4-12 months of age)

Ingredients	1	2	3	4
Ingredients, %				
Corn, cracked	78	—	—	50
Oats, rolled	20	35	—	27
Barley, rolled	—	50	76	—
Ground ear corn	—	—	5	—
Molasses, liquid	—	5	17	20
Soybean meal	—	8	—	1
Limestone	—	—	—	—
Dicalcium phosphate	1	1	1	1
Trace mineral salt	1	1	1	1
	100	100	100	100
Calculated analysis				
As-fed basis				
Crude protein, %	9.2	13.8	13.9	16.7
TDN, %	74.9	70.0	71.1	72.8
NE_m, Mcal/kg	1.87	1.71	1.84	1.82
NE_g, Mcal/kg	1.29	1.16	1.27	1.25
Calcium, %	0.25	0.33	0.35	0.68
Phosphorus, %	0.48	0.56	0.49	0.56
Dry matter, %	87.9	88.4	86.7	88.6
Dry matter basis				
Crude protein, %	10.5	15.6	16.0	18.8
TDN, %	85.2	79.2	82.0	82.2
NE_m, Mcal/kg	2.13	1.93	2.12	2.05
NE_g, Mcal/kg	1.47	1.31	1.46	1.41
Calcium, %	0.28	0.37	0.40	0.77
Phosphorus, %	0.55	0.63	0.56	0.63

Formulations are on an as-fed basis. Ration 1 is recommended to be fed with legume hay (14 – 17% CP). Rations 2 and 3 should be fed with legume-grass mixed hay (10 – 13% CP). Ration 4 is recommended to be fed with grass hay (6 – 9% CP).

Dairy calves should consume daily: 2.0 – 2.5% of their body weight as dry matter forage and 0.5 – 1.0% as dry matter grain mix.

Source: Jurgens, 2002.

2. 3. 2 From 12 Months of Age to Calving

Heifers should have sufficient rumen capacity to meet their nutrient needs from good quality forages.

- Should be gaining 1. 5 – 1. 8lb per day
- Feed grain mix only when/if forages are poor or limited in amount.
- Summer: Use pasture and (or) hay, and feed 2 – 8 lb of grain mix if necessary, depending on the body size.
- Winter: Use hay and silage, and feed 2 – 8 lb of grain mix if necessary (depending on body size).
- Provide minerals free-choice: Include Ca, P, and salt, and trace minerals if feeding poor forages.

To breed at 15 months, heifers should be weighing 550 (Jerseys) to 800 lb (Holstein and Brown Swiss). Heifers should gain about 1. 75 lb/day from birth. Growing heifers use available nutrients in an irreversible order: 1) daily maintenance, 2) growth, and 3) ovulation and conception. Avoid over-conditioning to prevent impairment of reproductive efficiency and also reduced milk production because of fatty deposits in the udder.

Management techniques for early conception:

- Flushing: Increase the intake of all the nutrients to heifers with appropriate age
- Bypass protein: Use during the first breeding period
- Proteinated trace minerals: May improve the breeding efficiency
- Ionophores: Not only reduce waste caused by methane production, but also spare intake protein by reducing ruminal ammonia production

2. 3. 3 Nutrition of Bred Heifers

Feeding to about 60 days before the expected calving date, you should aim for growth, yet avoid excess fat deposition, especially in the udder. In the last 60 days of gestation or transition period, start feeding a grain mix and increase gradually to adapt heifers to high grain intake, which will be necessary for lactation after calving, by doing so:

- Adjust the rumen population to increase microbes that ferment specific feeds in a lactation ration
- Increase nutrient intakes to increase body reserves necessary to support early lactation
- Provide for the increased demand for nutrients because of rapidly developing fetus

2. 3. 4 Feeding Bulls

Because of today's widespread use of artificial insemination, only a few dairy bull calves are raised for breeding purposes. Bull calves should be fed and handled much the same way as heifers, but bulls grow faster than heifers, thus should receive more feed.

Older bulls should be kept in thrifty, vigorous condition, but not too fat. Mature bulls can be maintained on forage with about 0. 5 lb of grain per 100 lb of body weight, if needed-The same grain ration as the one being fed to lactating cows.

Forage for young calves

Forages offered to calves are to be best quality. Palatable, fresh and leafy pasture grass is crucial to the calves for its proper rumen development and nutrients required for growth. Calves >2 weeks old must have access to fibrous food at least 100 g/ (day · calf), then increased to 250 g/ (day · calf) at 20 weeks old. Encouraging stable rumen conditions at an early age will help to avoid any nutritional disturbances, such as bloat or scouring. Forage provided to young calves is controversial, since too much reliance on low-quality forage reduces growth rates. Calves that rely too much on coarse forage develop a large rumen at the early age, which makes them look 'potbellied', and this may result in low weight gains. However, forage is important to create stable rumen conditions, particularly with high and relatively constant rumen pH. The addition of 15% straw to an otherwise all concentrate diet for calves will increase intake as they require fiber in their diet (USAID, 2012; Severidt et al., 2009).

2.3.5 Weaning

This is defined as the withdrawal of milk/milk replacer and the calf becomes fully dependent on other feeds. Traditionally, most dairy calves are weaned based on age, 12 weeks being the most common. Early weaning is possible if more milk is fed and calves introduced to pre-starter and starter early in life (Paul et al., 2001).

To minimize stress, weaning is practiced gradually; reduce milk feeding from twice to once/day, then once every other day to allow the calf's digestive system to adjust to the new diet. The criteria that have been used to determine weaning time include: age of the calf, attainment of twice the birth weight, and the amount of dry feed consumed by the calf (1.5% BW).

The age at weaning may determine the ability of the calf to grow adequately in its early life. Weaning at a too early age and without adequate high-quality solid food may lead to reduced growth rates for much of the calf's first year life (Berge, 1991).

However, the high cost of milk substitutes and whole milk to feed calves, as opposed to sale for human consumption, encourages farmers to transfer calves to solid feeds at an early age. In addition, calf rearing is time consuming, especially if they fed once/twice/day. Early weaning (5–8 weeks) may adopt to reduce the period of milk feeding and labor for calf-rearing. This needs a specific feeding program with low levels of milk and high energy-protein concentrates, preferably pelleted to stimulate rumen development. Liquid milk/replacer is reduced from 3 weeks old to encourage maximum intake of dry feeds by the calf (Rutgers and Grommers, 1988).

2.4 Calf Managements and Husbandry Routines

Housing of the young dairy calves is an important component of calf management. Claves are sheltered from adverse weather conditions (rain, cold, wind and severe heat), predators; and from reservoir animals and disease agents. The way calves are housing depends mainly on the climate and weather conditions. The colder the climate, the more attention is paid for housing. Since, in the first 10 days of calf's life the thermoregulatory mechanism is poor, causing death of the calf easily from heat stress or from wetting/chilling (Cook et al., 2003).

2.4.1 Housing and General Management Activities

A closed fixed shelter in hot-wet climate can be a serious source of diseases. Fixed shelters must have well-drained floors to be clean daily. Pens must be clean, dry, protect the calves against extreme temperature while providing adequate light and space, and prevent calf-to-calf contact, and devoid-off wet and filthy bedding materials. Good ventilation is necessary to remove excess CO_2 and NH_3. Calves are kept in an individual pen/hutch until they are 10 weeks old, and for the first several weeks after its purchase. Calves never housed together with adults in a cowshed.

Calf pens design

A badly designed house may increase the risk of disease. The risk of pneumonia is increased by poorly ventilated houses because of increased risk of cross infection and concentrated bacterial and viral agents. Each group of the young calves is sheltered in a separate individual shed, which is quite close to the shed/housing of the cow if possible. Each calf shed has an open paddock or exercise yard. For instance, 100 ft^2 areas for a group of 10 calves with a concomitant increase of 50 ft^2 for every additional calf will make a good paddock. Efficient management system classifies one-year old calves into 3 age groups viz., age of 3, 5 - 6, and > 6 months old with better allocation of the resting area (Albright et al., 1999). Table 6.5 shows the space requirement to each group.

Table 6.5 Space requirement recommended a group of calf

Age of calf (months)	Space (ft^2)
<3	20 - 25
3 - 6	25 - 30
6 - 12	30 - 40
>1 year	40 - 50

Source: Zollinger and Hanson, 2003.

Air space (400 - 500 ft^3/calf) is a good provision under tropical climatic conditions. Never neglect provision of water troughs inside each calf shed and exercise yard. Calf pen can be quite simple; constructed where possible from locally available materials. However, several features are considered in selecting and/or building individual pen like:

➢ Solid walls on 3 sides to prevent drafts and keep calves from sucking each other, and ease of cleaning.

- Approximately 2 m^2 (1.2 m×1.6 m) (24 ft^2) space/calf having a water cup, grain box and hay rack within the walls, and adequate ventilation.
- Well drained/bedded with material kept both clean and dry, in convenient location with an arrangement that will encourage quality care and observation.
- Distance between two calf hutches 1 m and 4 m apart from two rows of calf hutches.
- Strong to stand predator's invasion.

Calf hutches are best to use in tropical areas, consisting of two sections: a cell (covered area) bedded with wood shavings, sawdust or straw, or the cell could be covered with canvas or plastic sheet; and open exercise yard provided in front of the cell. The calf is provided with food and water on bucket (Fig. 32 and Fig. 33).

Advantages of calf hutch

- Low cost
- Easy to design and build
- Easy to clean and disinfect
- fewer disease problems
- Easy to move from contaminated areas
- Gets the calf used to existing weather
- Adequate natural ventilation

Management Activities

Management of dairy heifers 'calves-mean more intense management procedures. Some tips to optimize the care of young calves are to pay closer attention to feed quality like a high fat milk replacer (>10% fat), environment, and checking calves at least twice each day. Keeping on top of health care management during this stressful period will increase the chances of producing healthy calves (Albright et al., 1999).

2.4.2 Identification Systems of the Calves

Identification of the calf is an important activity to maintain a herd. Identification of each calf is positively carried-out immediately after birth or before the calf is removed from its dam to allow efficient and proper recording. This is required if calves are to be registered and it is essential for good breeding program records. A neck strap or chain with a number or a metal or plastic ear tag can be used for identification. The ear tag or registration numbers of calves, sires, dams, and the birth dates are to be entered in a permanent record book. Permanent identification methods include photographs/sketches, tattoos, and freeze branding as shown below:

Ear marking

It is the method of identification that includes:

- *Ear notching*: cut part of ear using an agreed code. This mark is permanent but exposes cow to infection.
- *Ear tattooing*: Uses a special tool to put inked numbers on an animal's skin. This method is done most commonly in the ear, though can be used on the lips or other locations. It is permanent, simple and relatively painless; however, it is hard to read from a distance, and does not work in dark animals.
- *Ear tagging*: Use special pliers/applicator to attach pieces with numbers on them. it is easy to read from the front view, but expensive, not permanent, and can be lost or removed (Fig. 34 – Fig. 36).

Ownership ID: Branding

This method comprises chemicals, cold-branding-freeze, and branding-hot iron.

1) Hot-iron branding

It is for a short time on the legs so as not to spoil the skin. Selection of the brand should be simple and ventilated with the face smooth.

Procedures include:

- Heat the irons to red-hot.
- Restrain the animal.
- Place brand on hide.
- Check it until the surface of the brand is a shoe brown color.

2) Fluid-branding

This method uses potash or acids, which can kill the hair follicles and there is more room for accidents and error to animals and humans. As a result of this it must be more careful because

3) Freeze-branding

It is permanent but not common in dairy cattle. It can be operated by using super cold iron, dry ice and alcohol or liquid nitrogen spray.

Practical procedures:

- Super-cold iron will burn.
- Dry-ice and alcohol or liquid nitrogen or spray can seal.
- Restrain the animal.
- Clip the brand area.
- Clean the area & place irons to get cold.
- Place iron firmly until the area is sealed and harden.
- Usually 50 seconds for dry-ice/alcohol method
- 25 – 30 seconds for liquid nitrogen

➢ 3 – 4 weeks hair will grow back.

2.4.3 Vaccination

Vaccination is another strategy to promote the immunity of calves. The pregnant cow can be vaccinated during late pregnancy to boost her immunity, which is passed onto the calf through feeding colostrum, or the calf can be vaccinated.

There is a diversity of diseases and vaccines available in different countries. An appropriate vaccination program requires local knowledge of endemic disease and vaccines available. In Australia, cattle are commonly administered vaccines to prevent clostridial diseases and leptospirosis, while other vaccines are available against diseases caused by *Moraxella bovis* (pinkeye), *Salmonella*, enterotoxigenic *E. coli*, *Pestivirus*, *Mannheimia haemolytica*, *Vibrio*, *Babesia bovis* and bovine ephemeral fever virus (John, 2011; James, 2001).

2.4.4 Infectious Disease Control

Since many of the calf disease agents are carried by the cows, those agents will be in the calf's environment when it is born, and in large doses, especially if all the cows are crowded in an area that also serves as the maternity area. Whether an agent causes a disease or not depends on how potent its disease-causing ability (pathogenicity) is, the number of organisms the calf is exposed to (dose), and the number of antibodies carried by the calf (strength of immunity). Infections through the navel (navel ill) by invasive *E. coli* occur at birth, especially under conditions of heavy contamination of wet muddy maternity areas. From the navel, the infections commonly spread to joints (joint ill), belly cavity, heart-sac, and brain. Clean calving areas and the practice of soaking the navel with a strong tincture of iodine soon after birth seem to be important and logical factors for preventing navel infections (John, 2011; Chiba. 2009).

One of the management objectives is to keep the environmental load of ever-present pathogens at the lowest possible level. The area where the calf is born is of concern because the time before suckling is when the calf is most susceptible. Therefore, the calving area should not be the area where (possible disease carrying) cows have been congregated before calving. The calving area should be chosen so that calves will not be born in muddy areas contaminated with feces and urine, and individual cows should be placed in this calving area only when calving is imminent (Chiba, 2009).

Since diseased calves shed vast quantities of infectious organisms, and calves may show diarrhoea as early as 2 – 3 days of age, a system of segregation should be designed to prevent exposure of young healthy calves to large doses of infectious organisms shed by unidentified carrier cows and sick calves. Ear tagging and dipping of the navel with a strong tincture of iodine at birth should be followed by moving the cow/calf pair from the calving area to a cow/calf area. Should any calf in this second area begin to show signs of illness, the pair should be moved to a sick pen/hospital area for thorough evaluation and treatment if indicated. The calving area and the second cow/calf area should be as free as possible of cross contamination with excretions of ill animals.

The location and design of the sick pen area should take into account the weather conditions and treatment ease. Adequate shelter, power for heat lamps, and dry bedding are minimal requirements. Water troughs should be low enough for calves, and loose salt (1/2 sodium chloride and 1/2 potassium chloride) should be accessible by the scouring calf. Low blood potassium is characteristic of some of the most depressed scouring calves.

This type of arrangement has been helpful in preventing or minimizing diarrhoea outbreaks, and in providing adequate supportive care for sick calves. With the water and salt available, some calves will treat themselves. If a calf is too dehydrated and depressed, the manager will have to supply adequate amounts of the right type of balanced electrolyte/fluid, perhaps by esophageal feeding (Chiba. 2009; John, 2001).

Of all the calf disease agents, *Salmonella* species, a bacterium, is probably the most fearsome. Fortunately, it is not as much a problem in beef herds as it is in dairy calf-raising facilities. When it occurs among beef calves, it can often be traced to a sale-yard dairy calf grafted on one of the beef cows. It is advisable to purchase calves for this purpose from dairies where there is good calf-rearing husbandry and little disease. There are no guarantees, however, since wild mice can carry the *Salmonella* bacteria. The grafted calf and cow should be kept separated from the rest of the cow-calf herd for at least 10 days (Chiba, 2009).

2.4.5 Castration

Castration is a component of the management activities, which means the processes of crushing or cutting the spermatic cords of the male. Skill is very important for proper castration, since bad castration is distressing and dangerous for animals. Castration is important to make animals less aggressive and stops them fighting each other. They are docile and easier to handle, it stops poor quality animals from breeding, and prevent males mating with immature or closely related females. Castrated animals grow faster and increase faster body weight gains. Castration may be performed on bull calves, except those being raised as veal calves. It is usually best to castrate animals when they are a few days old before they are weaned performed at the dry-season when there are not many flies. The best time for castration of calves is between 8 – 10 weeks of age. Young animals recover quickly (Zollinger and Hanson, 2003; Forse, 1999).

Castration can be done either by making an operation in the scrotum where the vas deferens are disconnected from the scrotum or by using a Burdizzo to crush the spermatic chords (Fig. 37) to destroy channels of nourishment.

Warning

Do not crush the spermatic cord or the testicles with a hammer or stones. This usually causes the animal much pain and distress; there are safer and better ways to castrate animals.

Burdizzo's castration method

The burdizzo is an instrument used to cut off the blood supply to the testicles, causing cell death of the testicular tissues resulting in degeneration of the testicles. The best time to apply the burdizzo is 2 – 3 months after birth when the spermatic cords can be felt. The bloodless operation leaves no open wound in which screw worm and other infections could develop. The burdizzo is applied to each spermatic cord separately in such a way that the blood supply to the testicles is damaged, while circulation to the scrotal sack remains intact. Gangrene can set in where blood circulation to the scrotum is lost.

The practical procedures:

➢ Hold the animal still using a nose ring.

➢ Squeeze one testicle to the end of the scrotum. Find the cord that comes from the scrotum with your hand and hold the cord close to the skin.

➢ Put the burdizzo jaws over the cord 2 cm above the testicles and close the jaws. Do the same thing again 1 cm higher. Then do the same with the other testicle.

➢ Make sure not to crush all the way across the scrotum in one line and be-careful not to crush the penis at that time.

➢ Check after three weeks that the testicles have become smaller or atrophied.

Surgical (open castration) method

It is also referred as knife/emasculator type, the only completely safe method to sterilize male animals and can be done at any age by a qualified veterinarian. With this method of castration there is always a danger that the wound can become infected and the necessary precautions must be taken. Equipment and chemicals like *iodine*, *alcohol*, *Savlon*, scissors and forceps, cotton gauze water, etc. are required.

The practical procedures include:

➢ Restrain the animal to be castrated.

➢ Examine the calf to see if testes have descended into the scrotum.

➢ Disinfect equipment, hands, and scrotum.

➢ Grab the end of the scrotum with one hand and pull down.

➢ Cut off the lower 1/3 of the scrotum.

➢ Pull testicles down one at a time/scrape if necessary.

➢ Trim excess fat and membrane carefully.

➢ Disinfect the area again.

➢ Incision method is performed as much the same way, but make an incision on each side of the scrotum first, and then pull down each testicle through the incision complete the process.

Elastrator-ring method

This is a bloodless castration method by which the rubber-ring is applied around the neck of the scrotal sack using the special instrument designed for this purpose. The testicles must be in the scrotal sack distal (away from the body of the calf) to the elastrator ring. To minimize pain when using the rubber ring method of castration, they must be applied within three days of birth. A strong rubber ring is placed around the top of the testicles thus cutting off the blood supply (Fig. 38). The testicles die off slowly.

The practical procedure includes:

- Restrain and hold the animal still.
- Squeeze both testicles down to the end of the scrotum.
- Put the rubber ring over the scrotum with the special tool and leave the rubber ring behind. Be-careful not to get the penis inside the rubber ring.
- Spread the rubber band and push both testes down through it.
- Check later for the scrotum to falloff after about two weeks, and disinfect (Fig. 39).

2.4.6 Dehorning

Dehorning is the process of removing or preventing the growth of horns. Disbudding means the arrest of horn growth at an early age when the horn-root is in the form of bud/button. The ideal time to dehorn calves is between 2 – 3 weeks old or as soon as the horn-button can be felt. Calves need to restrain to prevent the calves/operator from being hurt. Heavy-duty electric dehorners are one of the most humane, effective, and safest tools to use. When electrical dehorners are used correctly, a continuous copper-color ring will be displayed around the base of the horns. The surface of the iron should be cherry-red before it is touched to the horns. This results in a very minimum of pain to the calves and provides very rapid destruction of the horn buttons.

Advantages

- Dehorning increases spaces in the barn, shed.
- Reduce injuries or damages.
- It makes easier handling, and sometimes brings more money.

Disadvantages

- Sets the animal back due to stress.
- Cost and labor as well as equipment
- Loss due to death, bleeding, spreads disease, and scars may occur if not properly done.

Hot-iron method

Electric, gas or fire-heated iron is the most common in calves (4 – 6 weeks). When using this method, ensure that the killing of horn-bud is effective otherwise the horn will grow again.

Hot-iron dehorning can be done with ease-up to the age 3 months (while the dehorning-iron still fits over the bud comfortably); thereafter horn-growth is fairly rapid, making surgical removal necessary (Fig. 40).

Chemical method

This type of dehorning is operated by using the common chemicals such as caustic potash (KOH) or caustic soda (NaOH) is common chemicals used. Hold a caustic pencil carefully (Forse, 2003; Hanson, 2003).

The practical-procedures include:

- Prepare basic chemicals (KOH and NaOH), and equipment.
- Restrain the animal.
- Clip/remove hair around the horn-bud/button of calves, and cover the area with a ring of heavy grease or Vaseline to protect the eyes against the chemical.
- Scrape button so that it is raw.
- Put ring of mineral-oil around base of button.
- Rub the chemical over the button until blood appears.
- Protect the hand while doing so.
- Apply chemical on button in liquid, paste, or stick forms (Fig. 41).

Surgical method

This is undertaken by using saw or cutting wire: In older animals, surgical procedures must be used, especially if horns have grown to a length of 2 cm or more. The removal of larger horns causes a great deal of pain and anesthetics should be used with dehorning and steps taken to prevent bleeding. Blood attracts flies which cause serious problems in open wounds. Once horns have grown very large, removal of the horns exposes the hollows in the skull and these must be closed to prevent infection. Tools and equipment include: dehorning saw, and clippers, dehorning wire, hot-iron, and electric line/flaming.

The practical procedures to control hemorrhage during dehorning include:

- Perform dehorning early in the morning.
- Handle animals quietly as possible.
- Use forceps or hemostats to pull out the veins or use a hot iron to sear them.
- Stuff cotton in the hole.
- Use sterile materials and equipment.
- Never dehorn when cows are eating sweet clover.

2.4.7 Extra Teats Removing/clippings

Many dairy calves are born with more than the usual four teats. These supernumerary teats can grow and develop much like a normal teat. They detract from the general appearance of the animal and have the potential to disrupt the milking process later on and to become infected. For these reasons, it is a good practice to remove these extra teats as early as possible in the calf's life. If it is done immediately following birth at the same time as the navel is treated, the calf is easy to handle, and one qualified person can accomplish the task (Battaglia, 1998a). Because the calf is very young, the cut bleeds only slightly. Removal can be performed in the first 3 months of life with sharp scissors or a scalpel (Zollinger and Hanson, 2003).

The practical procedures best to remove an extra teat include:

- Disinfect the area around the teat.
- Clip the hairs with a pair of sterilized scissors or surgical blade.
- Usually there is no bleeding at all, if bleeding is considerable, holding a cotton pack over the wound for a few minutes will stop it.

2.5 Culling and Replacement

Culling policy is very important to make decision, to say how much genetic improvement, we can make by culling these animals, how much replacement we need to dairy farms cull or lose about 20% of cows from the milking herd each year through health, calving problems, mastitis, infertility, death or culling for age or production. Most of these herd exits are involuntary. These cows have to be replaced to maintain the same number of cows lactating each year. Good herd management will reduce involuntary losses (infertility, disease) to allow higher production culling and herd improvement through breeding.

Culling rate of 20 - 25% of its cows each year is enough in dairy cattle. For a herd of 100 cows, this would require 20 to 25 heifers to calve each year to supply the replacements within the herd. For high producing animals culling rate is high because of diseases such as mastitis and such amount of replacement of better animals is required.

Voluntary culling: Voluntary culling is the desirable culling. Excess animals and low producing animals are culled and give you the cash return. Culling these low producers will make genetic improvement. They are removed for the reasons of infertility, lameness, disease (mastitis, tuberculosis, and brucellosis), old age, death loss and low production and poor reproductive performance. The replacement stock should be with more genetic potentiality than the culled animals

Involuntary culling: Involuntary culling of dairy animals is practiced because it could be removal of genetic material due to: reproduction problem, mastitis, disease /injury, and miscellaneous problems.

Herd replacement

A herd replacement rate of 25 % with at least 10 % selective culling will make rapid genetic progress without sacrificing milk yield. By maintaining a younger herd, the farm can better meet requirements for high quality milk, as somatic cell counts, and mastitis incidence increase with age of the cow. To replace these milkers, farmers rear their own heifers, or cows or springing heifers are bought in. Most farms would rear their own herd replacements, with additional animals reared or purchased for herd expansion. Appropriate selection procedures should be applied in this replacement application.

3 HEIFER AND BULL REARING AND MANAGEMENT

The goal of a dairy heifer-rearing program is to provide a regimen, which will enable the heifer to develop her full lactation potential at the desired age and at a minimum of expense. Heifers' rearing is a financial investment that begins to bring dividends after the first-calving; therefore the goal is to make ensure proper growth rate at minimum costs to be inseminated on time in order to realize full lactation potential later in life.

Heifers rearing begin with a choice of a bull likely to produce animals with high genetic potential for milk. A well-managed dairy farm is believed to have as many calves born every year as there are cows in the herd. Most producers sell male-calves at an early age, while the females are reared as dairy replacement heifers for the herd or as heifers for sale. Rearing a high number of replacement heifers allows dairy producers to:

- ➢ Obtain the best replacement heifers through strict criteria from wide selection.
- ➢ Expand the dairy herd at low cost (without buying heifers/cows).
- ➢ Sell excess heifers to earn income.

3.1 Heifers Feeding and Management

According to James (2001), Banerjee (1998), and Wilson (1999) heifer rearing is the second largest expenditure in a farm after the milking herd, with feed costs takes the largest share. When feeding of the heifers, one should aim to:

- ➢ Reduce interval between weaning and first lactation. This will increase the number of calving per lifetime (more of lactations) and lead to faster genetic improvement.
- ➢ Minimize mortality.
- ➢ Achieve a growth rate of 0.5 - 0.7 kg/day.
- ➢ Feeding management should ensure heifers reach target live-weights for breeding at 14 - 16 months old for first calving at 22 - 24 months old to be achieved.

Combining both adequate development and early age at calving has several advantages as it:

- ➢ Reduce the risk of calving difficulty.
- ➢ Improve the lifetime of milk production.
- ➢ Reduce rearing costs (feed, labor, etc.).
- ➢ Reduce total number of heifers needed to maintain herd-size.

On most farms, heifers are normally the most neglected group in terms of feeding resulting in delayed calving. In pasture management systems, close supervision is required due to variation in pasture quality through the seasons which may affect heifer growth rates. Heifers can be reared on good quality pasture only as their nutrient requirement is low (growth and maintenance). Supplementation with concentrate should be at 1% of body weight. Generally, the amount of concentrate given to heifers should be 1 – 4 kg depending on its age/size and forage-quality (John, 2011; James, 2001).

According to FAO and IDF (2011) and Albright et al. (1999), while designing heifers feeding program, the points you need to consider include:

- Puberty is related to size rather than age. The consequences of poor feeding are manifested in delayed calving resulting in reduced milk production.
- Feeding heifers too much energy leads to deposition of fat in the mammary gland tissue displacing secretory tissue resulting in reduced milk yield. The key period in mammary gland development is between 3 – 9 months of age. During this period, mammary tissue is growing 3.5 times faster than body tissue.
- Heifers fed high-concentrate rations develop less milk-secretory tissue in the mammary gland than heifers raised on recommended rations.
- Heifers calving at 24 months old have higher lactation yield compared to calving at an older age.
- Size of animal is related to milk-yield. For twins of the same genetic makeup, the heavier one produces extra milk in a lactation.

3.1.1 Measuring Growth Rate (Weight) versus Age

A growth chart is a tool used to compare the height and weight of heifers to a standard curve and thereby determine whether feeding and other management practices are adequate or whether they must be adjusted during certain phases of the rearing period. Under most management systems (pasture, group feeding in confinement), it is difficult to assess heifer performance. The use of a growth chart allows the producer to monitor heifer growth rates (USAID, 2012).

Both under-and over-feeding heifers are undesirable during heifer rearing. Overfeeding may result in obesity, low conception-rate, difficult-calving and low milk-production, whereas underfeeding will result in a low conception rate, poor fetal growth, difficult calving and low first lactation milk yield. It is therefore important to monitor performance of heifers, particularly the body weight change and height at withers (Wattiaux, 1996).

Growth charts allow the producers to compare the height and weight of heifers to a standard curve that represents the average for the particular breed. This tool enables the farmer to monitor heifer performance to determine whether feeding and other management practices are adequate. Once the measurements are taken, they are then fitted into a growth chart which is breed-specific. If the body-weight falls below the band expected, then the heifer getting insufficient nutrients (energy) and vise-versa. Short heifers are an indication of low-protein in the diet (USAID, 2012) (Table 6.6).

Table 6.6 Heifers body condition scores (BCS) at various ages

Ages (months)	3	6	9	12	15	18	21	24
Body condition scores	2.2	2.3	2.4	2.8	2.9	3.2	3.4	3.5

Source: Patrick Hoffman, 1995.

Body weight and height at withers are three important measurements used to evaluate heifer growth.

3.1.2 Body weight, Withers Height and Body Condition Score

Body weight (BW) at a certain age is the most commonly used criterion to evaluate the growth of heifers. However, it should not be the only criterion that BW alone doesn't reflect the nutritional status of heifers. Heifer development is better evaluated when weight measurements are accompanied by a measurement of skeletal growth, such as wither height or body length. The height of the heifer reflects frame (skeletal growth) while BW reflects the growth of organs, muscles and adipose tissue (fat) (Zollinger and Hanson, 2003; NRC, 2001) (Fig. 42).

3.1.3 Measuring Body Weight

The most accurate method to determine BW is using a calibrated scale. However, time and labor involved in moving heifers around usually makes it impractical to use a scale when it is available on the farm.

Measurement of heart girth circumference can be used to predict BW accurately. A non-elastic measuring tape should be placed just behind the front legs and behind the shoulders of the heifer. Constructing charts specific to other breeds and/or localities around the world should be a part of any dairy improvement project (James, 2001).

To monitor the overall heifer-growth rate, measurements at birth and AFC are sufficient. However, multiple measurements of height, weight and body condition score at various points during the rearing-period allow a producer to monitor specific rearing phases (early calf-hood, weaning-period, pre-pubertal growth, etc.). A change in season usually leads to changes in housing and feeding practices that have strong effects on growth rate.

Heifers monitoring will be more successful if it is simple and requires little labor. There are two practical approaches to measure heifer's height and weight. The first is to do measurements when heifers are restrained or handled, which occurs most often when heifers are: born, moved from individual pens to group pens at weaning and/or when they are dehorned, restrained at the time of breeding, and placed in individual pens at first calving. The second method is a single time measurement. In this approach, measurements are not taken over time for the same heifer, but rather on all heifers present in the herd at one time (James, 2001; Roussel and Constable, 1999).

Clearly, the higher the number of heifers in each group, the more accurate the estimate of growth will be. In both approaches, average daily gain can be calculated, or data plotted and compared to a growth chart.

Explanation of the body condition scoring using a 0 – 6 scale as here under:

0=Very thin, emaciated, starving.

1= The individual sharp spines-processes to touch, no tail-head fat; and prominent hipbones and ribs.

2 = The individual processes can easily be felt, but feel rounded, rather than sharp. There is some tissue cover around the tail head. Individual ribs are no longer visually obvious.

3=The short-ribs can only be felt with firm thumb pressure. Areas either side of tail head have a fat cover which can be easily felt.

4=The processes cannot be felt and fat-cover around the tail-head is easily seen as slight mounds, soft to touch. Folds of fat are beginning to develop over the ribs and thighs.

5=The bone structure of the animal is no longer noticeable, and the tail-head is almost completely buried in fatty tissue.

6= Area around the tail base completely filled-out and the back is completely rounded.

The score can be varied half a score depending upon the amount of tail-head fat, e. g. if the short rib palpation (using the thumb) gives score-4 but the tail-head is a typical 3, the score would then be 3. 5 (USAID, 2012; James, 2001; Roussel and Constable, 1999) (Table 6. 7)

Table 6. 7 Nutrient requirements for large breed heifers at 800 g gain/day

Item	Unit	Heifer weight (kg)				
		150	250	350	450	550
Intake	kg/d	4. 2	6. 2	7. 9	10. 5	12. 2
Energy ME	Mcal/kg	2. 29	2. 27	2. 30	2. 33	2. 32
Protein CP	(% of DM)	15. 9	13. 1	11. 7	14. 2	13. 3
RUP	(% of CP)	39	26	17	30	26
RDP	(% of CP)	61	74	83	70	74
CP/ME	g CP/Mcal ME	69	57	51	60	57

Crude protein required only if the ration is balanced for RDP and RUP. ME is metabolizable energy; RUP is Rumen undegradable protein; RDP is rumen degradable protein.

Source: NRC, 2001.

3. 2 Feeding and Management of Bulls

- Wean the bull at 6 – 8 months of age
- Feed high energy rations for about 5 months old, and avoid fattening
- Allow full feed until spring, then put on pasture to complete growth
- Bulls will continue to grow slowly until about 4 years of age

Feeds of the bull

The feed resources for feeding of the bulls include:

- Forages, hay and grain
- Amount depends on type and quality of the diets.
- Minerals free choice
- Feed vitamin A if the ration is mostly corn silage or limited hay.
- The bulls may be self-fed or hand fed.
- When self-feeding uses plenty of roughage to keep bulls from getting too fat or going off their feed. Title and description do not match.

Rate of growth and needs

- Yearling bulls should be fed to gain 1.5 – 2 lb /day.
- 2 – 4 years old bulls need more energy and protein in the winter than cows and should be fed accordingly.
- Mature bulls in good condition may be fed the same as the cow herd.

After the breeding season

- The bull loses weight and it needs to feed to regain that weight.
- Give additional feed 6 – 8 weeks before the start of the next breeding season.
- Bulls that are too fat or too thin have poor fertility.
- They should be in medium flesh and have plenty of exercise.
- Keep bulls separate from cows; if no place to keep bulls it is safe to run them with steers.

Before the breeding season

- If necessary, trim hoofs several weeks before breeding season begins.
- Test semen for fertility and disease.

4 IDENTIFY COMMON HEALTH PROBLEMS OF DAIRY CALVES

Most health problems experienced by calves are due to poor management, environmental, and physiologic factors that can affect young calves making their life more difficult. The first place where good management can contribute to good health care is at calving. Then diligent feeding management and housing is essential to maintain calf health. Some of the common problems associated with management practices are pneumonia, diarrhoea and internal parasites.

4.1 Pneumonia

Research findings revealed that the immediate causes of pneumonia are bacteria and viruses, but of greater significance are the predisposing factors of dampness, drafts, chilling, and toxic gases. The accumulation of gases in confinement housing irritates the respiratory tract. Ammonia is one of the major irritants, and when its smell is noticeable, which probably exists a high risk of damage to the lung defense mechanisms of such confined animals (FAO, 2012).

At 6 – 8 weeks old, respiratory disorders seem to the largest problem and are often associated with high-population density and inadequate ventilation. Respiratory disorders are accentuated with high relative humidity at low environmental temperatures. Respiratory-diseases such as pneumonia tend to be worse during winter and early spring. Research has shown a high correlation between respiratory-disease and calf-hood morbidity, growth rate, as well as reproductive efficiency, and average age at first calving.

The other cause of pneumonia in young calves is fluids going to the lungs via the windpipe (trachea). The first feeding of colostrum can cause problems if the feeding rate is faster than swallowing rate. If colostrum is bottle fed it is important to use a nipple that matches the calf's ability to swallow. Greedy calves swallow large quantities of milk from the bucket, some of which may end up in the windpipe leading to pneumonia (FAO, 2012; Garg, 2009).

4.2 Calf Scour

Calf-scours, during the first month of life, is the most common cause of calf-sickness and death. Calves suffer from two major forms of diarrhoea or scours-viral-diarrhoea (Rotavirus and Coronavirus); and bacterial-diarrhoea (usually *E. coli*, *Salmonella*, *Coccidia*, and *Cryptosporidium*) (John, 2011; Zigler and Hanson, 2003).

Sources of infection have been discussed under the headings of calving-management, colostrum, milk and feeding management. Early introduction of solid-feed will help to reduce problems with milk feeding. The risk of scours is increased if calves are subjected to crowding, cold to stress, e. g. by a sudden change of diet, and less frequent manure removal (Payne and Wilson, 1999).

Symptoms

Calf scour (*white scours*) is manifested by severe diarrhoea with light colored, foul-smelling, watery and foamy feces. Calves with viral-diarrhoea can't reabsorb water in the gut because of villi damage, so dehydration is the major problem. Dehydration can be averted by recognizing the symptoms early and providing oral rehydration therapy. It should be accompanied by alkalinizing therapy with sodium bicarbonate as the diarrhoea is usually accompanied by acidosis (Owens et al., 1998), caused by poor renal excretion of hydrogen-ions. Reduced nutrient absorption accompanies a severe acidosis and milk should be withdrawn and replaced with a glucose solution for energy.

According to John (2011), calves with scours experience a number of medical problems that may contribute to death; these include dehydration, acidosis, bacterial infection, low blood glucose and hypothermia. Fluid therapy is the cornerstone of treating calf scours. Calves infected with bacterial pathogens and calves that are severely debilitated also benefit from antimicrobial therapy. Affected calves need a clean, dry, warm environment and nutritional support.

Treatment

Calf scours occur at any time up-to 4 weeks of life, at which the rumen is sufficiently inoculated with benign-bacteria to prevent it being colonized by bacterial pathogens. Calves can be vaccinated against certain forms of *E. coli* scours, but the most important means of protection is ensuring that the calf has adequate intake of colostrum in the first 6 hours of life.

Prevention and control

The best cure for scours is prevention. It is crucial for calves to feed correctly and housed in a clean environment. Calves with scours should be treated immediately with a homemade or commercial electrolyte solution to keep them from dehydrating. Milk should not be fed when scours occur because milk may encourage growth of bacteria in the intestine and further complicate the scours. Milk or milk replacer should be replaced with an electrolyte solution for no more than 24 – 48 hours (Payne and Wilson, 1999).

4.3 Internal-parasites

Coccidiosis is a protozoan parasitic infection becoming more of a problem in recent years in calves. Coccidia are single cell protozoan blood parasite which lives within the mesenteric cells of the digestive tract. After a coccidian infection has begun in the animal, the coccidian organisms spread to various locations within the intestines, where the organisms cause extensive damage of young animals. Calves often become infected during the age of 3 – 6 weeks and while confined in pens, although older-calves weaned in confined lots also show symptoms of coccidiosis.

Signs

Infected calves show signs of bloody-diarrhoea and may become dehydrated and die. Coccidiosis at the subclinical level (undetectable by usual clinical observations), reduces the growth rate of calves. With time and treatment, animals develop an immunity that keeps the numbers of these organisms at low levels.

Worms

Diagnosis of a worm problem is on clinical signs (being careful not to confuse with starvation) and on the number of eggs passed in the dung (fecal egg counts). The feces of at least 10 animals should be sampled and tested, either in a vet laboratory or by using a fecal egg counting kit available commercially. Care must be taken in interpreting results because worm burdens fluctuate rapidly. Also, a moderate burden under normal weather-conditions may be potentially lethal in a drought situation (USAID, 2012).

The most common internal parasites or worms of dairy-calves are tapeworms (*Moniezia expansa*, *Bunostomum*) and roundworms (*Ascaris*, *Haemonchus species*, etc.).

4.4 Fluids and Electrolytes in Calf Health and Disease

Fluids and electrolytes are necessary nutritional and functional components for all mammals and are required for normal cellular and organ function and for maintaining the acid: base balance with a blood pH 7.35 – 7.45. The normal animal maintains the balance of fluid, electrolytes, and acid: base (blood pH) within narrow limits by consuming water, minerals from supplements, feedstuffs, and salt. Many diseases cause fluid, electrolyte, and acid: base imbalances that can result in death. Appropriate fluid and electrolyte therapy (rehydration, electrolyte, and acid-base balance) is necessary to restore normal activity (Table 6.8) (Payne and Wilson, 1999).

Table 6.8 Goals and treatments for fluid therapy

Goals of fluid therapy	Treatments
Correct hydration and circulating blood volumes	Fluids
Correct acid: base balance to normal pH	Bicarbonate fluids
Correct mental depression	Fluids and electrolytes
Correct electrolyte imbalances	Electrolytes
Facilitate intestinal repair	Fluids and electrolytes
Restore suckle reflex	Glucose, fluids, electrolytes

4.4.1 Fluid and Electrolyte Requirement

Water represents the liquid portion of the fluid components of mammals and is one of the five major nutrients. Water provides the fluid medium in which the chemical reactions of the body take place. It also has an ability to absorb and give off heat with a relatively small change in its temperature; therefore, it is an ideal temperature-buffering system for the body.

Water is also the medium for transportation of nutrients and wastes within the body. Fluid requirement for maintenance of cattle is approximately 45 cc/(lb · day); therefore, a 100 pounds calf needs approximately 1 gallon of water a day, at 60 – 70℉, just to maintain normal bodily functions.

Electrolytes are dissolved in both intracellular and extracellular fluid compartments of the mammalian system. Electrolytes are required for normal cellular metabolic functions. The electrolytes of note in calf health are sodium (Na^+), potassium (K^+), hydrogen (H^+), chloride (Cl^-), and bicarbonate (HCO_3^-). Electrolyte needs are generally met through consumption of feed and salt and mineral supplements.

4.4.2 Causes of Fluid and Electrolyte Imbalances

These are characteristic of scours, intestinal blockage (LDA), kidney disease, blood loss, salivation (VSV/FMD), persistent fever, or water deprivation. One of the most common causes of fluid/electrolyte/acid: base imbalance is diarrhoea (scours). Fluid loss results in dehydration that results in decreased temperature, increased pulse and respiration, and other changes, such as sunken eyes and loss of skin elasticity. Loss of body fluid causes changes in the electrolyte and acid-base balance of the body.

Fluid loss routinely includes the loss of bicarbonate resulting in acidosis (blood pH $<$ 7.35). Clinically, dehydration, electrolyte imbalances, and acidosis are presented as weakness and downer animals. The body's mechanisms to correct dehydration can also result in electrolyte imbalances. Diseases such as scours can alter the integrity of the intestine resulting in further loss of fluids and electrolytes as well as decreasing the intestine's ability to absorb water and electrolytes.

Fluid and electrolyte deficits and imbalances require specific treatment protocols to correct imbalances. Oral and/or intravenous fluid therapy can be used quickly as discussed below.

4.5 Methods of Fluid and Electrolyte Therapy

The common methods of fluid administration are orally or intravenously. Oral administration of fluids is the safest in that it is more difficult to over treat an animal, but this method is most beneficial in treatment of fluid deficits in early disease or animals $<6\%$ dehydrated. Care must be used to avoid administration accidents, such as placing the fluids into the lungs or causing injury to the esophagus or trachea. Intravenous administration requires moderate surgical skills and increased cleanliness to avoid introducing infectious agents through the needle. IV administration is generally used in down or recumbent animals. IV therapy requires close monitoring as excess fluids and electrolytes can be fatal (Roussel and Constable, 1999).

Oral fluid therapy

Oral fluid/electrolyte solutions can be successfully administered by a nipple bottle for a calf that will suck or via orogastric intubation. Orogastric tube feeding systems consist of a bag or bottle reservoir attached to a rigid tube with protective bulbous end.

Place the tube into the mouth over the tongue and direct it to the left side. The calf will usually swallow the tube, which can be seen and felt passing down the esophagus into the stomach. The fluid is then allowed to flow via gravity into the stomach. Since oral therapy is effective in early or less severely dehydrated animals, placing fluid into the lungs is rarely a problem, however, when withdrawing the tube with any fluid remaining in the reservoir, crimp the tube to prohibit inhalation of fluid during withdrawal (Veterinary Anesthesia, 1996).

Intravenous fluid administration

Intravenous (IV) fluid administration in calves and cattle requires the placement of a catheter into the jugular vein. IV catheterization is a minor surgical procedure that your attending veterinarian may provide instruction in the procedure. Briefly, a thorough scrubbing—important to decrease contamination of the site and equipment—of the site is made with a skin cleansing detergent. An 18 ga×1.5-inch catheter is placed into the jugular vein and secured with suture or tissue (super) glue. The appropriate fluids are placed into an administration setup (Roussel and Constable, 1999; Veterinary Anesthesia, 1996) (Table 6.9).

Table 6.9 Selected fluid and oral electrolyte supplements

Name	Use	Components	Administration	Notes
Sodium bicarbonate	Restore fluid volume Correct acidosis	$NaHCO_3$ (baking soda)	Intravenous	(4 oz) + 1-gal distilled water
Calf Quencher	Correct acidosis	Dextrose, sodium and potassium	Oral therapy	1 qt/treatment every 4 to 6 hours
Entrolyte-HE	Correct acidosis and electrolytes, provide energy	Dextrose, glycine, sodium and potassium chloride, bicarbonate	Oral therapy	As above

Source: Western Beef Resource Committee, 2001.

4.6 General Practical Health Care Program for Dairy Calve-Heifers

According to research studies, health care of dairy-calves can be best described by age and state of the animal as set-out in the headings below:

Newborn calf at three days old

- Clean the newborn calf with a dry towel or dry-hay. This will stimulate respiration and blood circulation (USAID, 2012).
- Remove slime from the nose and mouth to assist breathing and holding up the rear legs of the calf, let the head hang down to release any water in the lungs, mouth or nose.
- If the navel is too long, cut it and leave two to three inches from the stalk, then dip the navel in tincture of iodine to prevent local infection. This procedure is important for prevention of navel-ill (omphalitis) and helps the umbilicus heal quickly.
- Feed the calf with colostrum within 1-2 hours after birth. The optimum time for absorption of antibodies through calf's small intestine is in the first 6-8 hours. It needs to provide the calf with colostrum 10-15% of its body weight during the first 12-24 hours to prevent early infection (FAO and IDF, 2011).
- In general, remove the calf from the dam after calving to the isolated dry-clean pen. Straw for bedding must be clean-dry and be changed regularly (Land O'Lakes, 2010).

Calves of three days to one-month age

Feeding with whole milk is expensive, so milk replacer is used for routine feeding, which is twice a day. Bucket feeding is commonly used; it should be cleaned well between uses to avoid digestive disorders due to poor hygiene (Garg, 2009; Wattiaux, 1996). Besides:

- Train the calf to take concentrate and roughage at about 1 week old. Solid food stimulates rumen development. Make available clean water at all times in the pen.
- Identify calves using an ear-tag/tattoo. Remove extra-teats in the first week.
- Common health problems during this period are omphalitis (navel-ill), diarrhoea (calf scours), respiratory infection (pneumonia) and arthritis.

Calves of one-month old to weaning (3-4/5 months)

- Calves are dehorned at 1 - 2 months of age.
- All female calves must be vaccinated against brucellosis at 3 - 8 months of age.
- Weaning should take place at 3 - 4 months of age or when the calf is able to eat roughage and concentrate of more than 1 kg/day or at a calf body weight between 80 - 90 kg (depending on the breed).
- De-worm the calf against internal parasites such as roundworm, tapeworm and flukes. Also, eliminate external parasites such as ticks by spraying of appropriate acaricide.
- In this period, problems to be aware of are parasites, bloat and arthritis.

Calves 4-12 months old

- Vaccinate against FMD (foot-and-mouth disease), hemorrhagic septicemia and/or anthrax every 6 months.
- De-worm against internal parasites such as roundworm, tapeworm, and flukes and, also eliminate external parasites such as ticks, by spraying.
- In this period, problems to be aware of include parasites, tick-fever, pneumonia, diarrhoea, bloat and arthritis.

Heifers 12-18 months of age

- Record the growth rate for which should not be<270 kg in crossbred or 300 kg in purebred cattle at first service.
- Take blood for brucellosis and perform a tuberculosis test.
- Vaccinate against foot-and-mouth disease (FMD) and hemorrhagic septicemia every 6 months; and black-quarter (black-leg) from six months of age.
- De-worming should be carried out every six months.
- Heat detection should be carried out to determine the right time for artificial insemination and use of selected semen in accordance with the breeding plan of the

region or farms.

- Heifers requiring repeated insemination (>3 times) check by a veterinarian.
- Check signs of estrus of heifers>18 months old and/or weighing >270 kg which haven't shown by a veterinarian.
- Perform pregnancy diagnosis on each animal at 45 – 60 days after the last insemination.

>>> SELF-CHECK QUESTIONS

Part 1. Multiple Choices.

1. Which of the following statements is true about calf rearing?
   A. The major goal of calf rearing is to have a strong and viable calf
   B. Calf is the foundation stock of the future dairy herd
   C. Calf rearing starts from when the cow is pregnant
   D. All are correct
2. The series of husbandry routines implemented to crop the futures foundation stocks of any dairy enterprise is known as
   A. Calf rearing　　B. Cow milking
   C. Cow's drying-off　　D. Calf marketing
3. Among the following ________ is commonly used to identify dairy-animals.
   A. Ear-tagging　　B. Ear notching
   C. Ear twisting　　D. Branding
4. Which of the following is a component of the calf's feeding activities responsible to protect the calf against infectious diseases?
   A. Whole-milk　　B. Colostrum
   C. Milk-powder　　D. Grain-feeding
5. The common chemicals used for dehorning of the calf are termed as
   A. NaOH & $CaCO_3$　　B. $Ca(OH)_2$ & KOH
   C. $CaCO_3$ & KOH　　D. All are correct
6. The appropriate recommended age for the calf to be castrated is
   A. 2 – 3 weeks　　B. 2 – 3 days
   C. 2 – 3 months　　D. 9 – 12 months
7. Which one of the following is true about the newborn calf?
   A. They are not to be fed with solid feed at 2 weeks old.
   B. They can be vaccinated after one month old.
   C. At an early age, they are recommended to be overfed.
   D. None of the above.
8. Weaning practices of the calf should take place when/at
   A. 3 – 4 months old

B. Calf eats roughage & concentrate >1 kg/day

C. Body weight of 80 - 90 kg as breed

D. All are correct

9. Which one of the following is the correct position of the fetus presentation during birth?

A. Forelegs with head come together B. Forelegs with head bend

C. Beck-tail first D. Breech presentation

10. The major goals of raising heifers is

A. provide replacements for cows culled

B. Improve genetics and production

C. Raise heifers economically

D. All of the above

11. The purpose of steaming-up for pregnant cow include all below except

A. Produce strong and healthy calf B. Restore the cow's body tissue lost

C. Prepare for good milk-yield D. None of the above

12. ________ is an age of the calf after calving recommended for calf put into solid food.

A. 2 - 4 days B. After 2 weeks

C. 2 - 4 months D. 3 - 4 months

13. Among below ________ is the major advantage of calf rearing in a dairy operation.

A. Produce strong calves B. Produce sound replacement heifers

C. Improve calf genetic potential D. All are correct

14. During heifer rearing over-feeding of heifers is undesirable. Since it results in

A. Obesity and low conception rate B. Difficult calving

C. Low milk production D. All are correct

15. When does measurements of heifers occurs most often?

A. At the time of born and moved to group pens at weaning

B. They are dehorned, and at the time of breeding

C. Placed in individual pens at first calving

D. All are correct

16. Under/over feeding of the heifers may result in one of the following

A. Low conception rate B. Difficult-calving

C. Low first lactation yield D. All are correct

17. Which one of the following is the most humane, effective, and safest tools to dehorn dairy calves?

A. Heavy-duty electric dehorners B. Chemical paste

C. Hot-iron D. Surgical-method

18. Which one is a temporary method of identification?

A. Ear-notching

B. Tattooing

C. Ear-tagging

D. Freeze-branding

19. The amount of supplementation with concentrate diets should be provided to growing heifers is at ________ body weight.

A. 10% to 15%
B. 1.5%
C. 20%
D. 5% to 10%

20. Which of the following seems to be the largest problem of calves at the age of 6 - 8 weeks often associated with high-population density and inadequate ventilation?

A. Parasitic infestation

B. Respiratory disorder

C. Metabolic disorder

D. Emaciation

21. The amount of concentrate diet given to heifers in general should be ________ per day depending on its age/size and forage-quality.

A. 8 - 10 kg
B. 5 - 10 kg
C. 1 - 4 kg
D. 6 - 8 kg

22. The management decision tool for dairy cows to be replaced for better production is

A. Feeding policy
B. Culling policy
C. Breeding program
D. milking policy

Part 2. True or false.

1. The calf is fed whole milk for 6 - 8 weeks from the 4th week of age with total amount of 114 - 136 kg of milk.

2. The first place where good management can contribute to good health care is at calving.

3. Dairy heifers to be bred at 15 months, should be weighing 800 lb (Jerseys) to 550 lb (Holstein and Brown Swiss).

4. From approximately 4 days of age, many calves are fed 'artificial' milk-milk replacer-for 5 - 10 weeks.

5. The calf needs to be provided with colostrum 10 - 15% of its body weight during the first 12 - 24 hours to prevent early infection.

6. Calf-scour is the most common cause of calf sickness and mortality in the first month of life.

7. The first step of calf rearing starts from the period at which calf is in the dam's womb, particularly from 6 - 8 weeks before parturition.

8. Sometimes it is recommended to crush the spermatic cord or testicles with a hammer.

9. Feeding management should ensure heifers to reach target live-weights for breeding at 14 - 16 months old for first calving at 22 - 24 months old to be achieved.

10. Growing heifers use available nutrients in an irreversible order: 1) Daily maintenance, 2) growth, and 3) ovulation and conception.

Part 3. Matching.

Match the words or phrases under column A with those best suited under column B given below.

A	B
1. Passive immunity	A. Steaming-up
2. White scour	B. Navel dipping
3. First feces of the newborn-calf	C. Ear tagging
4. Iodine tincture	D. Difficult-birth
5. Omphalitis	E. Electrolyte solution
6. Housing of cows giving birth	F. Last 6 - 8 weeks of pregnant
7. Individuals calf pen	G. Dehorning
8. Dystocia	H. Castration
9. Identification of the calf	I. Navel ill
10. Cows fed adequate, balanced diets ration	J. Maternity-pen
11. Reduce the barn or feeding space shortages	K. Maternal-antibodies
12. Starting point for care of the calf	L. Hutch
	M. Meconium
	N. Colostrum
	O. Calf's starter feed

A	B
1. Limited whole-milk	A. Fluid-therapy
2. Burdizzo	B. The first 5 days
3. Wash & sterilizing feeding utensils	C. 5 months
4. The dam's serum antibodies	D. 400 - 500 g/day
5. Create rumen-development	E. 1.5 kg concentrate diet
6. Period of colostrum feeding	F. 10% of its body-weight
7. Calf-weaning period	G. Preventing internal-parasites
8. Sawdust, hay and straw	H. Dry calf starter method
9. Treatments of calf-scour	I. IgG, IgM
10. De-worming of the calves	J. Testicles degeneration
11. Heifers weight gain/day	K. Prevent cross-contamination
12. Quantity of feed fed to heifer/day	L. Palatable, fresh & leafy pasture
13. Total amount of colostrum fed to calf/day	M. Bedding materials
	N. Bloodless castration

Part 4. Describe the following briefly and precisely.

1. Describe the calf's-rearing management activities undertaken before the near-calving.

2. Discuss management practices which must be undertaken before the calf is born.

3. The calf breathing should commence, what delivery practices undertaken? If breathing may not start, what should you do to start calf breathing?

4. What is the primary concern in rearing the newborn calf?

5. Mention and discuss the different phase of calf's-feeding management.

6. The newborn-calf at 3 - 5 days after birth is highly dependent upon colostrum feeding. Why?

7. Outline and discuss the different methods of calf's-feeding.

8. The common difference between single-suckling and foster/multiple suckling.

9. Illustrate the requirements that a calf pen should fulfil for the calf to be healthy.

10. Outline the general management-practices that must be implemented.

11. Define the term identification; and identify the most commonly used methods.

>>> REFERENCES

Blowey R W, 1993. A Veterinary Book for Dairy Farmers [M] . 2nd ed. Farming press, Ipswich, UK: 456.

Fox D G, van Amburgh M E, Tylutki T P, 1999. Predicting Requirements for Growth, Maturity, and Body Reserves in Dairy Cattle [J] . J. Dairy Sci, 82: 1968 - 1977.

Gachuiri C K, Lukuyu M N, Ahuya C, 2012. Dairy Farmers Training Manual [E] . http: //www. kdb. co. ke/press/publications/manuals/2 - dairy-farmers-training-manual/file.

Garg M R, 2012. Balanced Feeding for Improving Livestock Productivity-Increase in Milk Production and Nutrient Use Efficiency and Decrease in Methane Emission [E] . http: //www. fao. org/docrep/016/i3014e/i3014e00. pdf.

Heath E, Olusanya S, 1985. Anatomy and Physiology of Tropical Livestock [M] . Longman: 138.

IDF and FAO, 2004. Guide to Good Dairy Farming Practice [E] . http: //www. fao. org/docrep/006/Y5224E/Y5224E00. htm.

Land O'Lakes, Inc. , 2010. The Next Stage in Dairy Development for Ethiopia: Dairy Value Chains, End Markets and Food Security [E] . https: //www. usaid. gov/sites/default/files/documents/1860/Dairy Industry Development Assessment _ 0. pdf.

Matthewman R W, 1993. Dairying: the Tropical Agriculturalist [M] . Macmillan Ltd: 152.

Meijering A, 1984. Dystocia and Stillbirth in Cattle-a Review of Causes, Relations, and Implications [J] . Livestock Prod. Sci, 11: 143.

Reinaldo C, Aurora V, Charles E, 2008. Calving School Handbook [E] . http: //blogs. oregonstate. edu/beefcattle/files/2016/08/CalvingSchoolHandbook _ 000 - 2. pdf.

Robert E J, 2001. Growth Standards and Nutrient Requirements for Dairy Heifers-Weaning to Calving [J] . Advances in Dairy Technology, 13: 63.

Rousse A J, Smith G W, 2014. Fluid and Electrolyte Therapy [J] . Veterinary Clinics of North America:

Food Animal Practice, 30 (2): 295 - 486.

Senger P L, 2003. Pathways to Pregnancy and Parturition [M] . 2nd ed. Current Conceptions.

Severidt J, Hirst H, van Metre D, et al. , 2009. Calving and Calf Care on Dairy Farms [E] . http: //www. cvmbs. colostate. edu/ilm/proinfo/calving/notes/home. htm.

National Research Council, 2001. Nutrient Requirements of Dairy Cattle [E] . 7th ed. Washington, D C: The National Academies Press.

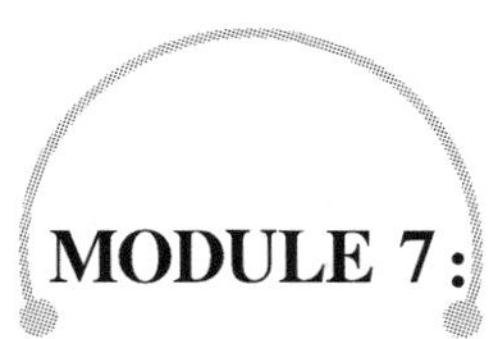

MODULE 7: CARRYING OUT MILKING OPERATIONS

>>> INTRODUCTION

Milk is secreted by the mammary gland of mammals to feed their offspring. It is manufactured from the raw materials in the blood by the alveolar cells and the storage of that milk in the cavity of the alveolus which is termed as milk secretion. Proper management of cows during and between each milking is required for maximum milk production and mastitis prevention. The economic loss from mastitis makes it the dairy industry's most important disease. The technologies to control and eradicate mastitis have been available for many years, yet bacteria still take thousands of cows out of production every year (Schroeder, 1997).

Understanding and following proper milking procedures is a critical step in maintaining maximum milk quality. The milking process should be consistent, and cows should be milked at the same time every day. The primary goals of the milking process should be to harvest large quantities of a high-quality end-product for consumers, minimize mastitis infections, milk clean, dry teats, and minimize stress on both cows and workers within the parlor. The first step to establishing a good milking procedure understands the milking process so that you can determine what procedures best fit your individual dairy operation. Write these procedures down and post them where everyone involved in the milking process. These standard operating procedures should be communicated to employees and periodically evaluated. The following are recommended best practices to consider applying in your milking procedure (Jeffery et al. , 2012)

Operating a dairy milking center involves managing a number of issues so that a satisfactory end result is accomplished. It involves labor management, work routine organization, mastitis control, cow physiology, Grade A milk production regulations, and agricultural economics (Schroeder, 1997).

Generally, this module generates useful information on milking operation with the objective to: 1) providing information on mammary system and milk let down phenomenon in milking cows, 2) identifying milk composition, and 3) providing guidance on the standard milking procedure in milking cows.

1 THE MAMMARY GLAND AND MILK LET DOWN PHENOMENA

1.1 Understanding Mammary System and Milk Let down Phenomenon in Milking Cows

Milk secretion is the process of manufacture of milk from the raw materials in the blood by the alveolar cells and the storage of that milk in the cavity of the alveolus. This is a continuous process which only stops when the alveolar cavity is full and the pressure on the alveolar cells inhibits further secretion. Milk secretion is controlled by hormones, unlike some other glands such as the salivary glands. The hormones of major importance are those secreted by the anterior lobe of the pituitary gland. Prolactin and pituitary growth hormone have a direct effect on the action of the alveolar cells. The hormones which control the thyroid and the adrenal glands are also important.

Milk ejection is the process through which milk is released from the alveoli and flows into the ducts, the gland cistern and the teat cistern, where it can be removed by the calf or the human milker. Unless the cow lets down milk, neither the calf nor the dairy farmer can remove the full yield. Milk secretion takes place continuously. As the cavities of the alveoli fill with milk some of it will pass into the duct system and move down into the gland and teat cisterns. This milk can easily be withdrawn. All the rest of the milk is held in the cavities of the alveoli, and unless forced out, it remains there.

1.2 Structure of Mammary Gland

The mammary gland is a modified sweat gland that nourishes the young. It consists of the mamma and the teat. Undeveloped in both male and female at birth, the female mammary gland begins to develop as a secondary sex characteristic at puberty. With the birth of the first young and first lactation, the mammary gland attains its full size and function. When suckling by the young stops, milk production ceases and the gland regresses. Shortly before the next and subsequent parturitions, the gland is stimulated by hormonal changes to produce milk.

Development of the mammary gland (plural = mammae): An ectodermic thickening develops along the ventral body wall extending from the thoracic to inguinal region-this is the mammary ridge. Cells aggregate, multiply and differentiate to form a chain of condensed mammary buds. Most mammary buds regress, but those that remain each develop into a mammary gland. A mammary gland is the secretary and duct system associated with one teat. Mammary buds grow into overlying mesenchyme, and primary epidermal sprouts grow out of the bud apex. The epidermal sprout branches extensively and develops a complete duct system. Mammary adipose tissue is derived from mesoderm. This is required for complete mammary development and is absent in the male. As a result, mammary development in the male is halted at the epidermal sprout stage.

The mamma is the glandular structure associated with a papilla (teat) and may contain

one or more duct systems.

The udder is a term designating all the mammae in the ruminant and the mare (sometimes also used for the sow). The lobes are the internal compartments of the mamma, separated by adipose tissue. The lobes are divided into lobules, consisting of connective tissue containing alveoli, which are clusters of milk secreting cells. The lactiferous ducts are large ducts, conveying milk from the alveoli to the lactiferous sinus. The openings of the lactiferous ducts convey milk formed in the alveolus to the gland sinus. The lactiferous sinus (milk sinus) is the milk storage cavity within the teat and glandular body. The gland sinus is part of the milk sinus within the glandular body and the teat sinus is part of the milk sinus within the teat.

Teat

It is the projecting part of the mammary gland containing part of the milk sinus. The papillary duct (teat canal) is the canal leading from the teat sinus to the teat opening and may be single or multiple. The teat opening is the opening of the papillary duct and the exit point for milk or entrance point for bacteria. The sphincter consists of muscular fibers surrounding the teat opening that prevent milk flow except during suckling or milking.

Suspensory apparatus

In species with large udders, especially in dairy cattle, there is a suspensory apparatus, which is organized into the lateral and medial laminae suspending the mammary gland from the ventral aspect of the trunk by their attachment to the pubic symphysis.

Lateral lamina

Lateral lamina consists of collagen fibers from the fascia of the pubic symphysis and the edge of the superficial inguinal ring. The medial lamina consists of elastic fibers from the tunica flava ventral to the pubic symphysis. The intra-mammary groove divides the left and right rows of mammae.

Arteries

The main blood supply to the inguinal mammary glands is from the external pudendal artery. This arises indirectly from the external iliac artery via the deep femoral artery. The external pudendal artery passes through the inguinal canal. In species which also have thoracic and abdominal mammary glands (bitch, queen, sow) additional blood supply is derived from the internal thoracic artery and its branches-cranial superficial epigastric arteries as well as from lateral thoracic and intercostal arteries.

Veins

In most species thoracic and cranial abdominal mammary glands drain via cranial superficial epigastric veins into the internal thoracic vein. Caudal abdominal and inguinal mammary glands drain via caudal superficial epigastric veins into the external pudendal vein.

In cattle a venous ring is formed between the base of the udder and the abdominal wall. During the first pregnancy, an anastomosis develops between cranial and caudal superficial epigastric veins forming the subcutaneous abdominal vein (milk vein). As a result, some drainage from venous ring passes in a cranial direction via this vessel, which then drains deeply through the abdominal wall (milk well) into the internal thoracic vein. Other drainage passes into the external pudendal veins or to perineal veins.

Innervations

Somatic innervations are via the ventral of the spinal nerves. In the cow, the ventral branches of L_1 and L_2 (iliohypogastric and ilioinguinal) supply the skin of the cranial glands. Mammary branches of the pudendal nerve supply the caudal aspect of the udder. There are sympathetic innervations to the blood vessels and teat sphincter smooth muscle. Mammary glands also have major influence from endocrine hormones.

Lymphatic

The more caudal mammary glands drain to the superficial inguinal lymph node and the more cranial mammary glands to the auxiliary or sternal lymph nodes.

Lymphatic drainage in the cow: The afferent lymphatic ducts pass dorso-caudally to reach the superficial inguinal (mammary) lymph nodes at the dorso-caudal side of the udder. These are usually palpable large, kidney-shaped nodes between the caudal side of the udder base and the thigh. The efferent lymphatic ducts pass into the abdomen through the inguinal canal to empty into the deep inguinal node (Fig. 43).

1.3 Desirable Appearances of Mammary Gland

The gross anatomy of the mammary gland differs a lot among species. However, the microscopic anatomy is very similar among species. The mammary gland development starts early in the fetal life. Already teat formation starts in the 2nd month and the development continues up to the 6th month of gestation. When the calf fetus is six months, the udder is almost fully developed with 4 separate glands and a median ligament, teat and gland cisterns. The developments of milk ducts and the milk secreting tissue take place between puberty and parturition. The udder continues to increase in cell size and cell numbers throughout the first five lactations, and the milk producing capacity increases correspondingly. This is not always fully utilized, since the productive lifetime of many cows today is as short as 2.5 lactations. The mammary gland of the dairy cow consists of four separate glands, each with a teat. Milk which is synthesized in one gland cannot pass over to any of the other glands. The right and left side of the udder are also separated by a median ligament, while the front and the hind quarters are more diffusely separated. The udder is a very big organ weighing, around 50 kg (including milk and blood). However, weights up to 100 kg have been reported.

1.4 Hormonal Action in Milk Let down Phenomenon

The "let-down" or milk ejection reflex is controlled by hormones. s. When the calf suckles her mother, or when the milker washes the udder or starts to milk the cow, messages are taken from the nerve endings in the teats to the brain. When the message reaches the brain the rear end of the pituitary gland, the posterior lobe, is set into action. "Let down" hormone (oxytocin) is released into the blood, it reaches the udder and causes the Myoepithelial cells, which surround the alveolus, to contract. Using the analogy of a bunch of grape, the contraction of the myoepithelial cells acts as a series of hands squeezing the grapes in turn "let down" is an involuntary process.

Oxytocin in the blood does not last for a very long time. Most of it will have disappeared in about two minutes. Once it has gone the myoepithelial cells relax, the alveoli resume their normal shape and as they do so, they draw the milk back from the ducts and cisterns into the cavity of the alveolus. From there it cannot be withdrawn until another release of oxytocin cause's the milk to be again squeezed from the alveoli. Some cows cannot release oxytocin for the second time for about fifteen minutes; others can do so almost immediately. The average is seven or eight minutes. It is therefore important that the removal of milk takes place immediately after milk ejection has occurred. If hand milking does not commence immediately after "let down", milking can turn into a long drawn out process, and if using a milking machine, it may cause injury to the udder by unnecessarily prolonging contact with the machine.

The neuro-hormonal reflex of milk ejection: Stimulus that a cow associates with milking causes a nerve impulse to travel via the inguinal nerve to the spinal cord and the brain. The brain causes the release of oxytocin from the posterior pituitary. Oxytocin is released into a branch of the jugular vein and travels to the heart and is then transported to all parts of the body by the arterial blood. The oxytocin reaching the udder leaves the heart by the aorta and enters the udder through the external pudic arteries. In the udder, it causes the myoepithelial cells to contract, resulting in milk ejection from the alveoli.

2 IDENTIFY CHEMICAL COMPOSITION OF THE MILK

2.1 Milk Composition

The major compositions of milk are water, fat, protein, lactose, ash or mineral matter, the minor constituents of milk are phospholipids, sterols, vitamins, enzymes, pigments, etc. Both milk yield and composition vary considerably among breeds of dairy cattle. Jersey and Guernsey breeds give milk with about 5% fat while the milk of Shorthorns and Friesians contains about 3.5% fat. Zebu cows can give milk containing up to 7%

fat. Milk of individual cows within a breed varies over a wide range both in yield and in the content of the various constituents (Table 7. 1).

Table 7. 1 The average milk composition in different breeds of cow

Breed	Fat	Protein	Lactose	Ash
Zebu	5. 6	3. 1	4. 6	0. 71
Ayrshire	3. 8	3. 4	4. 8	0. 70
Friesian	3. 4	3. 2	4. 6	0. 74
Guernsey	4. 9	3. 8	4. 8	0. 75
Jersey	5. 1	3. 8	4. 9	0. 75
Shorthorn	3. 6	3. 4	4. 8	0. 70

Source: International Livestock Research Institute, 1995.

The potential fat content of milk from an individual cow is determined genetically, as are protein and lactose levels. Thus, selection for breeding based on individual performance is effective in improving milk compositional quality. Herd recording of total milk yields and fat and solids-not-fat (SNF) percentages will indicate the most productive cows, and replacement stock should be bred from these.

2. 2 Factors Affecting Milk Composition

Milk differs widely in composition as breeds. All milks contain the same kind of constituents, but in varying amounts. Milk from individual cows shows greater variation than mixed herd milk. In general, milk fat shows the greatest daily variation, and then comes protein followed by ash and sugar.

- Species of animal: Each species of animal yield milk of a characteristic composition. Breed: The milk of some breeds of same species is comparatively higher in fat content than those of the other breeds e. g. milk of Red Sindhi cow contains higher fat than those of Holstein Friesian and Brown Swiss.
- Individual variation: Individuality of the animals is responsible for some of the greatest variations in the composition of milk.
- Stage of lactation: The period from the time the calf is born until the cow ceases to give milk is called period of lactation. The secretion of milk immediately after calving is known as colostrum. It may last from 3 to 6 days. It contains more protein and more total solids than those of normal milk.
- Age: The fat content in milk rises from 1st to 3rd lactation period, then it remains fairly constant in subsequent lactation periods, but later towards advancing age there is a slight reduction in the fat content of the milk.
- Seasonal Variation: Generally, fat and Solids-not fat content in milk show slight but

well-defined variations during the whole year.

- First and last milk: There is a considerable variation in the fat content of the fore milk, mid milk and stripping. The fore milk is very poor in fat and stripping are very rich in fat. This variation is more when the milk yield is high.
- Feeds of the animal: When the milk animals are given sufficient balanced ration, feed has no significant effect on composition. When the feeding is changed there is some variation in the composition of milk, but such variations are temporary.
- Milking Interval: In general, a longer interval is associated with more milk and lower fat test.
- Frequency of Milking: Whether a cow is milked two or three times a day, it has no great effect on the fat test.
- Physical condition of the animal: There may be a change in the composition of the milk if the animal is suffering from any disease viz. in the milk of cow or buffalo suffering from mastitis, there is a reduction in fat, protein, lactose content and a marked increase in chloride content.
- Environment at the time of milking: Anything which causes discontentment and uneasiness in the cow at the time of milking causes the cow to be nervous and leads to the holding up of her milk. As last portion of milk is rich in fat, hence it affects the composition.
- Administration of drugs and hormones: Certain drugs may affect temporary change in the fat%, injection or feeding of hormones results in increased milk yield and fat%.
- Milker: If the milker is not an efficient one and not able to draw milk completely from the udder, the fat content of the milk is reduced.
- Genetic factors: The genetic of Ethiopia's livestock have involved largely as a result of natural selection influenced by environmental factors. This has made the stock better conditioned to withstand feed and water shortages, disease challenges and harsh climates. But the capacity for the high level of production has remained low.
- Environmental factors: In tropical and subtropical countries, an animal may often be under heat stress. When the environmental temperature exceeds the upper critical level (18 to 24℃, depending on the species) there is usually a drop-in production or a reduced rate of gain. Furthermore, when the temperature falls outside the comfort zone, other climatic factors assume greater significance. Humidity becomes increasingly important, as do solar radiation and wind velocity. Dairy cattle show a reduced feed intake under heat stress, resulting in lower milk production and reduced growth. Reproduction is also adversely affected. There are, however, important differences between breeds. European cattle (*Bos taurus*) produce well at temperatures ranging from 4 to 24℃, even at high humidity. Much lower temperatures (−10℃) have little effect, provided that fluctuations are not too rapid or frequent. On the other hand, a drop-in milk production results when temperatures exceed 25℃. The drop may be as much as 50% at temperatures of 32℃ or higher. In

contrast, zebu cattle (Bos indicus), which are native to warm climates, have a comfort zone of 15 – 27℃ and milk production begins to drop only when temperatures rise above 35℃.

- Interval between milking: The fat content of milk varies considerably between the morning and evening milking because there is usually a much shorter interval between morning and evening milking than between evening and morning milking. If cows were milked at 12 hours intervals the variation in fat content between milking would be negligible, but this is not practicable on most farms. Normally, SNF content does not vary with the length of time between milking.
- Stage of lactation: The fat, lactose and protein contents of milk vary according to stage of lactation. Solids not fat content is usually highest during the first two to three weeks, after which it decreases slightly. Fat content is high immediately after calving but soon begins to fall, and continues to do so for 10 – 12 weeks, after which it tends to rise again until the end of the lactation. The high protein content of early lactation milk is due mainly to the high globulin content.
- Age and health: As cows grow older the fat content of their milk decreases by about 0.02% units per lactation while the fall in SNF content is about 0.04 percentage units. Both fat and SNF contents can be reduced by disease, particularly mastitis.
- Feeding regime: Underfeeding reduces both the fat and the SNF content of milk, although SNF content is more sensitive to feeding level. Fat content and fat composition are influenced more by roughage (fiber) intake. The SNF content may fall if the cow is fed a low-energy diet, but is not greatly influenced by protein deficiency, unless the deficiency is acute.
- Completeness of milking: The first milk drawn from the udder contains about 1.4% fat while the last milk (or stripping) contains about 8.7% fat. Thus, it is essential to milk the cow completely and thoroughly mix all the milk removed before taking a sample for analysis. The fat left in the udder at the end of a milking is usually picked up during subsequent milking, so there is no net loss of fat.

2.3 Major Composition of Milk

The quantities of the main milk constituents can vary considerably depending on the individual animal, its breed, stage of lactation, age and health status. Herd management practices and environmental conditions also influence milk composition. The average composition of cow milk is shown in Table 7.2.

Water is the main constituent of milk and milk processing is usually designed to remove water from milk or reduce the moisture content of the product.

Table 7.2 Composition of cow milk

Main milk composition	Range (%)	Mean (%)
Water	85.5 - 89.5	87.0
Total solids	10.5 - 14.5	13.0
Fat	2.5 - 6.0	4.0
Proteins	2.9 - 5.0	3.4
Lactose	3.6 - 5.5	4.8
Minerals	0.6 - 0.9	0.8

Source: International Livestock Research Institute, 1995.

2.3.1 Fat

If milk is left to stand, a layer of cream forms on the surface. The cream differs considerably in appearance from the lower layer of skim milk.

Cream consists of a large number of spherical microscopic globules of varying sizes floating in the milk. Each globule is surrounded by a thin skin the fat globule membrane which acts as the emulsifying agent for the fat suspended in milk. The membrane protects the fat from enzymes and prevents the globules coalescing into butter grains. The fat is present as oil in water emulsions that can be broken by mechanical action such as shaking.

Fats are partly solid at room temperature. Fats are partly solid at room temperature. The term oil is reserved for fats that are completely liquid at room temperature. Fats and oils are soluble in non-polar solvents, e.g. ether. The lipid content of milk is usually defined as the fraction which is extracted by organic solvents. Table 7.3 gives the main lipid classes of milk fat.

Table 7.3 Composition of lipids in whole bovine milk

Lipid	Weight (%)
Carotenoids + vitamin A	Trace
Cholesterol esters	0.02
Triglycerides	98.3
Diglycerides	0.3
Monoglycerides	0.03
Free fatty acids	0.1
Cholesterol	0.2 - 0.4
Phospholipids	0.2 - 1.0

Source: International Livestock Research Institute, 1995.

2.3.2 Proteins

Proteins are an extremely important class of naturally occurring compounds that are essential to all life processes. They perform a variety of functions in living organisms ranging from providing structure to reproduction. Milk proteins represent one of the greatest contributions of milk to human nutrition.

Proteins are polymers of amino acids. Only 20 different amino acids occur regularly in proteins. They have the following general chemical structure:

$$\begin{array}{c} NH_2 \\ | \\ R—C—COOH \\ | \\ H \end{array}$$

R represents the organic radical. Each amino acid has a different radical and this affects the properties of the acid. The content and sequence of amino acids in a protein therefore affect its properties. Some proteins contain substances other than amino acids and are called conjugated proteins. These include:

- Phosphoproteins in which phosphate is linked chemically to the protein, e. g. casein in milk and phosphoproteins in egg yolk.
- Lipoproteins which are combinations of lipid and protein and is excellent emulsifying agents. They are found in milk and egg yolk.
- Chromoproteins which have a colored prosthetic group and include hemoglobin and myoglobin.

Protein composition of milk

Cow milk contains about 32 g/L protein. Of this, about 26 g/L consists of caseins which are precipitated upon acidification to pH 4. 6 at temperatures above 20℃. The proteins (6 g/L) remaining in solution at pH4. 6 are called whey proteins and consist of a diverse group, including lactalbumin, β-lactoglobulin, blood serum albumin and immunoglobulin's.

1) Caseins

Casein was first separated from milk in 1830 by adding acid to milk, thus establishing its existence as a distinct protein. It was subsequently shown that casein is made up of a number of fractions and is therefore heterogeneous. In general, caseins are high in phosphorus, low in sulphur and are not significantly affected by moderate heat. All the major caseins associate with themselves and with each other. In the presence of calcium ions (Ca^{2+}) these associations lead to the formation of casein micelles. About 95% of the casein in milk exists as particles of colloidal dimensions known as micelles. The micelles are generally spherical in shape with diameters ranging from 40 to 300 nm (average about 100 nm) and molecular weight of about 10^8. Casein is easily separated from milk by acid precipitation at about pH 4. 6.

2) Whey proteins

When milk is brought to pH 4. 6, the caseins precipitate. The supernatant contains four principal proteins in the whey fraction, β-lactoglobulin, lactalbumin, blood serum albumin, immunoglobulin and a number of minor proteins, e. g. lactoferrin and enzymes. Most of the whey proteins are denatured by heat, i. e. they become less soluble if milk is heated.

β-lactoglobulin is the principal whey protein of the cow, goat and sheep, although there are slight interspecies differences. β-lactoglobulin accounts for about 50% of the total whey proteins or about 11% of the total protein in milk. Related but substantially different proteins occur in porcine milk. No β-lactoglobulin has been identified in human, camel or horse milk in which α-lactalbumin in the principal whey protein.

Denaturation of whey proteins and β-lactoglobulin, in particular, is of major technological significance-lacto globulin interacts with k-casein during heating and this reduces the heat stability of milk, slows down rennet clotting during cheese manufacture and gives a soft curd which tends to retain water.

α-lactalbumin represents about 20% of the protein of bovine whey (3.5% of the total milk protein) and is a relatively minor protein in terms of quantity. It functions as part of the enzyme system involved in lactose synthesis.

The immunoglobulins are antibodies which are present in high concentrations in colostrum. Infants and mammals are born without circulating antibodies and the main way in which they acquire these is by ingestion of colostrum.

3) Minor protein constituents

About 50 enzymes have been detected in bovine milk. The concentration of milk enzymes varies greatly among species. Some milk enzymes act on substrates present as normal constituents of milk and may play either beneficial or deleterious roles during milk processing.

- Catalase: This enzyme catalyses' the decomposition of hydrogen peroxide (H_2O_2) to H_2O and O_2. Its activity is higher in mastitis milk and colostrum's than in normal milk and increases with increase in bacterial numbers.
- Lactoperoxidase: This enzyme catalyses' oxidation of a range of substrates by H_2O_2. The enzyme catalyses' oxidation of thiocyanate to products that inhibit certain bacteria. It is relatively heat stable; it is not inactivated by pasteurization (72℃×15 seconds) but is destroyed when milk is heated above 80℃. The absence of lacto peroxides in milk indicates that the milk has been heated to at least 80℃. The test for the presence of lactoperoxidase is based on the oxidation of the substrate paraphenylenediamine in the presence of H_2O_2.
- Phosphates: Phosphates enzymes catalyze the hydrolysis of phosphate esters. Milk contains acid and alkaline phosphates. Alkaline phosphates have a pH optimum near 9 and are inactivated by heating milk to 72℃ for 15 seconds. Its absence indicates that milk has been properly pasteurized. If milk is inadequately pasteurized, the residual enzyme will catalyze the hydrolysis of added disodium para-nitrophenol phosphate liberating para-nitrophenol which is yellow in alkaline solution. Acid phosphates, which have a pH optimum of 4, are more heat stable than alkaline phosphates.
- Other milk enzymes: Milk also contains lipases proteases, amylases, xanthine oxidase, carbonic anhydrase and lysozyme.

2.3.3 Carbohydrates

The average lactose content of milk varies between 4.7% and 4.9%, although milk from individual cows may vary more. Mastitis reduces lactose secretion.

Lactose is a source of energy for the young calf and provides 4 calories/g of lactose metabolized. It is less soluble in water than sucrose and is also less sweet. It can be broken down to glucose and galactose by Bacteria that have the enzyme β-galactosidase. The glucose and galactose can then be fermented to lactic acid. This occurs when milk goes sour. Under controlled conditions they can also be fermented to other acids to give a desired flavor, such as prop ionic acid fermentation in Swiss-cheese manufacture.

Lactose is present in milk in molecular solution. In cheese making, almost all of the lactose remains in the whey fraction. It has been recovered from whey for use in the pharmaceutical industry, where its low solubility in water makes it suitable for coating tablets. It is also used to fortify baby-food formulations. Lactose can be sprayed on silage to increase the rate of acid development in silage fermentation. It can be converted into ethanol using certain strains of yeast; the yeast biomass is recovered and used as animal feed.

However, these processes are expensive, and a large throughput is necessary for them to be profitable. For smallholders, whey is best used as a food for humans and animals without any further processing.

Heating milk, above 100℃ causes lactose to combine irreversibly with the milk proteins. This reduces the nutritional value of the milk and also turns it brown.

Because lactose is not as soluble in water as sucrose, adding sucrose to milk forces lactose out of solution and it crystallizes. This causes sandiness in such products as ice cream. Special processing is required to crystallize lactose when manufacturing products such as instant skim milk powders.

In addition to lactose, milk contains traces of glucose and galactose. Carbohydrates are also present in association with protein. K-casein, which stabilizes the casein system, is a carbohydrate-containing protein.

2.3.4 Minor Milk Constituents

In addition to the major constituents already discussed, milk also contains a number of organic and inorganic compounds in small or trace amounts, some of which affect both the processing and nutritional properties of milk.

Minerals

The salts of milk are composed mainly of the chlorides, phosphates, citrates, carbonates and bicarbonates of sodium, potassium, calcium and magnesium. Approximately 20 other elements are found in milk in trace amounts. These include copper, iron, lead, boron, manganese and iodine. Milk salts are important in human nutrition, stability of milk lipids and in the processing of milk proteins.

The ash content of cow milk remains relatively constant at 0.7% to 0.8%, but the relative concentrations of the various ions vary considerably. The composition is influenced by a number of factors including breed, individuality of the cow, stage of lactation, feed, infection of the udder and season of the year. Certain milk salts, e. g. sodium and potassium chlorides are sufficiently soluble to be present almost in the dissolved phase. The content of others, in particular calcium phosphate, is greater than can be maintained in solution at the normal pH of milk. Consequently, these exist partly in soluble form and partly in insoluble or colloidal form.

Vitamins

Milk contains the fat-soluble vitamin A, vitamin D, vitamin E and vitamin K in association with the fat fraction and water-soluble vitamins B complex and vitamin C in association with the water phase. Vitamins are unstable, and processing can therefore reduce the effective vitamin content of milk.

2.4 Chemical and Physical Properties of Milk

Milk is the natural physiological secretion from normally functioning mammary gland of a mammal intended to nourish the young ones. All female mammals can secrete milk, the properties of which are similar to those of cow's milk in general, but there are considerable differences still existing with respect to their physical and chemical nature.

1) Color

Milk is a creamy-yellowish white liquid in color, which varies from bluish white to light yellow, depending upon the breed, the feed fed, and the butterfat quantity and other solids present in it. The yellow color of the milk is due to a pigment known as carotene, which is synthesized from the green feed fed to the cow conversion of carotenes into vitamin A chiefly occurs in liver. In cases of buffalo's milk, this change is complete and thus buffalo's milk is white. In case of cows this conversion of carotene into vitamin A is partial so cow's milk is yellow in color. The white color (adolescence) of milk is due to reflection of light by the fat globules and the colloidal protein, calcium-caseinate and phosphate. The bluish color of separated milk or whey is due to another pigment known as *Riboflavin* (vitamin B_2) or *Lacto-chrome*.

2) Taste

Milk is slightly sweet in taste. This is due to the presence of lactose (milk sugar). The sweet taste of lactose is balanced against the salty taste of chloride in milk.

3) Smell

Milk has got a characteristic odor of its own, when it is drawn from the udder.

4) Odor

Freshly drawn milk has a "cowey" odor, which disappears when kept exposed for some time. Milk has got the capacity to acquire odor from the surrounding and also from the feed etc. but these odors are abnormal. Milk has developed odor due to bacterial action and change in its chemical composition. Certain metals may have an adverse effect on the flavor of the milk which comes in contact with them the metals are like copper and copper alloys, as nickel, brass, bronze etc. Rusty cans or other rusty surfaces may prove harmful producing a metallic or tallow flavor.

5) Acid base equilibrium

Freshly drawn milk has got "Amphoteric Reaction" i. e. it changes red litmus blue and blue litmus red. Its average pH value is 6. 7 on titrating it with an alkali it is found to contain 0. 1% to 0. 17% acidity. This is not due to lactic-acid developed, but due to phosphates of milk proteins Citrates and carbon dioxide present in milk (Natural).

6) Specific gravity of milk

The specific gravity of freshly drawn milk is lower than specific gravity obtained, after an hour or later. The rise in specific gravity is regular, more rapid at low temperature than at higher ones and amounts on an average to 0. 001. This is called "Recknagel's phenomenon" and is attributed to change in the specific gravity of fat due to:

- Partial cooling and solidification
- Hydration of the proteins
- Loss of carbon dioxide
- Presence of air bubbles

7) Freezing point

Milk freezes at −0. 55 to −0. 56℃ (31 to 30. 96℉). Skim, whole milk or creams have same freezing point. Milk has a lower freezing point than water due to the presence of lactose and salts in aqueous phase. The freezing point is affected by:

- Increased acidity (decreased freezing point)
- Addition of preservatives (decreased freezing point)
- Addition of water

8) Boiling point

Milk is slightly heavier than water because of its solute content and boiling point of a liquid is influenced by factors responsible for its specific gravity. Milk boils at a temperature slightly higher water. Water boils at 212℉ (100℃) at sea level, while average milk boils at 212. 3℉ (100. 17℃).

9) Viscosity

It is the resistance to flow and is the reverse of fluidity. Viscosity is the property of all fluids. It can be expressed in only relative terms and for convenience the relative viscosity of any fluid is compared with water. Water flows with ease. Syrup and honey pour much more slowly and possess greater viscosity. Milk is 1. 5 − 1. 7 times more viscous than water owing to the presence of solids in milk.

Factors affecting viscosity:

- Temperature (At 0℃ milk has a fluidity of 0. 233 and water has a fluidity of 0. 558 and at 20℃, these values change to 0. 473 and 1. 00)
- Fat content

- Homogenization
- Souring ageing
- Microbial growth
- High heating followed by cooling. Heating the milk to pasteurization temperature or agitating it lowers the viscosity.

10) Adhesiveness of milk

A piece of paper moistened with milk sticks to a flat surface of wood, glass or metal. This property is undoubtedly due to casein, which is used in large quantities in the manufacture of casein glue, one of the strongest glues made.

11) Refractive index

Milk has a refractive index of about 1.35 that of water being 1.33. Addition of water would therefore lower the refractive index of milk, but since considerable variation is found in values for genuine milk; it is not possible to use this property alone as a criterion for the genuineness of milk samples.

12) Surface tension

Compared with water, the surface tension of milk is low (milk-50; water-72.75 dynes/cm 20℃). Somewhat higher values are shown by separated milk while cream has a lower surface tension. The colloidal constituents, particularly the proteins are responsible for this lowering of surface tension, have a tendency to get concentrated at liquid / air interface.

13) Cream rising

When whole milk is permitted to stand, the fat rises to the top and eventually forms a layer packed with fat globules called cream. The difference in specific gravity between the milk serum and milk fat is one of the most important factors responsible for cream rising. At least for the rapid and complete rising of the milk fat, the fat globules must aggregate or clump together. The rate of creaming, then, is dependent on the factors that affect Clumping. The fat particles are held together in the clumps by the musing-like material surrounding each fat particle.

14) Foaming

Milk has the property of foaming on agitation. Foam is due to the formation of a physical phase in which air becomes incorporated in the milk with thin layers of milk separating the air bubbles from one-another. The capacity for foaming is due to materials lowering the surface tension. The milk protein and fat reduce surface tension and therefore are the causes of the foaming capacity. Milk fat not only increases the foaming capacity, but also increases the stability of the foam. Milk foam is unstable and breaks down when allowed to stand.

15) Orbital rotational potential

Fresh milk exhibits a potential of +0.2 to +0.3V at noble metal electrodes Ascorbate, lactates, riboflavin are principal contributors.

16) Electrical conductivity

Considered a possible index of mastitis infection, added water, and added neutralize means of controlling solid concentration. The specific conductance of milk reflects its concentration and activity of ions and is of the order of 0.005 Siemens/ (ohms · cm) at 25℃, ranging from 0.0040 – 0.0055.

3 MILKING PROCEDURES AND REQUIREMENTS

Proper milking procedures, attention to detail, and a clean environment are important to prevent mastitis, maximize production of quality milk and for insuring complete milk removal from the udder. Mastitis can decrease total milk production by 15 – 20%. To minimize loss and achieve maximum milk yield, a practical milking management scheme should be followed. The term "milking management" includes care for the environment in which cows are housed or pastured. The dairy cow should have a clean, dry environment, which helps reduce the potential for mastitis and increases milking efficiency by reducing time and labor to clean udders before the milking process. Milking should be done by people who are responsible, trained, conscientious, and have a clear vision that they are harvesting food for human consumption. Research and experience indicate repeated regular training of milking technicians, whether family or non-family workers is a very important part of this process.

3.1 Selecting, Checking, Maintaining, and Using of Equipment and Materials

Maintenance is the upkeep of plant and machinery in proper working condition at all times. Preventive maintenance is the persistent and systematic procedure for the care of all production, control and auxiliary machinery in a dairy factory, including regular servicing, upkeep and overhaul, record keeping and stocking of essential spare parts for the purpose of preventing breakdowns and emergency shutdowns for repair.

Preventive maintenance is useful and necessary because it will prevent the loss of money and profits due to unnecessary machinery shut downs, shortened machine life, machine inefficiency, and reduced productivity.

The main purpose of maintenance is to:

- Increase the efficiency and improve the performance of all processing and service equipment.
- Increase the overall productivity of the entire plant by achieving coordinated and continuous operation of all plant equipment.
- Increase the certainty of meeting daily production schedules.
- Reduce unscheduled downtime.
- Extend the useful life of all plant equipment.
- Minimize property and personnel hazards.

Maintenance of milk cans

Great care should be observed in the handling of milk cans i.e. that they are not dented or damaged more than necessary. During cleaning of cans, the cleaning solution should be kept at the proper strength as alkali or acid cleaner of high concentration remove the tin and allow rusting. Thorough drying of cans will increase their life span and also improve on milk quality handled.

Maintenance of milk cooling equipment

Various types of refrigeration equipment ranging from surface coolers, immersion coolers, ice-bank and direct expansion refrigeration systems are in use throughout the dairy industry. Manufacturer's instructions on service and scheduled repairs should be followed very strictly. Special attention should be paid to lubrication of compressors and detection and timely repair of refrigerant gas leakages.

Where brine is used as a coolant, its corrosiveness to dairy equipment should also receive particular precautions during its circulation and handling.

Maintenance of milk separator

- The gears must be well lubricated; Follow manufacturer's instructions.
- The level of the lubricant must be kept constant; observe the oil level through the sight glass.
- The bowl must be carefully balanced.
- The bowl should be cleaned thoroughly immediately after use to ensure proper functioning of the separator and for hygiene.

Maintenance of butter churn

- The churn and butter making equipment should be washed as soon as possible, preferably while the wood is still damp in the case of wooden churns.
- Wash the inside of the churn thoroughly with hot water. Invert the churn with the lid on in order to clean the ventilator; this should be pressed a few times with the back of a scrubbing brush to allow water to pass through.
- Remove the rubber seal from the lid and scrub the groove. Scald the inside of the churn with boiling water or steam. Invert and leave to dry. Dry the outside and treat metal parts with food grade grease or vaseline to prevent rusting. The rubber seal should be placed in boiling water or dipping in warm water with disinfectant is enough.

Maintenance of Milking Machine

Milking machines are used 5 – 6 or more hours every day. Broken-down machines or machines operating inefficiently cost in reduced milk, time, damaged udders, and reduced milk quality. Regular service inspection will allow performance at high efficiency. There are certain maintenance checks that should be performed by the operator at every milking, and every 50 and 250 hours.

1) Daily

➢ Wash outside of milk line, receiver jar and trap, claws, and hoses. Empty trap.

➢ Check all rubber parts for holes, tears, or water in shells. Replace liners or short air tubes that have holes.

➢ Check vacuum level and vacuum recovery time (no more than 2 - 3 seconds after opening one milking unit).

➢ Check vacuum pumps for belt tension (within 1/2 inch from rest position) and oil reservoir.

➢ Make sure pulsators are working. Check with your thumb in the liner. Check each bleeder vent in every milking unit. A paper clip could be used on claw vents, but a finer wire is needed for vents in short milk tubes.

➢ Listen for air leaks.

2) Weekly

➢ Clean vacuum regulator and moisture drain valves.

➢ Check pulsator and vacuum regulator filters and clean or replace if necessary.

➢ Check short vacuum tubes with vacuum gauge to determine if each pulsator opens and closes fully.

➢ Check stall cocks for leaks and electrical connections for tightness.

➢ Break down receiver jars and weigh jars and clean fittings and gaskets.

3) Every 4 to 6 Weeks

➢ Disassemble pulsator and clean air ports and screens. Replace worn parts.

➢ Check each pulsator for proper operation with a vacuum gauge.

➢ Disassemble and clean vacuum regulators and replace air filters.

➢ Flush pulsator vacuum lines.

➢ Check condition of air tubes, vacuum hoses, and milk hoses.

➢ Wash trap inside and out, inspect float.

4) Every 6 Months

➢ Evaluate the entire milking system.

➢ Replace all pulsator rubber parts, hoses, and air tubes.

➢ Replace all milk hoses.

➢ Replace receiver jar gasket.

➢ Replace rubber hoses and rubber hose nozzles used to wash udders.

➢ Check belts and oiler on vacuum pump (s).

3.2 Sanitation and Hygiene, according to Codes of Production

The World Health Organization defines the term "sanitation" as follows: sanitation generally refers to the provision of facilities and services for the safe disposal of human urine and feces. The word "sanitation" also refers to the maintenance of hygienic conditions,

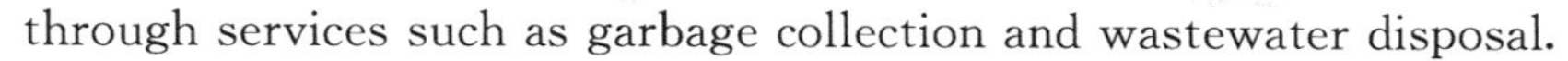

through services such as garbage collection and wastewater disposal.

Sanitation includes all four of these engineering infrastructure items: excreta management systems, waste water management systems (included here are wastewater treatment plants), and solid waste management systems.

There are some slight variations on the definition of sanitation in use. For example, for many organizations, hygiene promotion is seen as an integral part of sanitation. For this reason, the Water Supply and Sanitation Collaborative Council defines sanitation as collection, transport, treatment and disposal or reuse of human excreta, domestic wastewater and solid waste, and associated hygiene promotion (Evans et al. , 2009).

Despite the fact that sanitation includes wastewater treatment, the two terms are often used side by side as "sanitation and wastewater management" . The term sanitation has been connected to several descriptors so that the terms sustainable sanitation, improved sanitation, unimproved sanitation, environmental sanitation, on-site sanitation, ecological sanitation, dry sanitation are all in use today. Sanitation should be regarded with a systems approach in mind which includes collection/containment, conveyance/transport, treatment, disposal or reuse (Tilley, E. et al 2014).

The overall purposes of sanitation are to provide a healthy living environment for everyone, to protect the natural resources (such as surface water, groundwater, soil), and to provide safety, security and dignity for people when they defecate or urinate.

Effective sanitation systems provide barriers between excreta and humans in such a way as to break the disease transmission cycle (for example, in the case of fecal-borne diseases) (Thor Axel Stenström, 2005).

Hygiene is a set of practices performed for the preservation of health. According to the WHO, hygiene refers to conditions and practices that help to maintain health and prevent the spread of diseases. Whereas in popular culture and parlance it can often mean mere "cleanliness", hygiene in its fullest and original meaning goes much beyond that to include all circumstances and practices, lifestyle issues, premises and commodities that engender a safe and healthy environment.

3. 3 Milking Methods

Milking is the act of removing milk from the mammary glands of the cow. It may be done by hand or by machine.

3. 3. 1 Hand Milking

Cows are milked from the left or right side. After letting down of milk, the milker starts milking teats either crosswise or four quarters together and then hind quarters together or teats appearing most distended milked first few streams of fore milk from each teat be let on to a strip cup. This removes any dirt from the teat canal and gives the operator a chance to detect mastitis. There are types of hand milking techniques.

➢ Stripping: When you're ready to start milking the cow, grab a teat in each hand,

holding it between your thumb and forefinger. Start at the top of the teat, and then slide your thumb and forefinger down the length of the teat, compressing it as you slide. Squirt a few streams out of each teat in this manner into separate cups, and then check the milk for signs of mastitis, such as flakes or clumps in the milk. Never drink milk from an infected teat. After squirting out enough milk to inspect, stop the stripping technique. The friction from sliding your fingers down the teat can irritate the cow's skin. At the end of your milking session, use the stripping technique for a few squirts; the last dribbles of milk often are richer than the rest.

- Full-hand milking: The proper technique for the majority of your milking time is using your full hand. Grasp the top of the teat by wrapping your thumb and forefinger around it, then compress the teat and wrap your other fingers around it as you squeeze. These forces milk already in the teat out in a stream without sliding your hand up and down the teat. Release the compression without fully letting go of the teat. This allows more milk to drop from the udder into the teat, so you can compress your fingers from top to bottom and squeeze out the milk. Because your hand isn't sliding on the teat, it reduces the chance of irritation to the cow.
- Pinch Milking Techniques: is done by pinching the teat between the first and middle fingers, and stripping downwards (Fig. 7. 1)

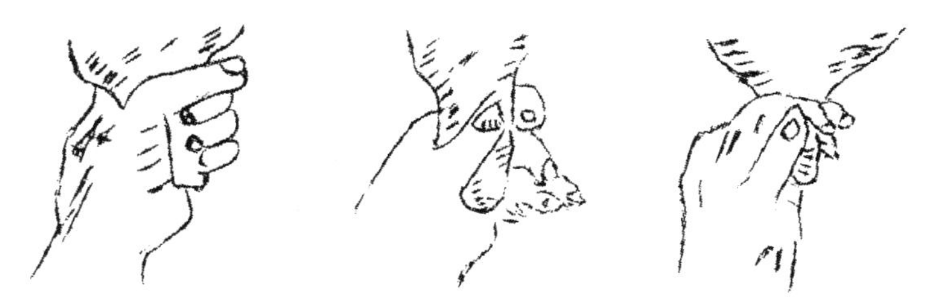

Fig. 7. 1 Hand milking techniques

3. 3. 2 Machine Milking

Modern milking machines are capable of milking cows quickly and efficiently, without injuring the udder and the teats, if they are properly installed, maintained in excellent operating conditions, and used properly; however, it must be remembered that this machine is one of the few devices which has direct contact with living animal tissue.

The milking machine performs two basic functions:

- It imposes a controlled vacuum at the teat end, which opens the teat orifice and provides a differential pressure necessary for milk flow.
- It massages the teat at regular interval, which prevents congestion of blood in the teat end and assists in the stimulation of milk let-down.

A milking operation which results in discomfort to the cow and is caused by faulty milking equipment or techniques may lead to injury or mastitis. Consequently, before a person attempts to milk cows he/she should thoroughly understand the basic operation of the milking equipment and fully realize the significance of maintaining the equipment in good condition at all times and of employing good milking skills and techniques.

The advantages of milking machine are manifold. It is easy to operate, costs low, saves time as it milks 1.5 – 2 liters/minute. It is also very hygienic and energy conserving as electricity is not required. All the milk from the udder can be removed. The machine is also easily adaptable and gives a suckling feeling to the cow and avoids pain in the udder as well as leakage of milk.

All milker units operate in basically the same way and consist of the following components:

- Pulsator
- Teat cup shells and liners (inflations)
- Milk receptacle: Bucket and teat-cup claw (attached to a floor pail milker or to a pipeline)

1) Pulsator

It causes the chamber between the shell and the liner to alternate regularly from vacuum to air source. Keep in mind that the inside of the teat-cup liner is under a milking vacuum at all times. Thus, when air is admitted between the shell and liner the line collapses around the cow's teat. The pressure of the collapse liner is applied to the teat giving a massaging action. This is called the rest or massage phase. Milk does not flow from the teat during this phase.

During the milk phase, the space between the liner and the shell is exposed to the vacuum by way of the pulsator. The fact that there is now equal pressure on both sides of the liner causes it to open. The end of the cow's teat exposed to the vacuum and the influence of internal milk pressure within the cow's udder causes the milk to be drawn out through the teat opening.

2) Teat-cup shells and liners (inflations)

Many types of teat-cup shell and liner combinations are available. Make sure that the shell and liner are compatible. For instance, make sure that the liner has enough room inside the shell so that it can fully collapse without hitting the inside walls.

Choose a liner that has a mouthpiece which helps prevent downward slippage or riding-up action on the base of the udder. To reduce teat and udder irritation, the use of narrow to intermediate bore liners is recommended. If a herd is presently being milked with a wide-bore style liner and one wishes to change to a narrow-bore liner, be conscious of the following facts:

- Many drop-offs may be experienced at first until you relearn how to handle the milker units, such as putting less tension on the units.
- Slower milking may be experienced on some cows at first.

- Some long, flabby teat cows may never be able to adapt to the narrow-bore style.
- The liners (inflations) should be replaced as recommended by the manufacturer. A liner should be replaced immediately if it becomes damaged.

3) Milk receptacle

Milker units may be classified either as being the suspension bucket or claw type. The claw-type milker is attached to a floor pail milker or to a milk pipeline system via a hose used to transfer the milk. Air is needed in the claw to help lift the milk up into the pail milker or into the pipeline. This air is let into the claw through the air admission inlet located in the claw or in the individual teat-cup liners. Care must be taken so that the air admission hole (s) is kept clean, but not enlarged to the extent that excessive agitation could cause rancidity problems with the milk. An alternative to air admission is a two-vacuum level system within the milker unit which pushes the milk as a column up the hose and into the pipeline.

4) Milk Movement-Pipeline to Bulk Tank

Once in the pipeline, milk flows by gravity into a receiver jar. A milk pump removes the milk from the receiver jar and it passes through a filter as it makes its way to the bulk tank for cooling and storage.

Effective machine milking technique

- Clusters should be attached to cows in batches as soon as possible after preparation
- The inverted cluster should be held level in the hand such that vacuum is cut off from the cups and air leakage is minimized.
- Attach each teat cup in sequence and as quickly as possible.
- If right-handed, apply teat cups in a right-direction sequence, leaving the nearest teat cup (to operator) to the last; if left-handed, apply teat cups in a left-direction sequence leaving the nearest teat cup to the last.
- Air leaks in the milking system during milking should be corrected immediately
- When animals with three active quarters are being milked, a clean dummy teat should be used for the remaining milking cup and a mechanism to easily identify the cow, such as leg band and udder spray should be in place.
- Where manual cluster take-off is practiced, clusters should be removed when milk flow ceases in the claw-piece by shutting off vacuum at the claw-piece before removing the cluster and the four Teat cups should be removed together (automatic cluster removers).
- Is an effective method by which to avoid over or under milking, particularly if there are different operators in the parlor.
- It is most important that a set of procedures is chosen that requires each milker to be absolutely consistent at every milking.

Hygiene considered during milking by machine

- Avoid splashes or sprays of milk on hands or on clusters.
- Use running water and disinfectant solution to remove infected milk from gloves, clusters or other surfaces.
- Dirty clusters should be washed externally between milking.
- Dirty clusters should not be washed /cleaned while still attached to the cow.
- Standing areas should be washed only when cows have exited.
- All used gloves and towels, disinfectant wipes and used antibiotic tubes should be assembled as waste.
- Washing of yards should not be done while cows are still present.
- Avoid walking cows through a foot bath after milking when not required (Fig. 44 – Fig. 46).

3.4 Milking Procedure

Good milking management is critical in the production of quality milk. During the milking process, bacteria can be transferred from the environment into the milk. Furthermore, there is a high risk of spread of contagious mastitis from infected to uninfected cows during the milking process unless proper control measures are taken. The milking process may be defined as the overall procedure associated with three distinct activities, i. e. cow preparation for milking, milking technique and post milking teat disinfection.

3.4.1 Hand Milking Procedure

Step 1: Prepares milking items

- Milking pail Graduated jug of one litter, plastic pails.
- Milk strainer/sieve/cloth
- Milk can/bulk container
- Towels
- Dip cup
- Strip bowels
- Surgical glove
- Disinfectant
- Waste basket
- Milker seat
- Rope

Step 2: Apply milking order

Before milking your cow; observe the udder and teats for any signs of mastitis, such as swollenness or fever. Mastitis is a disease in dairy cows that can be caused by a number of different pathogens and that can affect the production and quality of milk. Often mastitis symptoms aren't readily apparent through visual or tactile observation, but it's best practice to know what looks normal for your cow and what could be symptoms of disease. Start milking with first-calf heifers, follow by uninfected cows, segregate chronic mastitis cows, and milk them last.

Step 3: Clean teats for milking

Restrain the cow properly in a clean, dry, stress free environment and clean teats for milking.

➢ Wear clean surgical gloves.

➢ Wipe dirt and debris from the udder.

➢ Remove dry dirt and beddings in milking room.

➢ Use a clean towel to dry the teat and udder. Use more than one if necessary.

➢ Do not put soiled towels into the disinfectant.

➢ Wring excess water from towels.

➢ Completely clean teat from base of udder to end of teat.

➢ Disinfect hands.

➢ Pay special attention to the tip of the teat while milk and clean (Fig. 47).

Step 4: Fore strip

Strip out each teat by performing three to four squirts each. Stripping allows you to remove dirt and bacteria from the teat as well look for signs of infection, such as clumps, clear milk and blood, any of which could indicate mastitis. Mastitis checking procedure by California mastitis test (CMT).

➢ Squirt milk onto the black surface of the strip cup (2 times).

➢ Add equal amount of CMT reagent into the paddles of strip cup.

➢ Shake and mix for 10 – 30 seconds.

➢ Color and textural change will be observed if the milk is abnormal.

➢ Abnormal milk may appear watery, bloody, have flakes or clots.

➢ If any abnormal milk is observed, stop milking and refer to animal health professional (Fig. 48).

Step 5: Teat pre-dip

Apply a pre-milking disinfectant to the teats to help prevent mastitis.

➢ Immerse at least 3/4 of the teat into pre-dip disinfectant.

➢ Leave the disinfectant on the teats for approximately 30 seconds and dry with an individual towel.

➢ Disinfect hands and dry.

➢ Wipe the pre-milking disinfectant off of each teat with a single-use towel.

Step 6: Milking

Properly hold the teat by wrapping your thumb and forefinger around the base of the teat. This captures the milk in the teat.

1) Start milking in 60 – 90 seconds after pre-dip teats.

2) Place a clean milk bucket/pail under the cow's udder or between your legs.

- Each milker should have his own milk pail.
- Use milk pails only for milking.

3) Grasp the base of each teat just below the udder floor with your thumb, index and middle finger.

4) Expel the milk by using all three fingers, your middle fingers, following the index finger in gentle squeezing.

5) Complete milking within 5 – 7 minutes.

Step 7: Post teat-dip

- Immerse at least 3/4 of the teat into post-dip disinfectant
- Allow 30 seconds contact time.

After dipping, keep cows standing and let them out for exercise. If the cow is feeding a calf, you can return the cow to pasture, as the calf will finish cleaning and disinfecting the teats. However, if your cow doesn't have a calf, immediately apply a post-milking disinfectant to the teats. This solution performs a different function from the pre-milking disinfectant and each should be clearly labelled. Use of a post-milking disinfectant reduces the rate of mastitis by 50%, according to the National Mastitis Council.

Step 8

Weigh the milk and add in to milk can through using sieve.

Step 9

Disinfect and dry hands between cows.

Step 10

Apply the same procedures for the rest of cows.

Step 11

Clean all used utensils promptly and disinfect properly. Finally, return back to store equipment and distribute the milk for user.

Important recommendations

- Milkers should keep their personal hygiene and trim their nails.
- Milkers should keep their hand dry and lubricated.
- Apply regular milking times (12 – hour intervals) and avoid noise during milking.
- Use clean water for all purposes and use at least amount as possible.
- Towels must always clean and kept in a clean area hanging.
- Have a specific milking area and kept it clean.
- Keep barn clean and should have good ventilation.
- Wash udder twice a week (after morning milking).
- Address the cause of dirt, if the cows are persistently dirty.
- Clip hairs from udder.
- Each milker should have a pair of gloves.
- For post dip 1% iodine is preferable.

- If the recommended strip bowel is not available black color plastic or clay bowel or indicator paper could be used.
- Milker should be tested for tuberculosis annually.
- Milker should be trained on proper milking procedures.
- Farm manager should monitor the application of this milking procedure.

3.4.2 Machine Milking Procedure

Milker preparation

The hands of a person milking cows can become contaminated with mastitis-causing pathogens, either from handling dirty equipment or from contact with contaminated milk from infected cows. Some microorganisms prefer living and growing on skin, whether it is the cow's teat skin of the milker's hands. Today, most milking operations will have the milkers wear disposable latex gloves. These are replaced periodically through the milking process.

Clean the teats

The teats are prepared by thoroughly cleaning the teat and teat-ends with some solution that removes dirt and provides some sanitation to the teat skin. Many people now use a pre-milking germicide dip solution (for example, at the UIUC farm they use a 1% iodine solution) called a pre-dip. This wets the teat, provides sufficient moisture to wipe off the teat and get it clean, and sanitizes the teat skin. The act of massaging the teats while wiping them off also is stimulating the oxytocin release that will cause milk ejection. It is important to avoid getting the udder wet. Use of spray hoses (drop hoses) to spray germicide onto the teats can get the udder hair wet, where the contaminated fluid then can drain down the teat to the teat end even after wiping off the teat. Long udder hair is not desirable, and it is usual for many dairy producers to remove the hair from the udder, especially during winter months. This is done by clipping udders or by singeing the hair with a flame. If done properly, the latter method is very effective with no effect on the cow.

Dry the teats

Use a separate dry towel (usually paper or cloth) to wipe-off and dry the teats thoroughly. It is particularly important to get the entire teat and tip of the teat clean.

When a pre-dip is used, wiping off the teat will remove most of the iodine solution resulting in negligible contamination of milk with the iodine. Typically, milkers will dip teats on several cows and then return to the first cow, wipe off the teat and go to step three. The use of sponges is discouraged. Sponges can harbor mastitis-causing pathogens, even when soaked in germicide. Use of individual towels so that each cow is separately dried is highly recommended. Reuse of a towel from one cow to the next can spread mastitis-causing pathogens from cow-to-cow.

Foremilk stripping

Several squirts of milk are removed from each quarter. This is done into a strip cup, where the white flakes or clots in the milk will be collected and show up against the black screen of the strip cup top. Alternatively, milk is stripped onto the floor under the cow and observed for flakes or clots. The latter approach is most commonly used, although using the strip cup is the preferred means of identifying flakes or clots. Cows with flakes or clots in their milk probably have some form of mastitis. This is the most common means of identifying clinical mastitis. Typically, the milk that was furthest down in the gland at the start of milking, that is closest to the teat end, is high in somatic cells. Eliminating this by stripping, results in lowered overall somatic cells in the milk that is harvested.

Application of the machine

The milking machine should be applied within one minute of the initial wiping of the teats to take maximum advantage of the milk letdown response. The milker holds the claw in hand, the vacuum is turned on and four teat cups are applied as efficiently as possible, with minimal sucking of air when teat cups are turned up to place on the teat ends. Milk should start flowing immediately. Adjust the machine so that it hangs straight down from the cow. Teat cups that ride-up excessively high on a teat should be adjusted. This situation can potentially cause irritation to the teat lining.

Machine-on time

Maximal intra-mammary pressure caused by milk letdown occurs at about one minute after udder preparation begins and continues for about 5 minutes. Shortly after that, the milk flow will drop to a point where the automatic take-offs will detach the milking machine. Most cows will milk out in 5 to 7 minutes. Some cows are slow to milk out. This may occur because they produce more milk than can be removed in 5 minutes, even with maximal removal efficiency. Or, cows may have structural problems with the teat end or inside the udder that makes them milk out slowly. In the latter case, because the machine is on the cow repeatedly for long periods, the cows may be exposed to more chances of contacting mastitis-causing pathogens.

Detaching the machine at the end of milking

The vacuum must be turned off before the machine is removed. Otherwise, pulling on the teat cups while the vacuum is still on may cause trauma to the teat ends, weakening the sphincter muscles that keep the streak canal closed. Normally it takes about one hour after milking for the streak canal to re-close. Any teat end trauma may compromise the ability the sphincter muscles to close the canal and prolong the exposure of the teat end to mastitis-causing pathogens post-milking. Most people milking cows tend to over-milk the udder. To remove all of the milk, they will physically push down on the claw or pull down on one or more teat cups. This is called machine stripping, and while it does result in removal of more milk from the quarters, it also results in over milking and more stress on the gland. The purpose of the automatic take-off (ATO) is to prevent this over milking. The milking system detects flow rate of milk coming from the gland. When that flow rate drops to a specified level, the vacuum is turned off and a mechanical arm or chain retracts and pulls the machine from the cow's udder.

Post-milking teat germicide dipping

As indicated above in #5, the streak canal stays open for about an hour after milking. If a cow's teat then comes in contact with mastitis-causing pathogens, they may easily enter the teat and cause an infection. One of the most effective means of controlling mastitis is post-milking teat dipping with a germicide. This protects the teat end for a period after milking, kills pathogens that may be on the teat skin, and minimizes the potential passage of those pathogens from one cow to the other at the next milking. Post-milking teat dipping can reduce new infections by 50%. However, teat dipping must be done routinely at each milking. Only dipping teats for selected periods of time is not effective.

Post-milking cow management

Because the streak canal stays open for about an hour after milking, often producers will make feed (often hay or silage) available to the cows after they are done milking. Cows will remain standing while eating. This reduces the chances of the cow lying in manure that may contaminate the teat end before streak canal closure.

3.5 Environmental and OHS Hazard Control

Prevention and control systems should be designed to protect both workers' health and the general environment. Environmental consequences include the effect of fine particles on atmospheric visibility, damage to buildings, effects on vegetation and animals, and health effects on people outside the plant. As in the workplace, the first priority is to prevent the generation of waste, and, if generation cannot be prevented, then secondly, its removal. Measures that minimize waste generation should be given priority, and any inevitable waste disposal should be so planned as to avoid environmental damage.

Occupational safety and health (OHS) also commonly referred to as occupational health and safety in the workplace is an area concerned with the safety, health and welfare of people engaged in work or employment. The goals of occupational safety and health programs include fostering a safe and healthy work environment. OSH may also protect co-workers, family members, employers, customers, and many others who might be affected by the workplace environment. As defined by the World Health Organization

(WHO) "occupational health deals with all aspects of health and safety in the workplace and has a strong focus on primary prevention of hazards. Health has been defined as " a state of complete physical, mental and social well-being and not merely the absence of disease or infirmity. Occupational health is a multidisciplinary field of healthcare concerned with enabling an individual to undertake their occupation, in the way that causes least harm to their health. Health has been defined as it contrasts, for example, with the promotion of health and safety at work, which is concerned with preventing harm from any incidental hazards, arising in the workplace. The main focus of occupational health is on three different objectives:

- The maintenance and promotion of workers' health and working capacity.
- The improvement of working environment and work to become conducive to safety and health.
- Development of work organizations and working cultures in a direction which supports health and safety at work and in doing so also promotes a positive social climate and smooth operation and may enhance productivity of the undertakings.

Generally, at any time:

- All farm machinery must be regularly maintained according to manufacturers' instructions, and all controls must be clearly marked. Do not use faulty machinery.
- Observe appropriate hygiene practices: no smoking, eating or drinking around the dairy; wash hand with detergents following contact with effluent and manure.
- Provide appropriate clothing and protective equipment such as gloves, aprons, rubber boots, goggles and other skin protection, and ensure that it is worn by staffs that come in contact with animal effluent and manure.
- Maintain or replace all personal protective equipment regularly.
- Avoid inhalation of aerosols during reuse of effluent for yard or alley washing or spray irrigation.
- Follow effluent and manure management guidelines; poor practices increase the health risks associated with flies and insects.
- Whenever chemicals are used, read and understand the material safety data sheet for the chemical involved and follow the safety precautions prescribed.
- Extinguishers must be properly maintained to ensure that they will work when needed, and that they are safe to use. A carbon dioxide extinguisher, for example, can build up a high static charge if it is used when there is a breakdown of the insulation around the discharge horn. This can cause electric shock. Adequate maintenance of extinguishers consists of regular inspections, recharging as needed, and a complete annual check-up and servicing.
- Records must be kept of all maintenance work carried out, including inspections. Testing and servicing is usually carried out by a service agency.

>>> SELF-CHECK QUESTIONS

Part 1. Multiple Choices.

1. Which one of the following factor affects milk composition?
   A. Breed
   B. Age
   C. Environmental factor
   D. All of the above
2. From the following main constituent of milk, which one holds a higher portion of milk?
   A. Water
   B. Protein
   C. Lactose
   D. Fat
3. The freezing point of milk affected by:
   A. Increased acidity
   B. Addition of preservatives
   C. Addition of water
   D. None of the above
4. Which one of the following is not Factors affecting viscosity?
   A. Homogenization.
   B. Souring ageing.
   C. Temperature.
   D. Low heating followed by cooling
5. Which one of the following hormone play role in milk let down?
   A. Progesterone
   B. Oxytocin
   C. Adrenaline
   D. Androgen
6. Which one of the following milking machine parts attached to the udder of cows during milking?
   A. Soft rubber liner
   B. Claw
   C. Pulsator
   D. Connecting tube
7. What is the goal of occupational health and safety (OHS)?
   A. Affect co-workers, family members, employers, customers, in workplace environment.
   B. Fostering a safe and healthy work environment.
   C. Minimize a safe and healthy work environment.
   D. All of the Above

8. Which one of the following hormone causes the myoepithelial cells, which surround the alveolus, to contract?

   A. Adrenaline

   B. Tyrosine

   C. Oxytocin

   D. Progesterone

9. The main objective of proper milking procedure is

   A. To prevent mastitis.

   B. Maximize production of quality milk and for insuring complete milk removal from the udder.

   C. To increase quantity of milk production.

   D. A & B

Part 2. Fill the following blank.

1. ________ is a pale liquid produced by the mammary glands of mammals.

2. ________ is a modified sweat gland that nourishes the young.

3. ________ is the process of manufacture of milk from the raw materials in the blood by the alveolar cells and the storage of that milk in the cavity of the alveolus.

4. ________ is the process through which milk is released from the alveoli and flows into the ducts, the gland and the teat cistern, where it can be removed by the calf or the human milker.

5. ________ is an area concerned with the safety, health and welfare of people engaged in work or employment.

Part 3. Discuss and describe in detail.

1. Write the difference between milk ejection and milk secretion.
2. Discuss and describe in detail the function of each milking machine unit.
3. Write major composition of milk.
4. List factors affecting milking composition.
5. Write standard milking procedure.
6. Write and discuss in detail the main objective of proper milking procedure

>>> REFERENCES

Abdolreza A, 2012. Food Outlook: Global Market Analysis [E]. http: //www.fao.org/docrep/015/al989e/al989e00.pdf.

Belachew H, Ahmed M, Haileleul T, et al., 1994. Dairy Products Marketing Survey in Addis Ababa and the Surrounding Regions [M]. DDE, Addis Ababa.

Evans B, van der Voorden C, Peal A, 2009. Public Funding for Sanitation-The Many Faces of Sanitation Subsidies [M]. Water Supply and Sanitation Collaborative Council, Geneva, Switzerland: 35.

Pehrsson P R, Haytowitz D B, Holden J M, et al., 2000. USDA's National Food and Nutrient Analysis Program: Food Sampling [J]. Journal of Food Composition and Analysis, 13 (4): 379 - 389.

SuSan A, 2008. Towards More Sustainable Sanitation Solutions-SuSan A Vision Document [M]. Sustainable Sanitation Alliance (SuSan A).

Tegegne A, Gebre-Wold A, 1998. Prospects for Peri-urban Dairy Development in Ethiopia [C] // Ethiopian Society of Animal Production. Fifth National Conference of Ethiopian Society of Animal Production, Addis Ababa, Ethiopia: 28 - 39.

Tilley E, Ulrich L, Lüthi C, et al., 2014. Compendium of Sanitation Systems and Technologies [M]. 2nd ed. Swiss Federal Institute of Aquatic Science and Technology, Duebendorf, Switzerland.

MODULE 8: PERFORMING MILK HANDLING AND PROCESSING OPERATIONS

>>> INTRODUCTION

Milk is a valuable nutritious food that has a short shelf-life and requires careful handling. Milk is highly perishable because it is an excellent medium for the growth of microorganisms-particularly bacterial pathogens-that can cause spoilage and diseases in consumers. Milk processing allows the preservation of milk for days, weeks or months and helps to reduce food-borne illness.

The usable life of milk can be extended for several days through techniques such as cooling (which is the factor most likely to influence the quality of raw milk) or fermentation. Pasteurization is a heat treatment process that extends the usable life of milk and reduces the numbers of possible pathogenic microorganisms to levels at which they do not represent a significant health hazard. Milk can be processed further to convert it into high-value, concentrated and easily transportable dairy products with long shelf-lives, such as butter, cheese and ghee, powder milk.

Processing of dairy products gives small-scale dairy producers higher cash incomes than selling raw milk and offers better opportunities to reach regional and urban markets. Milk processing can also help to deal with seasonal fluctuations in milk supply. The transformation of raw milk into processed milk and products can benefit entire communities by generating off-farm jobs in milk collection, transportation, processing and marketing. Most of the milk in developing countries is processed by small-scale processors who produce a variety of milk products. The type of processing employed may vary from country to country and region to region, depending on local tastes, dietary habits, culinary traditions and market demand (FAO, 2016).

Small-scale processing in developing countries typically involves traditional or semi-traditional technologies, but the products must compete with those of the organized dairy sector (large-scale manufacturers or multinationals). Most small-scale processors do not have access to training, and learn by seeing, hearing and doing. As these businesses are run without formal skills they incur high risks and are of limited cost-effectiveness. Small-scale equipment for pasteurization, packaging in plastic sachets, and butter, yoghurt and cheese making exists, but small-scale processors often lack access to the equipment they need (FAO, 2016).

Milk micro-flora includes spoilage and pathogenic microorganisms. Many milk borne diseases such as tuberculosis, brucellosis and typhoid fever are known (Goff and Horst, 1995). Milk is spoiled by a wide range of microorganisms, some of which are pathogenic and are responsible for milk borne diseases. The milk is very easily contaminated if collected unhygienically and handled carelessly leading to quick spoilage (Prajapati, 1995; Chatterjee et al., 2006) and is often contaminated by *Escherichia coli* bacteria under poor sanitary conditions which can affect public health.

This module covers milk handling and quality control, milk preservation, processing, packing, transportation activities, marketing of dairy products and animals and is aimed to equip students and other users with knowledge and skills.

1 MILK HANDLING AND QUALITY CONTROL

Dairy producers know that the quality of milk and dairy products that consumers purchase depends in large part on the quality of milk they produce. While there can be handling problems by collectors, processors, distributors, and retailers, the quality of the milk will never be better than it was when produced at the farm. To monitor the milk quality Farm-to-Table (includes all steps involved in the production, storage, handling, processing, distribution and preparation of a dairy product), regulatory agencies check periodically for somatic cell and bacteria counts, among other things. In addition, milk marketing cooperatives/handlers also check the milk for various things.

Aseptically drawn milk has the following characteristics; i. e. a yellowish-white non-transparent color, has a pleasant soft / sweet taste and carries hardly any smell. Consumer acceptance of milk is greatly affected by its flavor; this indicates that milk always needs attentive handling. There are several factors which may produce off-flavors and/or odors in milk. Some of the more common causes of flavor and odor problems are (Rhea Fernandes, 2008):

- Feed and weed flavors: Caused due to feed type and time of feeding-milking interval. Example: Strong flavored feedstuffs such as poor-quality silage, Strong smelling plants (like wild onion or garlic).
- Cow-barn flavors: From dung, etc. These are found when milk is obtained from a dirty or poorly ventilated environment or from improperly cleaned milking equipment.
- Rancid flavors: These are caused by excessive agitation of milk during collection and/or transport. Damage of the fat globules in the milk results in the presence of free fatty acids.

➢ Oxidized flavors: From contact with copper or exposure to sunlight

➢ Other such types of flavors from the use of chlorine fly sprays, medications, etc.

Hygienic milk when it emerges from a healthy udder contains only a very few useful bacteria's just like lactic acid bacteria *Streptococcus lactis*. However, milk is a perishable product. It is an ideal medium for micro-organisms and as it is a liquid, it is very easily contaminated and invaded by bacteria. Almost all bacteria in milk originate from the air, dirt, dung, hairs and other extraneous substances. In other words, milk is mainly contaminated with bacteria during milking. It is possible to milk animals in such a clean way that the raw milk contains only 500 to 1,000 bacteria per milliliter (G. S. Pandey, & G. C. J. Voskuil, 2011).

Usually the total bacteria count after milking is up to 50,000 per milliliter. However, counts may reach several million bacteria per milliliter. That indicates a very poor hygienic standard during milking and the handling of the milk or milk of a diseased animal with i. e. mastitis. Raw milk is one of the most suitable media for the growth of a wide variety of bacteria. Especially immediately after milking when it is almost at body temperature. However, milk contains a natural inhibitory system which prevents a significant rise in the bacteria count during the first 2 - 3 hours. If milk is cooled within this period to 4℃, it maintains nearly its original quality. Timely cooling ensures that the quality of the milk remains good for processing and consumption. The bacterial load in fresh raw milk should be less than 50,000 per milliliter when it reaches the collection point or processing plant. Pathogens can be present in the environment in the dry blending area. Product is usually exposed during blending. Ingredients may become contaminated by equipment that is unclean or uncleanable (Lund et al., 2000).

To prevent a too high multiplication of bacteria, the milk has to be produced as hygienic as possible and should be cooled or heated at the earliest. Hygienic milk only originates from mastitis free, clean and healthy animals. Cows suffering from a disease may secrete the pathogenic bacteria, which cause their disease, in the milk they produce. Consumption of raw milk therefore might be dangerous to the consumer. Some of these diseases, like tuberculosis, brucellosis and anthrax, can be transmitted to the consumer. Whatever the milk is used for during processing, the hygienic standard of the produced milk at farm-level forms the basis of the quality of the ultimate milk products too! Milk of standard quality meets requirements set by the Regulations (Table 8. 1). In terms of fat and protein content and number of microorganisms and SCC, such milk contains (Z. Topless et al., 2006):

Table 8.1 Chemical composition and physical characteristics of milk quality standard

Indicator of fresh milk quality	Cow milk	Ewe milk	Goat milk
Chemical composition			
Fat (%)	3.2	4.0	2.8
Protein (%)	3.0	3.8	2.5
Carbohydrate (%)	8.5	9.5	7.5
Water (%)	87.7	83.7	85.8
Physical properties of milk			
Density (g/cm^3)	1.028 - 1.034	1.034 - 1.042	1.024 - 1.040
Degree of Acidity (SH)	6.6°- 6.8°	8.0°- 12.0°	6.5°- 8.0°
pH value	6.5 - 6.7	6.5 - 6.8	6.4 - 6.7
Freezing point	<-0.517℃	<-0.56℃	<-0.54℃
Alcohol test with 72% ethyl alcohol	Negative	—	—
microbial count	≤ 100,000	≤ 1,500,000	≤ 1,500,000
Somatic Cells Counts (SCC)	≤ 400,000		—

Source: Dairy microbiology manual.

Quality control and hazard analysis and critical control point (HACCP)

HACCP are the two essential quality assurance tools in order to meet the customers' expectation and further milk consumption interest. The overall system of technical activities that measures the attributes and performance of the process, item, or service against defined standards to verify that they meet the stated requirements established by the customer; operational techniques and activities that are used to fulfil requirements for quality is called Quality Control. Quality Control is the reactive tool rather than proactive; which means doing identification of deviation of product quality through such testing method.

On the other hand, HACCP is a tool to assess hazards and establish control systems that focus on prevention rather than relying mainly on end-product testing. This is a scientific and systematic way of enhancing the safety of foods from primary production to final consumption through the identification and evaluation of specific hazards and measures for their control to ensure the safety of food. It has seven important principles (CFSAN, 2006) including:

- Establish hazard analysis
- Determine the critical control points (CCPs)
- Establish critical limit(s)
- Establish a monitoring system
- Establish corrective actions
- Establish verification procedures
- Establish documentation

1.1 Factors Affecting Milk Quality

Milk quality is affected by several factors, some of which are categorized as:

Poor dairy handling at the farm contributes to poor quality of milk (the milker, the dairy equipment, the cow, the milking parlor and the storage areas). Nature of the milk and the udder: Mastitis may cause an alteration in fat, lactose and protein content in milk (Nielsen et al., 2005). Declining fat content during mastitis is due to the reduced synthetic and secretory capacity of the mammary-gland. Free-fatty-acids in mastitis milk may increase because of inflammation, probably caused by increased activity of the enzyme lipase. Lactose decreases because of reduced synthetic capacity and losses in circulation, but also as a way to maintain the osmotic pressure, since mastitis causes an increase in ion content (Auldist & Hubble, 1998).

Environmental factors (temperature, sunlight, heat and cold-stress, estrus-condition, disease, vaccination and drug administration) (Barkema et al., 1998) may affect the SCC of individual cows. Stress may increase the number of leucocytes in blood (Lam, 2011).

The major group of bacteria in milk is the lactic acid bacteria such as *Streptococcus lactis* that can use the milk lactose and convert it into lactic acid. These bacteria multiply and grow very fast when the milk is kept at ambient temperatures after milking. The produced lactic-acid causes the natural souring of milk. The primary source of these bacteria is the environment: air, dust, dirty equipment and operators. How soon the milk turns sour depends on the degree of contamination and on the temperature of the milk. Therefore, proper hygiene and sanitation procedures are essential to control the quality of milk.

The types of micro-organisms which make use of other milk components, like the proteins and milk fat. All these microbial activities deteriorate the milk quality. Therefore, only fresh milk of tested quality should be used as raw material to enable processing into high quality milk products. For this reason, the dairy industry strictly controls the quality of the incoming milk from the dairy farmers. If the milk quality does not fulfill the set minimum quality standards, it is rejected. This means an economic loss to the farmer.

Most countries have implemented special laws and regulations concerning the composition and hygienic quality of milk and milk products to protect both the consumers and the public health. World milk producers have grades to the received raw milk of the highest hygienic quality as "Grade A". This grade allows a maximum of 50,000 bacteria per milliliter in raw bulk milk. Chronic offenders of this limit risk losing their license to sell milk to the Grade A market. Most dairy societies are able to deliver "Grade B" milk, which maintains bacteria counts between 50,000 and 200,000 per milliliter (Pandey & Voskuil, 2011). When high counts become a problem, it is generally due to one or more of the following reasons:

- Improper cleaning of milking equipment (the most common cause of high bacteria counts in milk)

- Improper cooling of milk
- Occasionally, a herd experiencing a high prevalence of bacterial infection (Table 8. 2)

Table 8. 2 Common hazards and their corrective actions which may happen within milk processing plant

Ingredient or process	Potential hazard	Type of hazard	Hazard rationale	Hazard management or controls
Receiving-materials shipped by bulk tanker, e. g. fluid milk and milk products	Biological	Contamination with vegetative pathogens	The truck unloading area has the potential to contaminate liquid milk products. These products are normally transmitted through equipment that if unclean (or uncleanable) can result in bacterial contamination	Truck unloading should be constructed to protect the milk (at a minimum overhead protection and concrete, or equivalent surface under the truck that is properly drained). Maintain the truck unloading area and equipment clean. Protect the milk that is being unloaded by closing in the unloading area or using filters over the vent /personnel access port area. Using equipment meeting sanitary design guidelines
	Chemical	Cleaning & sanitizing residues	Free of foreign material which constitute food safety hazards	Maintain proper separation or a physical break between circuits containing cleaning solutions and vessels and lines used to contain or conduct product
	Physical	Extraneous materials	Equipment in poor repair or improperly assembled may contaminate product with foreign material	Use a filter, screen or other appropriate device at some point in the system
Materials shipped by common carrier, e. g. dry ingredients, and packaging materials	Biological	Contamination with vegetative pathogens	Product may become contaminated if products containers are damaged during shipment	Inspect product during unloading operations for damage
	Chemical	Toxic chemicals	Delivery trucks may have been used to transport toxic chemicals prior to food products or packaging materials	Inspect vehicles prior to unloading for evidence of unsanitary conditions, spilled chemicals, off odors, of evidence that might Indicate the delivered product may have been contaminated
	Physical	Extraneous materials	Vehicles may have not been maintained in good repair or have been used to carry metal or wood articles	Inspect vehicles prior to unloading for evidence of foreign materials that may have contaminated the product

(Continued)

Ingredient or process	Potential hazard	Type of hazard	Hazard rationale	Hazard management or controls
Raw milk storage	Biological	Contamination with vegetative pathogens	These products are normally stored in vessels that, if unclean (or unclean), can result in bacterial contamination	Verify that storage vessels and associated lines and valves Similar appurtenances are constructed in such a way that they can be cleaned. Pipeline openings (e. g. flow control panels) and outlet valves are capped when not in use, other openings are closed with tight fitting covers
		Growth of vegetative pathogens	Without proper temperature and time controls, vegetative pathogens can multiply to levels that may be capable of overwhelming the pasteurization process without proper temperature and time controls	Maintain the temperature sufficiently low to minimize the growth of pathogens. Clean the storage vessels and associated lines and valves similar appurtenances at frequencies that do not allow for bacterial growth of pathogens in the product at the product temperature used
	Chemical	Cleaning & sanitizing solution residues	Without proper separation between cleaning & sanitizing solutions and product there could be product contamination	Maintain proper separation or physical break between circuits containing cleaning solution and vessels and lines used to contain product
	Physical	None	—	—
Storage, blending & addition of ingredients	Biological	Contamination with vegetative pathogens	Pathogens can be present in the environment in the dry blending area. Product is usually exposed during blending. Ingredients may become contaminated by equipment that is unclean or uncleanable	Verify that blending equipment and associated lines and valves similar appurtenances are constructed in such a way that they can be cleaned. Maintain records that they are cleaned as needed but at least each day used
	Chemical	Cleaning & sanitizing solution residues	Without proper separation between cleaning & sanitizing solutions and product there could be product contamination	Maintain proper separation or physical break between circuits containing cleaning solution and vessels and lines used to contain product
		Allergens being mixed with products that are not labeled as containing allergy	Foods which contain undeclared allergens may cause life threatening reactions in sensitive individuals	These documented and monitored programs need, at a minimum to include requirements and procedures to assure: separation and identification of such allergens during storage. Addition only of those products that are properly labeled must be monitored and documented

(Continued)

Ingredient or process	Potential hazard	Type of hazard	Hazard rationale	Hazard management or controls
Pasteurized milk & milk product storage (except dry products)	Biological	Contamination with vegetative pathogens	Human illness outbreaks have been linked to post-pasteurization contamination of milk and milk products	Openings and outlet valves are capped when not in use; other openings are closed with tight fitting covers. Verify that storage vessels and associated lines and valves and similar appurtenances are constructed in such a way they can be cleaned. Maintain records storage vessels are cleaned after each use
		Growth of vegetative pathogens	Human illness outbreaks have been linked to post-pasteurization contamination of milk and milk products	Maintain the temperature sufficiently low to minimize the growth of pathogens. Clean the storage vessels and associated lines and valves and similar appurtenances at frequencies that do not allow for bacterial growth of pathogens in the product at the product temperature used
	Chemical	Cleaning & sanitizing solution residues	Without proper separation between cleaning & sanitizing solutions and product there could be product contamination	Maintain proper separation or physical break between circuits containing cleaning solution and vessels and lines used to contain product
	Physical	None	—	—
Pasteurized milk & milk product packaging (except dry products)	Biological	Contamination with vegetative pathogens	Human illness outbreaks have been linked to post-pasteurization contamination of milk and milk products	Packaging may come from an IMS listed source with associated letters of guarantee, or the milk plant may perform tests to verify the ongoing safety of the packaging. After receipt, single service containers and other single service items must be protected from re-contamination. Filling equipment and appurtenances must be cleanable and inspectable and must be constructed and operated to protect the product being packaged from contamination

Proper handling has been used as a mechanism for protect the milk against deteriorations that may retard efficiency of processing and processing task also called works for value addition.

Cooling: to a temperature of 4℃ makes the bacteria inactive and prevents them to grow and produce the lactic acid. Once again, the milk should be produced as clean as possible in the first place, but after that it should be cooled soonest.

Heating: it is another method to prevent the *S. lactis* to produce too much lactic acid and make the milk sour. In the dairy plant this is usually done in the form of pasteurization. During this process the milk is heated to 72℃ for a period of 15 seconds. After pasteurization we are sure that all pathogenic bacteria, the one causing tuberculosis, and at the same time most lactic acid bacteria are destroyed. When the milk is cooled after pasteurization it can be kept for approximately 5 days by the consumer without spoilage. However, certain organisms are capable of surviving pasteurization and continue to multiply during refrigeration. These bacteria are an important source of concern because they reduce the product shelf-life. To eliminate these bacteria, milk can be boiled at 100℃ or sterilized at temperatures of 120 – 140℃. Sterilized milk will keep its quality for a long time without cooling. However, at these high temperatures the taste of the milk is affected.

1.2 Milk Quality Tests

The following platform tests can be applied during milk collection and reception at the collection center and/or the milk plant (Bashir et al., 2013) (For the execution of those tests see an operation sheet / laboratory manual for details):

1.2.1 Organoleptic Tests

In these tests the milk quality is judged using a person's senses view, smell, and taste. The organoleptic tests are always used for the first screening of the incoming raw milk. Not any equipment is required (Lore et al., 2006). The person carrying out the tests should be experienced for reliable results. On arrival the appearance of the milk and of the lid of the milk can or container is observed and inspected instantly after the lid is removed. The tester smells the milk, observes the appearance, checks the can for cleanliness, looks for sediment, flies, etc. and tastes if necessary. To classify the milk according to cleanliness, he/she needs to filter it with a special milk filter. If doubts arise about the quality of the milk after this simple examination, other tests can be carried out in the laboratory to determine the quality.

1.2.2 Lactometer/Adulteration/Specific Gravity Test

If during the organoleptic inspection the milk appears to be too thin and watery and its color is "blue thin", it is suspected that the milk contains added water. The lactometer test serves as a quick method to determine adulteration of milk by adding water, adding alkaline or skimming off. The test is based on the fact that the specific gravity of whole milk, skim milk and water differ from each other (Fig. 49).

Lactometer is used to measure the specific density of milk. At 15℃ the normal density of the milk ranges from 1.028 to 1.033 g/mL, whereas water has a density of 1.0 g/mL (Lore et al., 2006). So, when the lactometer reads a value closer to 1.0, probably water has been added to the milk. If possible, the lactometer reading can be combined with the fat test. The density of fat is lower than that of milk. So, in case the results of the fat test are low and the found density is still high (e.g. 1.035), then the milk might have been skimmed off. If the results of the fat test are low and the density is low (e.g. 1.025), then water might have been added to the milk. Always read the temperature of the milk first; the lactometer reading varies according to temperature.

Make sure you adjust readings as indicated in Table 8.3:

Table 8.3 Temperature adjustments for lactometer readings

Temp (℃)	17	18	19	20	21	22	23	24
Correction	−0.07	−0.05	−0.03	0.00	0.03	0.05	0.08	1.1

Determine the specific gravity through the following simple formula (IRIL):

$$\text{Specific gravity (sp. gr)} = (L_c/1000) + 1$$

$$L_c = L_r + C_f$$

L_c—Corrected lactometer;

L_r—Lactometer reading;

C_f—Correction factor.

According to a result from the calculation, if the value of specific gravity ranges from 1.028 to 1.033 g/mL, the milk is normal; if below the given range, foreign solids have been added in the milk; if above the range, the milk has been adulterated through the addition of water.

1.2.3 Clot-on-boiling Test

Clot-on-boiling test is performed simply by heating a small amount (5 mL) of milk in a test tube over a flame or by immersing it in boiling water for four minutes. The result can be seen immediately. If the milk is sour or if the milk is abnormal (colostrum or mastitis milk) the milk will clot and not pass this test. Heating will precipitate the proteins in the milk if it is sour. So, milk what clots should be rejected. This test is not very sensitive to slightly sour milk, but still very useful. If no coagulation occurs the milk can stand heating operations at the time of testing (Fig. 8.1).

1.2.4 Alcohol Test

In case there is any reason to suspect that milk is sour, the alcohol test is used for rapid determination of an elevated acidity of milk. The test is carried out by mixing equal quantities (2 mL) of milk and of a 75% ethanol solution (made by mixing 78 mL of 96% alcohol with 22 mL distilled water) in a test tube. If the milk contains more than 0.21% acid, this result in coagulation of the milk proteins and the milk is sour. The milk will clot and is not fit for any process which involves heating, like pasteurization. For the above reasons, it is

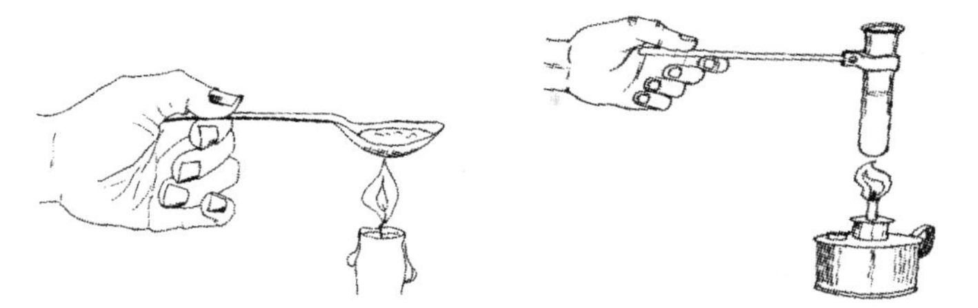

Fig. 8. 1 Carrying out clot-on-boiling test

Source: A training guide for farm-level workers and milk handlers in Eastern Africa.

recommended that the alcohol test is applied to each and every incoming milk can and container, whenever the milk is to be pasteurized. If the result of the alcohol test indicates a too high acidity, a milk sample can be taken to the laboratory for a more detailed testing by the titratable acidity test.

Also, so many common tests at small scale dairy plants, which will suffice the requirements of most milk quality control laboratories of small scale *processing units*, are briefly described here.

Titratable acidity test

This test measures the concentration of lactic acid in the milk. If the acidity is higher than 0. 18%, the milk quality is poor, and it cannot be heated and processed. For this test, add a sodium hydroxide solution to the milk by titration. The more sodium hydroxide you have to add before the milk is neutralized, the higher the acidity of the milk.

1. 2. 5 Gerber Test to Determine Fat Content

This test is used to determine the fat content of the milk. Some milk is added to a butyrometer together with sulphuric acid and amyl-alcohol. Some expensive equipment, like a special centrifuge and a water bath is needed for this test (Saiqa Bashir et al. , 2013).

1. 2. 6 Bacteriological Counting

Murphy from Cornell University described in a paper he wrote a few years ago what these two bacteria tests indicate. His comments are summarized below (Voskuil, 2011).

Standard plate count (SPC)

Standard Plate Count (SPC) of raw milk gives an indication of the total number of aerobic bacteria present in the milk at the time of pickup at the farm. Milk samples are plated onto a nutrient medium, incubated for 48 hours at 90℉, and then the number of bacteria colonies is counted. The value is reported as the number of colony forming units per milliliter of milk. The legal limit of the number that can be in milk is 100,000, but most producers usually have values below 10,000. The most frequent cause of a high SPC is poor cleaning of the milk system (milking units, lines, bulk tank). Another cause frequently found is failing to rapidly cool milk to less than 40℉. Sometimes milking cows with dirty teats and maintaining unclean milking and housing facilities can be the cause.

Preliminary incubation count (PIC)

Preliminary Incubation Count (PIC) reflects milk production practices. PICs are generally higher than SPCs, with values more than 3 - 4 times the SPCs being considered worthy of seeking corrective measures. Values more than 50, 000 should be of concern regardless of the SPC values. To obtain PICs, milk samples are held at 55℉ for 18 hours prior to plating and counting the bacteria colonies. This process encourages the growth of bacteria that grow well at cool temperatures. High PICs are usually associated with milking cows that have not been properly cleaned prior to milking or using milking equipment that is not properly cleaned and sanitized. Bacteria that are considered to be natural flora of the cow, including those that cause mastitis, are not thought to grow significantly at the PI temperature. Marginal cooling or prolonged storage times may also result in unacceptable PIC levels. PICs equal to or slightly higher than SPCs greater than 50, 000 may suggest that the high SPC is possibly due to mastitis. The PIC of a raw milk supply does not usually indicate the potential quality of a pasteurized product made from that milk.

1.2.7 Somatic Cells in Milk

Somatic cell counts represent another important milk quality parameter. The word somatic means body and thus a somatic cell is a body cell. Most important in milk are the leukocytes (white blood cells). Milk originating from an infected udder contains a high concentration of leukocytes. Consequently, somatic cell counts are an important indicator of udder health, in particular of mastitis (Fig. 50).

Fresh milk from healthy cows has a somatic cell count of less than 200, 000 cells per milliliter of milk. Cell counts from herd bulk milk, which are consistently more than 500,000 per milliliter, are an indication of a high prevalence of mastitis in the herd. At the farm the cows can be checked easily for mastitis by the California Mastitis Test (CMT). The relation between the results of the CMT and the somatic cell count of individual cows is shown in Table 8.4.

Table 8.4 Relation between California mastitis test and somatic cell count in milk of individual cows

CMT Score	Average Somatic Cell Count in Milk
Negative (no mastitis)	100, 000
Trace (of mastitis)	300, 000
Class 1 (light mastitis)	900, 000
Class 2 (severe mastitis)	2,700,000
Class 3 (very severe mastitis)	8,100,000

Milk with a high somatic cell concentration can be harmful to human health and contains less casein. This results in lower cheese yields. In addition, milk with a high cell count generally contains an increased amount of enzymes, which have an effect on the quality of the protein and the fat in milk. The presence of these enzymes in milk increases the potential for off-flavors and odors. Because the somatic cell content of raw milk is important for the shelf-life, flavor and the yields (particularly of cheese), milk processors strive to obtain raw milk of the highest hygienic quality from their producers.

1.2.8 Methylene Blue Test

Methylene blue test for the assessment of mastitis was performed according to the procedure described by Awan and Rahman (2005). The test is used to diagnose mastitis, the ability of bacteria to reduce the color of methylene blue dye from the milk sample. Dye reduction time is inversely proportional to the presence of total number of bacteria in sample; hence the greater the bacterial population, the shorter is the dye reduction time (Saiqa Bashir et al., 2013). Methylene blue reduction test is done to detect pasteurized milk samples containing bacteria. Tenfold serial dilution of samples was made up to 10^{-6} in Nutrient broth and MacConkey broth. Samples were plated in duplicate using pour plate technique. 0.5 mL of the diluted samples was delivered by pipette into 19.5 mL of enriched agar. Plates were incubated inverted in an incubator at 37℃ for 24 – 48 hours (J Okpalugo et al., 2008).

Summary

In summary, these two bacteria tests (SPCs and PICs), as well as the other milk quality tests done on raw milk, serve as monitors for both regulators and producers to use as they attempt to produce the highest quality milk possible for consumers. For more information on production practices to follow to keep SPCs and PICs low, producers should contact their milk handler field representative or other qualified advisor. The most crucial aspects to maintain high quality fresh milk are summarized below:

- Hygienic milking and milk handling from farm up to table
- Good health status of the animals
- Cooling of milk as soon as possible after milking
- Transport of milk to the collection center and/or processing plant within 2 – 3 hours after milking
- Good quality and well cleaned milk equipment

Maintaining a good quality of milk is benefiting all; farmers, milk processing plants and consumers. It will result in the milk producer receiving the full value for his produce, in minimal losses during processing at the milk plant and in a reliable quality of the milk products for the consumer.

1.3 Checking, Maintaining, and Using Processing and Storage Utensils

Processing and storage utensils are prepared based on the amount of milk to be processed. Milk processing utensil includes: PPE, heat source, sauce pan, thermometer, spoons, cream separator, churner, milk bucket, knife, skimmer, cheese mould, cheese presser, refrigerator, pasteurizer, homogenizer, weighing scale, jar, pail, milk-can and so on (Fig. 8. 2).

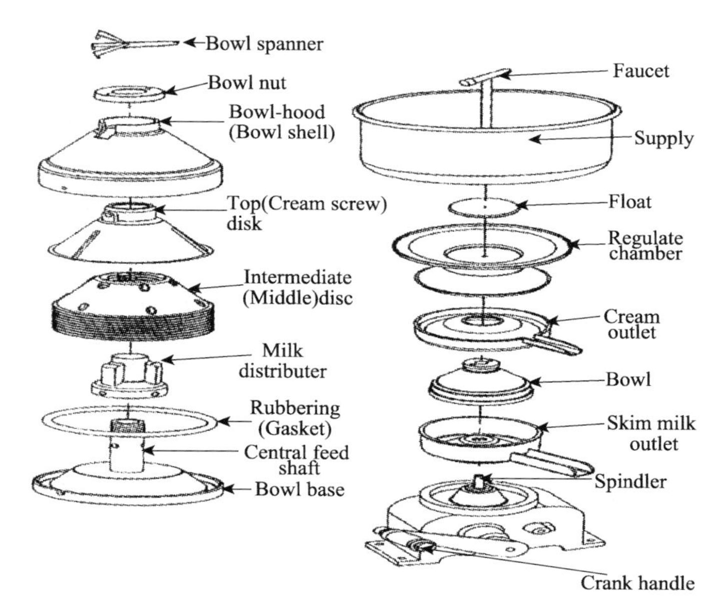

Fig. 8. 2 Dismantled cream separator

Source: Agarfa college, Ethiopia.

1. 3. 1 Utensils Maintenance

Work that is done regularly to keep a machine, building, or piece of equipment in good condition and working order for different objectives such as; to prevent breakdowns, to prolong life of equipment, to reduce breakdown time to a minimum, to keep repair costs to a minimum and to keep records on each piece equipment, as a guide to working with equipment manufacturers and to deciding on future purchases.

General maintenance

General maintenance is classified into preventive machine maintenance & breakdown maintenance.

- Preventive machine maintenance: A definite schedule of running and shut down inspections for each man who is to inspect, proper lubrication form.
- Breakdown maintenance: Written work order form, prompt and thorough correction, immediate inspections by production supervisor, to see if job is done properly. Broken-down machines or machines operating inefficiently result in reduced milk yield, increased milking time, damaged udders, reduced milk quality and regular inspection enhances performances.

Milking machines are used 5 - 6 hours or more daily. For example, cream separator maintenance (Abebe Tessema and Markos Tibbo, 2009) is important to maintain the cream separator properly. Always remember: The cream screw regulates the ratio of skim milk to cream. It should be adjusted so that the fat content of the cream is about 33%. If cream is too thin (i. e. fat content is too low), it reduces the quantity of separated milk available for other uses and increases the volume of cream to be handled. Low-fat cream is also more difficult to churn efficiently, i. e. butter production is reduced.

Cleaning

- Flush the separator with warm skim milk or warm water.
- Flush the bowl with clean water until the discharge from the skim milk spout is clean.
- Dismantle the bowl and all parts (bowl, bowl cover, discharge spouts, float, supply tank and buckets) and wash with a brush, hot water and detergent (Bleach 4%).
- Rinse with very hot water and allow the parts to drain in a clean place protected from dirt and flies. This should be done after each use.

Oiling

- Frequent oiling is not necessary.
- When oiling is necessary, remove screw in gear cover.
- Put enough oil in oil holes, also in screw hole.
- Do not use too much oil, or it will just be wasted.
- The gears must be well lubricated. Follow the directions of the manufacturer.
- The level of the lubricant must be kept constant; observe the oil level through the sight glass.
- The bowl must be perfectly balanced.
- The bowl should be cleaned thoroughly immediately after use to ensure proper functioning of the separator and for hygiene.

1.3.2 Cleaning, Sanitizing and Sterilizing Dairy Equipment

Thorough cleaning and sanitizing of all dairy equipment is an essential part of all milk processing. Water alone is inadequate for cleaning and sanitizing and, therefore, chemical agents must be used. Cleaning and sanitizing dairy equipment are necessary to prevent:

- Accumulation of undesirable micro-organisms in the equipment
- Development of bad smells in the equipment which pass on to the product
- Loss of efficiency
- Possible corrosion of metal parts due to lactic acid
- Mould growth on wooden surfaces leading to mould contamination in the product and discoloration of the churn surface and contamination of the product with pathogens

Processing equipment should be clean and look neat. The processing room should also be well lit and ventilated, clean and neat. However, sanitation of processing equipment means more than having the equipment looking clean and neat. The cleanliness of equipment can be classified at four levels:

- Physical cleanliness: all visible dirt has been removed
- Chemical cleanliness: in addition to all visible dirt, microscopically small residues have been removed
- Sanitation: in addition to being chemically clean, the equipment has been treated in such a manner as to remove most of the micro-organisms present on its surface
- Sterilization: in addition to being sanitized, the equipment has been treated in such a manner as to destroy all micro-organisms present on the equipment

Sanitation and sterilization are easier to achieve if the equipment is initially at least physically clean. Therefore, the equipment is normally cleaned before sanitation or sterilization. Let's see dirty milking machines versus udder damage as an example; most new udder infections are caused by factors that increase the number of bacteria present on the teat end or aid in the passage of bacteria through the teat canal. A keratin lining located along the streak canal has antibacterial properties and serves as a seal or barrier against bacterial invasion. Injury and local inflammation of the teat end reduces the ability of this barrier to repel or destroy a small number of bacteria. The milking machine may influence development of mastitis in the following ways:

- It transmits bacteria from cow to cow through contaminated milking equipment surfaces
- It damages or removes the teat canal's natural protective barrier (keratin) resulting in injury to teat ends and internal linings
- It permits back flows of air and milk droplets against the teat end into the teat canal during the milking process

Bacteria can penetrate the teat canal by accompanying the impacted droplets. These organisms can originate from other quarter or from external contamination through teat cup liners or unclean or wet teat or udder surfaces. Wash the inside of the churn thoroughly with hot water. Invert the churn with the lid on in order to clean the ventilator; this should be pressed a few times with the back of a scrubbing brush to allow water to pass through.

Notice: The ventilator should be dismantled occasionally for complete cleansing.

Remove the rubber band from the lid and scrub the groove. Scald the inside of the churn with boiling water. This step is very important. Invert and leave to air. Dry the outside and treat steel parts with Vaseline to prevent rusting. The rubber band should not be placed in boiling water; dipping in warm water is sufficient.

1.3.3 Chemicals Used for Cleaning

Detergents are chemical agents that assist in the cleaning process by solubilizing the deposited dirt, thereby making its removal easier. Sodium salts are the commonest and cheapest detergents. Sodium hydroxide, sodium carbonate and sodium tripolyphosphate are commonly used. Synthetic detergents, such as alkyl benzyl sulphate, and biological detergents are also used.

Sterilizers

Chemical sterilizers are agents which, when added to water at a specific concentration, reduce the number of microorganisms on previously cleaned surfaces to very low levels. The active sterilizing ingredient is usually iodine, chlorine, nitric acid or quaternary ammonium compounds. Organic sterilizers such as chloramine-T, hyaline and isocyanuric acids are also used.

Cleaning procedure

Before using any detergent or sterilizer, remove as much of the product as possible from the surface of the equipment. Product not removed before washing is wasted.

Prewash the equipment with clean, cold water. This removes much of the dirt and should be carried out immediately after the product has been removed. After washing with cold water, wash the equipment with warm water (50℃) to remove fatty material. If the equipment is washed thoroughly with water, much less detergent is required in later stages.

Wash the equipment with a detergent solution, following the manufacturer's instructions. The equipment should be cleaned thoroughly at this stage to ensure that it is chemically clean. If the equipment is cleaned by hand it should be scrubbed thoroughly using the detergent solution. Detergent cleaning also reduces bacterial numbers on the equipment.

Drain the detergent solution. It may be retained for washing other items of equipment, provided its strength is maintained. After draining the solution, rinse the equipment at least three times with cold water to remove all traces of the detergent. If not removed, traces of detergent may be incorporated in subsequent batches of product. Rinsing three times with small volumes of water removes detergent residues much more effectively than rinsing once with a large volume of water.

Sanitize the equipment using one of the compounds mentioned above. Chlorine compounds are particularly corrosive and should only be used in accordance with the manufacturer's instructions. After sanitizing, rinse the equipment again with clean water to remove all residues of the sanitizing agent. In the absence of a suitable chemical sterilizer, the equipment can be scalded with water at 80℃.

Once washed and rinsed the equipment should be stored in a clean, dry, dust-free area.

Remark

- Detergents and sterilizers are normally chemically active compounds and great care is required in their use and handling to avoid injury to personnel.
- Detergent sterilizers are compounds formulated to clean and sterilize equipment at the same time. They are generally expensive, but reduce the overall time required for

cleaning and sterilizing equipment and also reduce the amount of water needed for rinsing as only one set of risings are required.

2 MILK PROCESSING, PACKAGING, AND PRESERVATION

2.1 Milk Processing

Raw milk means milk produced by the secretion of the mammary gland of farmed animals that has not been heated to more than 40℃ or undergone any treatment that has an equivalent effect. Dairy products' means processed products resulting from the processing of raw milk or from the further processing of such processed products, which includes cream, yoghurt, butter, ghee, cheese (it has a lot of varieties), whey and other special products. Milk is brought from the farm to the dairy for processing.

Milk is nutritious and essential food for human beings and also serves as a good medium for microbial growth and contamination 240 raw milk samples and 72 pasteurized milk samples from different places of Warangal District for a period of six months were analyzed for microbial quality. Among the raw milk samples only 19.1% of samples were good quality and 28.3% are very poor quality. In the pasteurized milk samples 81.9% of samples were good for human consumption. The bacteria isolated from milk samples include *Lactobacilli*, *Staphylococcus aureus*, *Escherichia coli*, *Bacillus subtilis*, *Salmonella typhi*, and fecal coliforms (Srujana et al., 2011). When received at the dairy, the following information on the milk is required:

- Physical qualities: Before weighing the milk, its quality should be checked. Taste and smell are good preliminary indicators of milk quality, and visual observation can also be useful. If the person receiving the milk suspects that it is of poor quality, he or she can carry out one of the following tests: acidity, pH, alcohol and clot-on-boiling. These will determine the quality of the milk. Once the person receiving the milk is satisfied with its quality, it can be weighed, and the weight recorded.
- Weight: The quantity of milk received can be estimated either volumetrically or gravimetrically. Milk processors usually base payments for milk on its solids content, and hence it is more appropriate to use weight to estimate the quantity of milk being tendered. In a small-scale processing center, a spring balance and a stainless-steel bucket can be used to weigh milk. The milk weight must be recorded accurately as losses can be incurred or underpayments made to suppliers if care is not taken at this stage.
- Composition and presence of additives (neutralizers', preservatives etc.): A dairy engaged in butter-making will need to base its payments on the butterfat content of the milk. The milk received will have to be sampled for butterfat analysis. Spot checks can also be carried out to test for added water and the presence of neutralizers if malpractice is suspected.

➢ Presence of added water: This is simply the addition of table water for the matter of dodger profit through maximizes a volume.

The advantages of milk processing

➢ Provide regular income.
➢ Improve nutrition.
➢ Selling processed milk products is more profitable than selling fresh milk.
➢ Generate employment.
➢ Improves quality and safety (Abebe Tessema and Markos Tibbo, 2009).

In modern dairy farming system, milk processing tasks includes milk separation/cream separation, standardization, pasteurization, homogenization, churning and so on.

2. 1. 1 Cream Separation

Milk may be sold in many forms, including whole milk, skim milk, low-fat milk, flavored milk and other modified milks. Some of these products require the removal of the fat portion as cream. Cream is defined as "a milk product comparatively rich in fat, in the form of an emulsion of fat-in-skim milk, which can be obtained by separation from milk" . Cream is produced from whole milk by skimming or other separation means. Milk is subjected to a range of processing operations before being sold.

Typical processes include standardization or formulation of milk, which may include: separation steps such as filtration, centrifugation, and sometimes clarification; homogenization; and various forms of heat treatment such as thermization, pasteurization, sterilization and UHT (ultra-high temperature) processing.

Efficiency of separation is influenced by the following four factors (O'Connor, 1995):

➢ Speed of the separator: Reducing the speed of the separator to 12 r/min less than the recommended speed results in high fat losses, with up to 12% of the fat present remaining in the skim milk.
➢ Residence time in the separator: Overloading the separator reduces the time that the milk spends in the separator and consequently reduces skimming efficiency. However, operating the separator below capacity gives no special advantage, it does not increase the skimming efficiency appreciably, but increases the time needed to separate a given quantity of milk.
➢ Effect of temperature: Freshly drawn, uncooled milk is ideal for exhaustive skimming. Such milk is relatively fluid, and the fat is still in the form of liquid butterfat. If the temperature of the milk falls below 22℃ skimming efficiency is seriously reduced. Milk must therefore be heated to liquefy the fat. Heating milk to 50℃ gives the optimum skimming efficiency.

- Effect of the position of the cream screw: The cream screw regulates the ratio of skim milk to cream. Most separators permit a rather wide range of fat content of cream (18 – 50%) without adversely affecting skimming efficiency. However, production of cream containing less than 18% or more than 50% fat results in less efficient separation.

Methods of cream separation

1) Gravity separation

Fat globules in milk are lighter than the plasma phase, and hence rise to form a cream layer. The rate of rise (V) of the individual fat globule can be estimated using Stokes' Law, which defines the rate of settling of spherical particles in a liquid:

$$V = [r^2 (d_1 - d_2) g] / 9\eta$$

r^2—r is the radius of fat globules. As temperature increases, fat expands and therefore r^2 increases. Since the sedimentation velocity of the particle increases in proportion to the square of the particle diameter, a particle of radius 2 ($r^2 = 4$) will settle four times as fast as a particle of radius 1 ($r^2 = 1$). Thus, heating increases sedimentation velocity.

$d_1 - d_2$—Sedimentation rate increases as the difference between d_1 and d_2 increases. Between 20 and 50℃, milk fat expands faster than the liquid phase on heating. Therefore, the difference between d_1 and d_2 increases with increasing temperature.

g—Acceleration due to gravity is constant. This will be considered when discussing centrifugal separation.

η—Serum viscosity decreases with increasing temperature. Calculation of the sedimentation velocity of a fat globule reveals that it rises very slowly, as shown in the equation; the velocity of rise is directly proportional to the square of the radius of the globule. Larger globules overtake smaller ones quickly. When a large globule meets a small globule, the two join and rise together even faster, primarily because of their greater effective radius. As they rise, they come in contact with other globules, forming clusters of considerable size. These clusters rise much faster than individual globules. However, they do not behave strictly in accordance with Stokes' Law because they have an irregular shape and contain some milk serum.

2) Centrifugal separation

Gravity separation is slow and inefficient. Centrifugal separation is quicker and more efficient, leaving less than 0.1% fat in the separated milk, compared with 0.5 – 0.6% after gravity separation. The centrifugal separator was invented in 1897. By the turn of the century it had altered the dairy industry by making centralized dairy processing possible for the first time. The separation of cream from milk in the centrifugal separator is based on the fact that when liquids of different specific gravities revolve around the same center at the same distance with the same angular velocity, a greater centrifugal force is exerted on the heavier liquid than on the lighter one. Milk can be regarded as two liquids of different specific gravities, the serum and the fat.

Milk enters the rapidly revolving bowl at the top, the middle or the bottom of the bowl. When the bowl is revolving rapidly, the force of gravity is overcome by the centrifugal force, which is 5,000 to 10,000 times greater than the gravitational force. Every particle in the rotating vessel is subjected to a force which is determined by the distance of the particle from the axis of rotation and its angular velocity. If we substitute centrifugal acceleration expressed as $r_1\omega^2$ (where r_1 is the radial distance of the particle from the center of rotation and ω^2 is a measurement of the angular velocity) for acceleration due to gravity (g), we obtain:

$$V = [r^2 (d_1 - d_2) r_1\omega^2] / 9\eta$$

Thus, sedimentation rate is affected by $r_1\omega^2$. In gravity separation, the acceleration due to gravity is constant. In centrifugal separation, the centrifugal force acting on the particle can be altered by altering the speed of rotation of the separator bowl.

In separation, milk is introduced into separation channels at the outer edge of the disc stack and flows inwards. On the way through the channels, solid impurities are separated from the milk and thrown back along the undersides of the discs to the periphery of the separator bowl, where they collect in the sediment space. As the milk passes along the full radial width of the discs, the time passage allows even small particles to be separated. The cream, i. e. fat globules, is less dense than the skim milk and therefore settles inwards in the channels towards the axis of rotation and passes to an axial outlet. The skim milk moves outwards to the space outside the disc stack and then through a channel between the top of the disc stack and the conical hood of the separator bowl.

2. 1. 2 Standardization of Milk

A dairy plant receives milk from various dairy farms and chilling centers in varying fat content and quantity. In such cases like ice-cream, cheese or milk powder, it is necessary to standardize the milk to a certain fat and SNF ratio to ensure the desired composition of the products.

Standardization may be defined as the adjustment of one or more of the milk constituents to a nominated level. In the market milk industry this normally involves reducing the butterfat content by the addition of skim milk or the removal of cream these adjustments are carried out for the following reasons:

- To provide the consumer with a more uniform product
- To effect economies in production (addition of skim milk increases the volume of milk available for sale and removal of cream allows the production of table cream, butter, or other high fat products.)

The usual method of making standardization calculations is the Pearson's Square technique. To make this calculation, draw a square and write the desired fat percentage in the standardized product at its center and write the fat percentage of the materials to be mixed at the upper and lower left-hand corners. Subtract diagonally across the square the smaller from the larger figure and place the remainders on the diagonally opposite corners. The figures on the right-hand corners indicate the ratio in which the materials should be mixed to obtain the desired fat percentage.

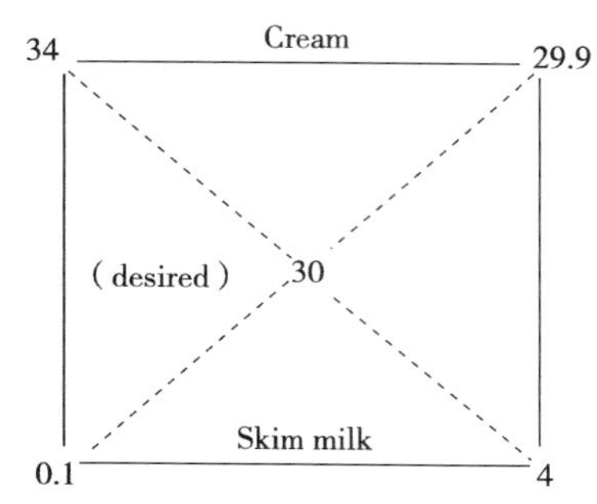

How much skim milk containing 0.1% fat is needed to reduce the percentage fat in 200 kg of cream from 34% to 30%? If 29.9 parts of cream require 4 parts of skim milk, 200 parts of cream require x parts of skim milk.

$$\textit{Weight of skim milk needed} = x = (200 \times 4) / 29.9 = 26.75 \text{ kg}$$

2.1.3 Homogenization

Homogenization is an entirely separate process that occurs after pasteurization in most cases, to emulsify the fat particles in milk or cream in order to give it an even consistency and prevent cream from separating from the rest of the milk.

The purpose of homogenization is to break down fat molecules in milk so that they resist separation. Without homogenization, fat molecules in milk will rise to the top and form a layer of cream. Homogenizing milk prevents this separation from occurring by breaking the molecules down to such a small size that they remain suspended evenly throughout the milk instead of rising to the top.

Homogenization is a mechanical process and doesn't involve any additives. Like pasteurization, arguments exist for and against it. It's advantageous for large-scale dairy farms to homogenize milk because the process allows them to mix milk from different herds without issue. By preventing cream from rising to the top, homogenization also leads to a longer shelf life of milk that will be most attractive to consumers who favor milk without the cream layer. This allows large farms to ship greater distances and do business with more retailers.

Finally, homogenization makes it easier for dairies to filtrate out the fat and create 2%, 1% and skim milk. It is also possible to achieve these different fat contents by skimming cream from the top, homogenization makes the process more precise. Some people worry, however, that by reducing the size of fat molecules, homogenization makes fat easier to absorb.

2.1.4 Churning

In churning, cream is agitated in a partly filled chamber. This incorporates a large amount of air into the cream as bubbles. The resultant whipped cream occupies a larger

volume than the original cream. As agitation continues the whipped cream becomes coarser. Eventually the fat forms semi-solid butter granules, which rapidly increase in size and separate sharply from the liquid buttermilk. The remainder of the butter-making process consists of removing the buttermilk, kneading the butter granules into a coherent mass and adjusting the water and salt contents to the levels desired. Butter is produced from cream by churning or an equivalent process. Butter spreads are based on vegetable fats, a blend of vegetable and butterfat, or butterfat alone (light butter) (Ryser, 1999). The main steps in the manufacturing butter are illustrated in Fig. 8. 3.

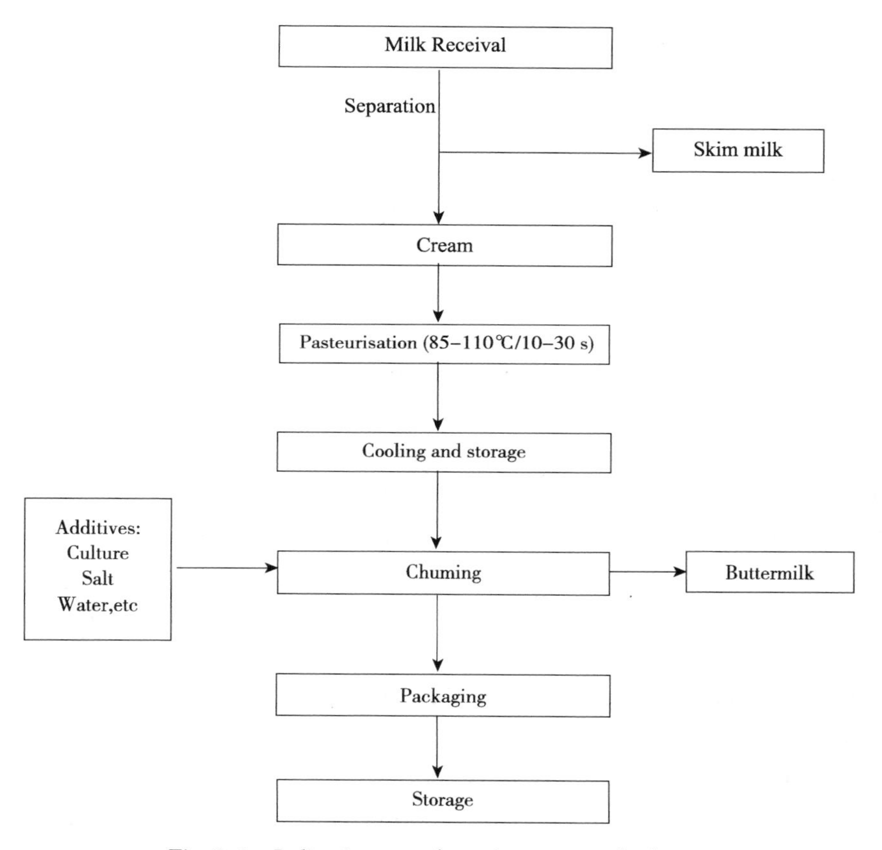

Fig. 8. 3 Indicative manufacturing process for butter

Types of churner

1) Electric butter churner

Electric butter churners are shown in Fig. 51.

2) Manual butter churner

Manual butter churners are shown in Fig. 52.

3) Hand paddle butter churner

Hand paddle butter churner is shown in Fig. 53.

4) Traditional butter churner and improved traditional butter churner

Traditional butter churner and improved traditional butter churner are shown in Fig. 8. 4.

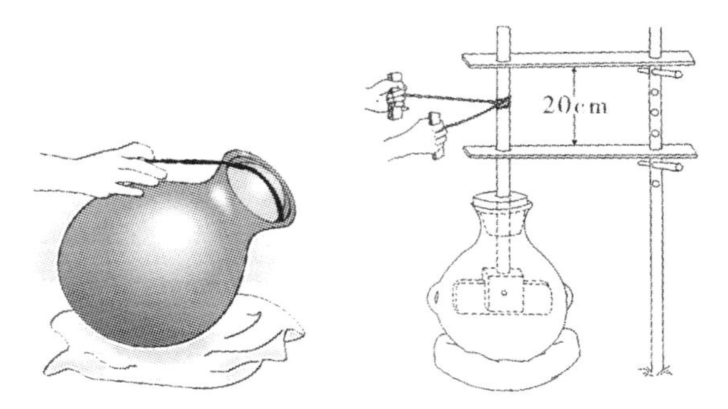

Fig. 8. 4 Traditional and improved traditional butter churner

In considering the mechanism of churning, the following factors must be considered:

- The function of air
- The release of stabilizing material from the fat globule surface into the buttermilk
- The differences in structure between butter and cream
- The temperature dependence of the process

Air is thought to be necessary for the process, but some workers have demonstrated that milk or cream can be churned in the absence of air, although it takes longer. Butterfat can be recovered from milk and converted to a number of products, the most common of which is butter. Butter is an emulsion of water in oil and has the following approximate composition: fat, 80%; moisture, 16%; salt, 2%; and milk SNF, 2%.

In good butter the moisture is evenly dispersed throughout the butter in tiny droplets. In most dairying countries legislation defines the composition of butter; and butter makers conform to these standards insofar as is possible. The butter is then either rolled out 8 to 10 times or ridged with the spatulas to remove excess moisture. If the butter is to be heavily salted, it must be worked more in proportion to the amount of salt used, as uneven distribution of the salt causes uneven color. The butter should be worked until it seems dry and solid, but it must not be worked too much, or it will become greasy and streaky.

The butter is then weighed and packed for storage. It should be packed in polythene-lined wooden or cardboard cartons and stored in a cool, dry place. The butter should be firm and of uniform color. Surplus good-quality butter can be stored but should contain more salt than usual—at least 30 g/kg. Low moisture content is desirable.

The butter must be packed in clean containers, such as seasoned boxes or glazed crocks, and stored in a cold room or in a cold, airy place. If a box is used, it should be lined with good-quality polythene. The container should be filled to capacity from one churning. The more firmly butter is packed, the better; it may be covered with a layer of salt, but this is not essential. The container should be securely covered with a lid or a sheet

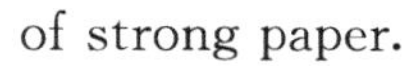

of strong paper.

An enterprise engaged in butter-making must be able to measure the efficiency of the process, i. e. by measuring the yield of butter from the butterfat purchased. First, the theoretical yield of butter has to be estimated. Butter contains an average of 80% butterfat. Thus, for every 80 kg of butterfat purchased 100 kg of butter should be produced, or for every 100 kg of butterfat purchased 125 kg of butter should be produced. The difference between the number of kilograms of butterfat churned and the number of kilograms of butter made is known as the overrun. This difference is due to the fact that butter contains non-fatty constituents such as moisture, salt, curd and small amounts of lactic acid and ash in addition to butterfat. Overrun is affected by:

- Accuracy of weighing milk received
- Accuracy of sampling and testing milk for fat

For example, if careless sampling and testing results in a reading of 3.6% butterfat against an actual content of 3.2% butterfat, what will be the effect on the overrun from 100 kg of milk?

- Fat paid for = 100 × 0.036 = 3.6 kg of butterfat.
- Maximum theoretical yield of butter = 3.6 × 1.25 kg = 4.5 kg
- Fat received = 100 × 0.032 = 3.2 kg
- Maximum theoretical yield = 3.2 × 1.25 = 4 kg
- our overrun therefore is Butter made = 4 kg
- Butterfat paid for = 3.6 kg
- Overrun = 4/3.6 =1.11=11%

2.1.5 Cheese Making

The term cheese covers over 1, 000 varieties of fermented dairy products with significant variations in their flavor, texture and appearance. The process of converting liquid milk into cheese involves a series of steps that are modified to produce a cheese of the desired characteristics (ICMSF, 1998). Cheeses manufactured from the milk of all species, including bovine, ovine and caprine animals undergo similar processing steps. The production of all cheese varieties generally follows a similar process comprising the following general stages (Fig. 8.5):

Cream cheese is a soft, fresh acid-coagulated cheese product, which is acidified by mesophilic lactic acid starter culture, i. e. *Lactococcus* and *Leuconostoc*. Cream cheese products are categorized into two main types based on the different fat content in the initial mix and the final composition. These are double-cream cheese with at least 9 - 11% fat content in the initial mix, and single-cream cheese with 4.5 - 5% fat content in the initial mix. Cream cheese was first made by using the cooked-curd method, which was developed in the early twenties, and the cold-pack and hot-pack methods were developed and are still used at present. The products with high quality should have a uniform white to light cream color with a lightly lactic acid and cultured diacetyl flavor and aroma (Phadungath, 2005) (Table 8.5).

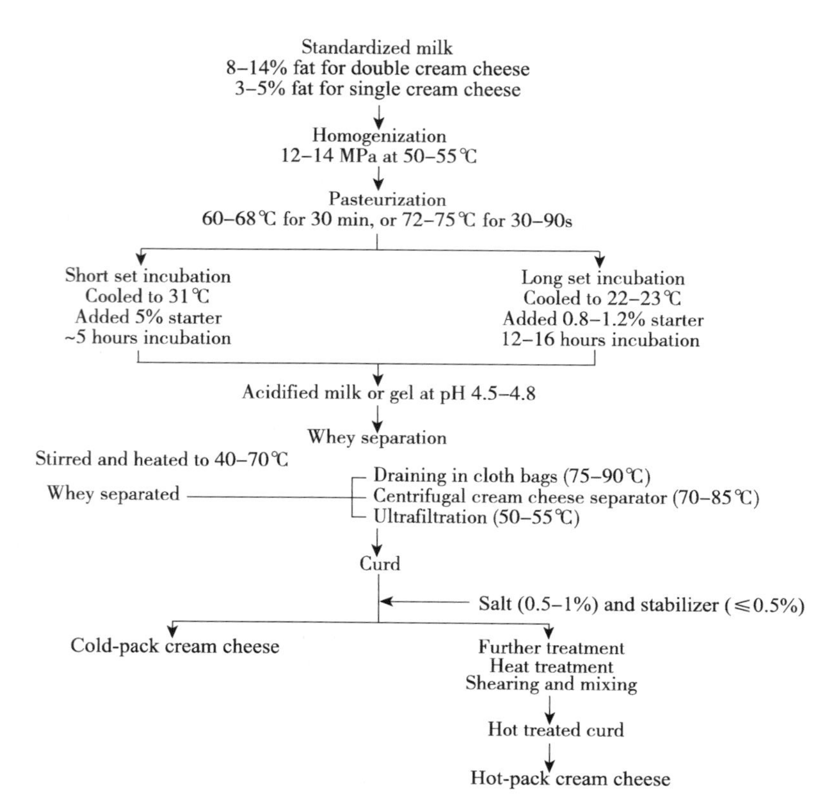

Fig. 8. 5 Processing steps for cream cheese making

Source: Singh and Tewari, 1990.

Table 8. 5 Classification of cheeses by type and moisture content

Cheese Type	Moisture	Description/Style
Very hard	<36%	Ripened by bacteria, e. g. Parmesan, Asiago, Romano
Hard	<42%	Ripened by bacteria, without eyes, e. g. Cheddar
		Ripened by bacteria, with eyes, e. g. Emmental, Gruyère
Semi-soft	43 - 55%	Ripened principally by bacteria, e. g. Gouda, Edam, Provolone
		Ripened by bacteria and surface microorganisms, e. g. Limburger
		Ripened principally by interior, blue mould, e. g. Roquefort, Stilton, Danablu
Soft	>55%	Ripened by surface mould, e. g. Brie, Camembert
		Unripened (also referred as fresh cheese), e. g. Cottage cheese, Quark, Cream cheese
		Salt cured or pickled, e. g. Feta

2. 1. 6 Casein, Whey Products and Other Functional Milk Derivatives

An increasing awareness of the nutritional and health benefits of dairy products has driven the development of markets for a wide array of functional and nutritional ingredients derived from milk. Improvements in fractionation technologies have allowed the manufacture of these on a commercial basis from surplus milk and other dairy by-products. The process typically uses pasteurized milk as a starting material. Separation of milk into cream and skim milk leads to processes for the enrichment of components derived from the fat-and protein-enriched fractions, respectively.

The further by-products of whey fractionation are milk salts, recovered from whey protein, ultra filtrates and electrodialysis (Horton 1995; Wit 2003).

2.1.7 Dried Milk Powders

Whole milk, skim milk, whey, buttermilk, cheese and cream may be dried into powders by the application of heat. The fluid is initially concentrated by evaporation, then spray dried to form a powder.

2.1.8 Ice-cream

Ice-cream is a frozen aerated emulsion made from cream or milk products or both, and other food components. Manufacture of ice-cream involves the preparation of an ingredient mix comprising milk fat, milk solids, sweetener, water and other ingredients which are pasteurized and homogenized, aged, then whipped with incorporated air while being frozen. The final product is then packaged and hardened during frozen storage prior to distribution (Goff, 2003). Other types of ice cream are available in many forms, flavors and packages. Different products prepared both from edible fats and milk or milk products; include gelatin, soft serve, stick ice creams and confections, etc.

2.1.9 Ghee, Butter Oil and Dry Butterfat

These products are almost entirely butterfat and contain practically no water or milk SNF. Which are made from butter through heating? Ghee is made in eastern tropical countries, usually from buffalo milk. An identical product called *Samn* is made in Sudan and *Nitr* made in Ethiopia. Much of the typical flavor comes from the burned milk SNF remaining in the product. Butter oil or anhydrous milk fat is a refined product made by centrifuging melted butter or by separating milk fat from high-fat cream.

Ghee is a more convenient product than butter in the tropics because it keeps better under warm conditions. It has low moisture and milk SNF contents, which inhibits bacterial growth. A considerable amount of moisture and milk SNF can be removed prior to boiling by melting the butter in hot water (80℃) and separating the fat layer.

The fat can be separated either by gravity or using a hand separator. The fat phase yields a product containing 1.5% moisture and little fat is lost in the aqueous phase. Alternatively, the mixture can be allowed to settle in a vessel similar to that used in the deep-setting method for separating whole milk. Once the fat has solidified the aqueous phase is drained. The fat is then removed and heated to evaporate residual moisture. Products made using these methods exhibited excellent keeping qualities over a 5 months test period.

2.2 Products Packing

Proper packing is recommended ever for the issue of quality product. So, the storage equipment's should be considered:

- Always use certified food grade containers, e.g. aluminum, stainless steel or food grade plastic jerry cans.
- Equipment should not be made from copper or any copper alloy or any toxic material.
- Equipment should have a smooth finish, without cracks or rust stains.
- Equipment should be used only for milk, not for anything else.
- Equipment used for storing milk should be clean, pest-free, well ventilated with light, & protect from dust, rain & direct sunlight.

2.3 Milk and Milk Products Preservation

Food business operators must ensure that, upon acceptance at a processing establishment, milk is quickly cooled to not more than 6℃ and kept at that temperature until processed. However, food business operators may keep milk and colostrum at a higher temperature if: 1) processing begins immediately after milking, or within four hours of acceptance at the processing establishment; or 2) the competent authority authorizes a higher temperature for technological reasons concerning the manufacture of certain dairy or colostrum-based products.

Milk preservation techniques include cooling milk, pasteurization, removal of water, fermented milk, boiling milk, and microfiltration.

2.3.1 Cooling Milk

Milk should be cooled immediately after milking. This is essentials to ensure the quality the product. The best temperature for cooling is 4℃. Methods of cooling milk:

- By keeping the milk in the shade
- By keeping milk in a well-ventilated place
- By using cold water (put the milk in cold water bath)
- By using different cooling equipment such as refrigerator, milk cooling tank

If chilled milk is collected, try to maintain the temperature as cool as possible by using insulated containers or boxes, or by placing ice around the container.

2.3.2 Pasteurization

For public health aspect, the purpose of pasteurization is to make milk & milk product safe for human consumption by destroying all harmful bacteria, for keeping quality aspect, to improve keeping quality of milk & milk product and for increasing shelf life of milk & milk product.

Pasteurization is the process of exposing milk to a high temperature that destroys all pathogenic bacteria, but neither reduces the nutritional value of milk nor causes it to curdle, or it is the process of heating every particle of milk or milk product to a specific temperature for specified time. Pasteurization used since the early 1900s (heating raw milk to 161℉ for 15 minutes) is expected to remove microorganisms from milk. The microbiology of pasteurized milk can be determined by dye reaction test.

Methylene blue when added to milk which is incubated at 37℃ will be chemically reduced if there is microbial activity in the milk, but do not indicate anything about the kind of bacteria in the milk.

The time it takes for the methylene blue to become colorless is the methylene blue reduction time (Okpalugo et al., 2008). Pasteurized cream should be cooled as soon as possible after heat treatment to a temperature of 5℃ or less, to prevent growth of thermoduric organisms, and then be packaged quickly. Most cream for retail sale is now packed in plastic pots sealed with metal foil lids. This type of packaging generally carries very low-level of microbial contamination. However, as with pasteurized milk, the hygienic operation of the filling process is very important to prevent post pasteurization contamination (Fernandes, 2008).

Pasteurization is achieved by a treatment involving:

- High temperature for a short time (at least 72℃ for 15 seconds)
- Low temperature for a long time (at least 63℃ for 30 minutes)
- Any other combination of time-temperature conditions to obtain an equivalent effect, such that the products show, where applicable, a negative reaction to an alkaline phosphatase test immediately after such treatment.

2.3.3 Sterilization of Raw Milk

Sterilized milk is defined as milk, which has been heated to a temperature of 108 - 111℃ for 25 to 30 min. Such type of milk is safe for human consumption for at least 7 days at room temperature.

Aim of sterilization

- To inactivate heat-resistant spores micro-organism
- To extend shelf life of milk.
- To make milk free from harmful micro-organism.

2.3.4 Removal of Water from Milk

By evaporated milk: About 60% of the water is removed from milk.

- Milk without added sugar
- It is a process of preheating to stabilize protein and then removing about 60% of water.
- It is sealed in a container, and then heat-treated to sterilization its content. It is no require refrigeration until open.

Concentrated or condensed milk: This also has water removed, but not subjected to further heat treatment to prevent spoilage (milk with added sugar for long life) (Fig. 8.6).

2.3.5 Fermented Milk

Fermented milk products are dairy products that have been fermented with lactic acid bacteria such as *Lactobacillus*, *Lactococcus* and *Leuconostoc*. It is known as cultured dairy products or cultured milk produced.

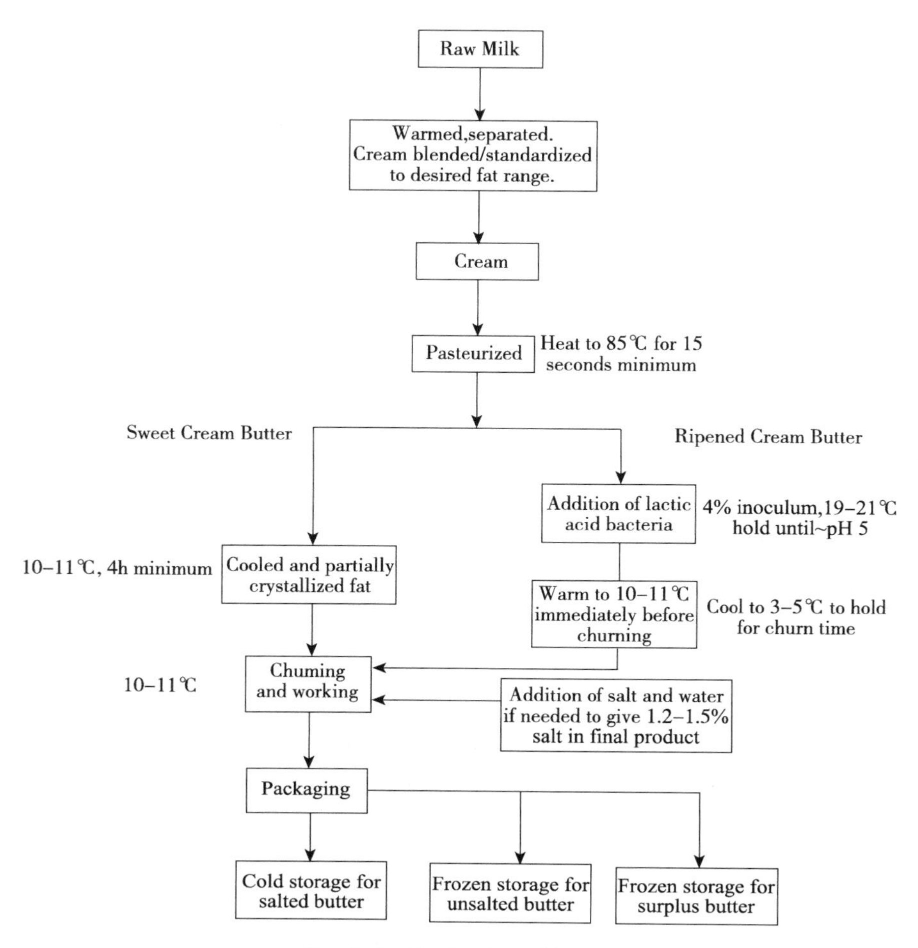

Fig. 8. 6 Production of butter

Importance of fermented milk

- Increase the shelf-life of the milk
- Add taste & flavor
- Improve digestibility of milk

Lactoperoxidase system of milk preservation

This method is a safe milk preservation method that can be used in situation where transportation from farm to processing unit takes a long time & where no cooling facility is available or affordable. Lactoperoxidase delays bacteria growth, prevents souring of the milk and helps to maintain the milk in a condition as healthy as it was when drawn from udder.

- Lactoperoxidase is an enzyme that exists naturally in milk & slows the growth of

spoilage bacteria.

➢ This system is only applicable if refrigeration is not available or practical.
➢ The effect of lactoperoxidase depends on temperature at 30℃; it can prevent souring of the milk for 7 – 8 hours (if the initial hygienic quality of the milk is reasonably good).

2.3.6 Boiling of Milk

Boiling of milk will depend a lot on the type of milk. For example, skim milk has a lower boiling point than whole milk.

➢ The exact boiling point depends on type of milk & what altitude it's at.
➢ Boiling point of milk will close to that of water boiling point which is around 100℃ (The exact number depends on atmospheric pressure).
➢ Milk boils at a temperature higher than the boiling point of pure water at the same atmospheric pressure.
➢ Factors affecting boiling point of milk include fat in the milk, sugars & salt in the milk, and dissolved protein in the milk.

3 MILK AND MILK BYPRODUCTS TRANSPORTATION AND RECEPTION

Don't expose milk to light, especially while transporting, because milk is very sensitive to light. If exposed to direct sunlight butter fat & some vitamins get oxidized & the milk develops bad, oxidized flavor. Also, don't delay above time and minimize excessive agitation because the time from milk collection to cooling is very crucial. Bacteria multiplication is very slow during the first hours. After that bacteria will multiply very fast during every 20 – 30 minutes. If you are transporting unchilled milk, make sure the milk reaches its final destination within 2 hours from the time of milking. When milk is agitated, the milk fat is destabilized & tends to oxidize easily. One big reason for agitation is transporting half-full cans. Try and avoid this wherever possible.

Milk transport equipment and time

Ideal milk transport equipment would be made from stainless steel or aluminum. Dirty vessel or entry of dust particles will increase the bacteria load in milk & reduce its shelf life. All dairy products have a shelf life that varies according to how an item is processed, packaged and stored. Example: yogurt, sour cream is sold in date-coded cartons that indicate a product's peak freshness. The shelf life of cheese depends on its type (hard or soft) and its form (cut or wax-coated). Type of container used can alter the freshness period. Refrigerator temperatures should be between 32°F and 38°F. Because the shelf life of milk and dairy products is shortened by a full 50% for every 5 degrees rise in temperature over 38°F. The higher temperature, the faster will bacteria growth, & cause the milk to become sour. High level of hygiene milk, speedy transport milk, and careful handling milk minimizes losses due to spillage, avoid contamination of milk by pathogens & increase the profits from your milk transportation business. During transport the cold chain must be maintained and, on arrival at the establishment of destination, the temperature of the milk must not be more than 10℃.

Transportation tank should be:

- Clean, intact, odorless
- Disapproved quality milk must not be filled into the tank
- Prohibited to use it for other purposes
- Compartments of the tank should be emptied separately
- Hose (cleaning, disinfection)
- Workers' personal hygiene
- Bulk tank is divided into 2 - 5 parts
- Double layer

Requirements of bulk tank:

- Temperature and amount of milk can be measured
- Sample taken at reception
- Documentation
- Totally empty
- Cells, tubes, surfaces disinfection and cleaning

Then, an expert for milk reception can be done the following duties:

- Quantitative reception: Volume or weight
- Qualitative reception: Physical-, chemical-, and microbiological parameters include temperature, organoleptic properties, pH value or acid degree, physical purity, density, inner substances (fat, protein, solids—nonfat content), antibiotic residues, freezing point, somatic cell counting, and microbe counting.

Plant and raw milk qualifying laboratories:

- Arriving at the plant, the temperature of milk $\leqslant$ 10℃
- Organoleptic inspection: Sourness; curdling; taste-, color-, odor-defects of milk

4 MARKETING OF DAIRY PRODUCTS AND ANIMALS

4.1 Marketing Milk and Milk Products

Consumption pattern and marketing of dairy products produced at home varied depending upon the amount of milk produced per household, dairy production system, market access, and season of the year, fasting period, and culture of the society. Rural dairy farmers have very little access to market fluid milk and milk is often processed into butter. The major dairy products commonly marketed include fresh milk, butter, ergo (fermented whole milk), cottage cheese and buttermilk. The dominant milk products marketed across all the PLWs with the exception of urban and peri-urban system is butter followed by cottage cheese. In areas (e. g. rural highlands) where milk marketing is practiced, the amount marketed is very small due to lack of surplus production, the desire to process into milk products and lack of access to market. In market-oriented urban and peri-urban system fluid milk marketing is dominant being higher in urban than peri-urban system. Although both formal and informal milk marketing systems do exist, the latter is the dominant system across all the production systems.

The marketing of surplus milk, after the requirements of the family and the farm are met, improves the farm income and creates employment in processing and distribution. So, the milk does not only contribute to the food security in the rural community, but it adds value as well. Marketing of milk is particularly difficult for small scale producers, who live, scattered in rural areas.

Products and technology must be chosen, which suit the scale and location of the operation. But at the same time the price, taste, and packaging must meet the local requirement. In urban markets in many developing countries the sale of raw milk by informal traders is one of the major outlets for milk. However, this is associated with health risks. These must be addressed, and steps have to be taken to minimize these risks. An example is the 'village milk processing' approach, which is promoted by FAO.

Higher incomes, larger urban populations and continued population growth will fuel higher demand for dairy products. However, the Ethiopian consumption rate during the last four decades ranged from 16 – 19 liters per capita (FAOSTAT), about half of the average African consumption rates, and well below the world average.

4. 1. 1 Informal Traditional Markets Model

Smallholder milk producers sell milk directly to consumers or milk supplier/middlemen at local markets (Fig. 8. 7). The middlemen cater to the demand of sweetmeat shops, bakeries, consumers, more distant markets and vendors. They pay producers up to 50% less for their milk than other models, such as those described in the following sections. In many cases, the middlemen provide loans to smallholders with interest rates of up to 20% per month. Ethiopian milk marketing is dominated by an informal marketing system, although the dairy cooperative system is getting a good footage. During the fasting periods, there is a serious market problem for milk, and producers lost a considerable amount of return from their product due to price reduction or wastage.

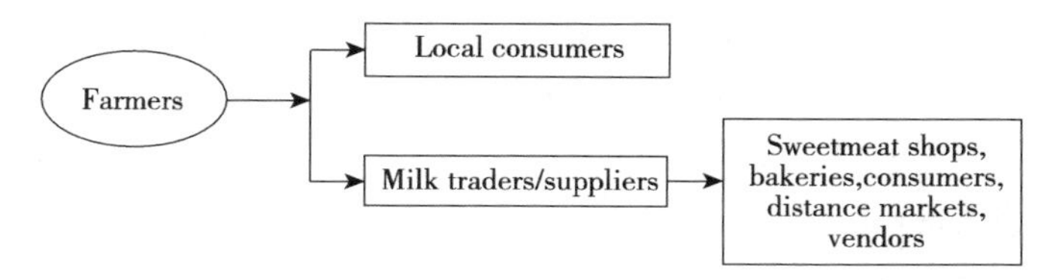

Fig. 8. 7 Traditional milk trader model

4.1.2 Milk Cooperative Model

The Milk (not clear, give synonyms in bracket or take it out) Cooperative model was adapted from the world-renowned Indian Dairy cooperative development model. It modestly started in the mid 1970's by providing 4,300 very poor, often landless households in remote rural areas with a complete package of milk production-enhancing technologies, organizational skills and a milk collection-processing-marketing system.

A novel aspect of the "Milk Vita" operation is its urban distributor cooperatives. These use locally fabricated "milks haws" -an insulated box mounted on a traditional three-wheeled-cycle rickshaw chassis-to deliver affordable pasteurized milk and dairy products to urban shops and consumers. For better marketing of milk and milk products, dairy farmers should be organized into dairy cooperative to sell huge volume of milk to dairy processing company/ plant in order to earn high price from their dairy products (Verschuur, 2014) (Fig. 8. 8).

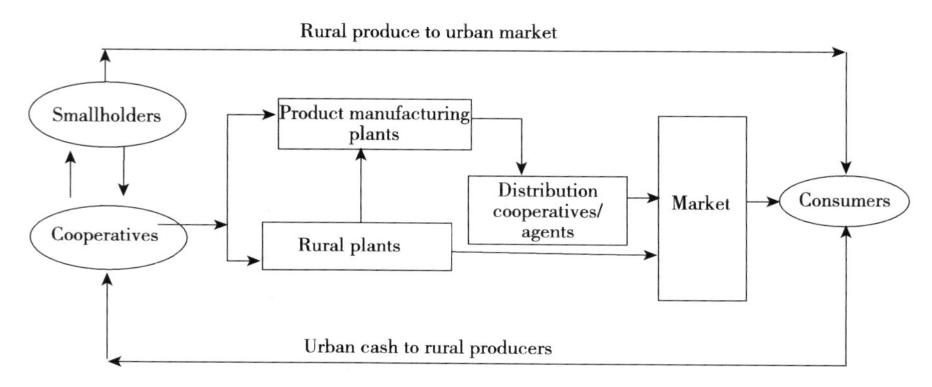

Fig. 8. 8 Milk Vita Cooperative model

The benefits of the Milk Vita cooperative model:

- The model is a holistic, cow-to-consumer model.
- Milk production and productivity increase.
- Household nutrition and incomes increase.
- Communities are empowered through the participation of poor farmers' in the organized cooperatives and through accountability of the Milk Vita board and management to its milk producer members.
- Quantity of affordable and safely processed milk and dairy products for urban consumers is increased while the quality is enhanced.
- Substantial off-farm employment is generated.

Milk Vita continues to be a flourishing venture and has many recent imitators that have

set up similar enterprises to process and market 70 million liters of milk annually.

4. 1. 3 Private Entrepreneur Model

Private dairies, some owned by non-government organizations (NGOs), such as the Bangladesh Rural Advancement Committee (BRAC), usually operate through milk supplier/middlemen (known as *ghoshes* or *dudhwalas*) in place of rural groups or cooperatives (Fig. 8. 9). They collect milk for a specific dairy. However, smallholders involved in the system do not receive any value-added benefit-only the basic price for their milk.

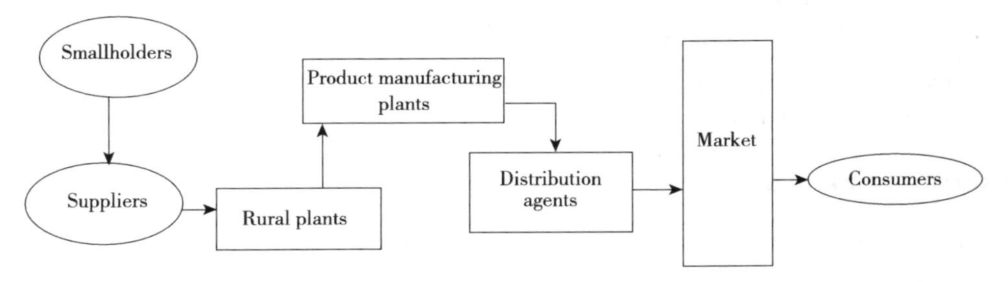

Fig. 8. 9 Private entrepreneur model

4. 2 Assess Seasonal Price Trends

In non-urban areas, milk consumption varies by season and cow productivity. Consumed milk-preference is for whole, raw milk, is either from milk production at home or by neighbors. Processed milk is not available in rural markets. An estimated 40% is converted to butter, and 9% to cheese. Traditional butter can be kept for more than a year. Milk and milk products are an important source of food security. Women often play a large role in decision-making regarding the processing and marketing of milk (Fig. 8. 10).

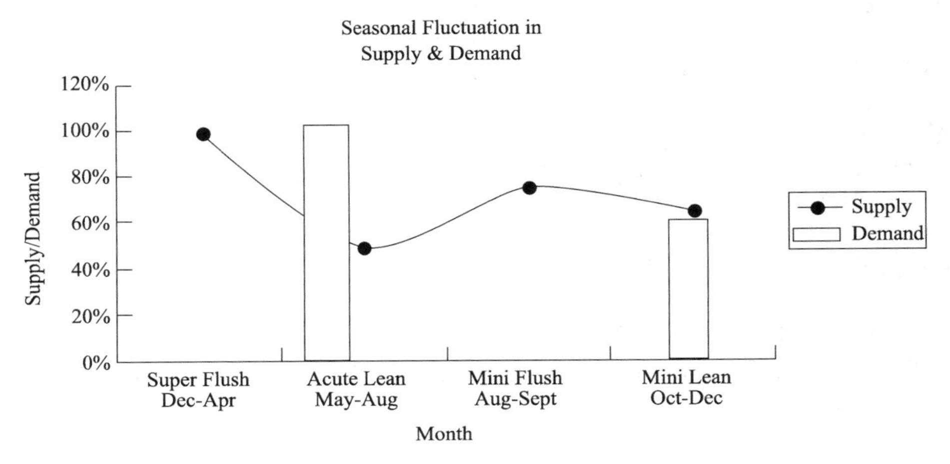

Fig. 8. 10 Seasonal fluctuation in supply and demand

Source: Umm, 2006.

>>> OPERATION SHEET

Regular quality control of milk is essential. Routine chemical analyses are carried out in the laboratory to check milk composition of fat and total solids (TS).

Part 1. Sampling

Laboratory samples should be kept in closed containers, sealed and labelled indicating the nature of the product, identification number, batch number of the sample and date of sampling. The method of sampling depends on the nature of the product (liquid or solid) and whether chemical or bacteriological analyses are required. The size of the samples is regulated by the size of the product to evaluate.

Technical instructions

Clean and dry all equipment for chemical purposes. Equipment used in sampling for bacteriological purposes should be clean and treated by one of the following methods:

- Expose to hot air at 170℃ for 2 h.
- Autoclave 15 - 20 min at 121℃.
- Expose to steam for 1 hour at 100℃ for immediate use.
- Immerse in 70% alcohol and flame for immediate use.

Liquid sample

Use clean and dry containers: glass, stainless metal or plastic with a suitable shape and capacity for material to be sampled. Containers should be closed with a suitable plastic stopper or screw cap.

Note: Containers should be filled well to prevent churning during transportation.

Solid and semi-solid samples

Use clean and dry containers of suitable waterproof material like glass, stainless metal or plastic material. Avoid narrow neck containers and close well. Airtight plastic bags can be used too.

Preservation of laboratory samples

Preservatives have to be added to liquid samples or cheese intended for chemical analysis. The preservative should not affect analysis. Sodium azide or potassium dichromate is the preferred preservatives.

Transport of laboratory samples

Transport samples to the laboratory as fast as possible. Take precautions to prevent exposure to direct sunlight, or to temperatures above 10℃ in case of perishable products. For laboratory samples intended for bacteriological examination, maintain temperature at 0 – 5℃. Maintain samples of cheese under the same conditions to avoid separation of fat and moisture.

Part 2. Sample Preparation

Principle

Homogenizing of the milk before analysis is essential. Milk fat tends to separate and to flow on the surface.

Procedure

- Put the milk sample in a water bath at 38℃.
- Mix until homogenization.
- Cool samples to approximate 20℃ before analysis.

Part 3. Determination of Fat (Gerber Method)

Principle

Milk is mixed with sulphuric-acid in a special glass tube called a Gerber tube. The temperature of the mixture becomes hot due to the dilution of a strong acid. Therefore, milk proteins are digested, and fat is set free from the fat globules. It can be measured in the glass scale

Procedure

- Pipette 10 mL of sulphuric-acid into the butyrometer.
- Slowly pipette 10.75 mL milk. Milk should not be mixed with the sulphuric-acid, it should appear as a separated layer.
- Add 1 mL of amyl alcohol.
- Close with Gerber tube stopper.
- Centrifuge in Gerber centrifuge for 5 min on (1100±100) r/min.
- Place the butyrometer for 5 min in a water bath, at 65℃.
- Read the percentage of fat.

Caution

- Dry neck of butyrometer before and shake the butyrometer until all the milk is digested.
- Glass becomes very hot.

Calculation

The reading represents the percentage of fat content in milk.

Reagents/solutions

- Sulphuric-acid density 1.81

- Amyl alcohol

Equipment

- Water bath
- Gerber milk butyrometer
- Standard rubber stopper
- Volumetric pipette (H_2SO_4) with safety bulb, 10 mL
- Volumetric pipette (milk), 10.75 mL
- Volumetric pipette (amyl alcohol), 1 mL
- Gerber centrifuge [(1100±100) r/min, diameter 45 - 50 cm] .

Part 4. Total Solids

Principle

Milk is dried under constant temperature in the oven until weight is stabilized.

Procedure

- Weigh a clean pre-heated crucible.
- Add 3 - 4 g of milk and record weight.
- Dry the sample in an air forced oven at 105℃ overnight to stabilize weight.
- Weigh the crucible with the dried sample.

$$\%TS = [(\text{Weight of dried milk}) / (\text{weight of milk})] \times 100\%$$

Equipment

Air-forced oven, crucibles, desiccators, pipette

Part 5. Fat, Protein Total Solids and Lactose Analysis

Principle

The analysis of milk with the Milkoscan device is based on absorption of near infrared energy at specific wavelength by:

- Carbon-hydrogen groups (3.48 μm) and carbonyl groups of fat (5.723 μm)
- Peptide bonds of protein molecules (6.465 μm)
- Hydroxy groups in lactose molecules (9.61 μm)

The Total Solids (TS) are set by the factory based on other solid milk components (e.g. protein, fat, lactose).

Procedures

- Clean and calibrate of the Milk scan.
- Warm Triton 1% solution and Stella 5% solution to 40℃.
- Purge the device with Stella.

- Purge the device with Triton.
- Calibrate the device with Triton 1%.

Measurement

- Warm milk samples to 40℃ in water bath.
- Insert sample in aspirator device of milk scan.

Equipment

Milk scan 133B device, and water bath

Part 6. Milk Density

Principle

Based on specific gravity a special device is used to determine the milk density.

Procedure

- Inject approximately 5 mL of milk to ensure that the milk line is well flushed with the sample.
- Read the density on the device.

Note: Ensure that there are no bubbles in the milk line, bubbles cause an error.

Part 7. Milk Conductivity

Principle

Mastitis causes an increase of salt concentration. Conductivity is measured with an electronic conductivity meter by immersing a probe directly into the milk sample.

Procedure

- Immerse the conductivity probe into the milk sample.
- Read the result on the device.

Equipment

Electric conductivity meter

Part 8. Milk pH

Principle

pH is the negative logarithm of the hydrogen ion concentration. Measurement of pH is particularly important in the dairy industry since it provides, in many cases, a more meaningful measurement than titratable acidity. pH values from 0 – 7 are acidic, while those from 7 – 14 are alkaline. pH measures the level of the acidity.

Procedure

- Pour the sample into a clean, dry beaker.
- Carefully press electrodes into the milk-containing beaker and determine the pH.

Equipment

pH meter

Part 9. Milk Acidity

Principle

The pH value of fresh milk is normally about 6.6 at 25℃, which indicates a natural acidity. The acidity is determined by titration with alkali NaOH, in the presence of phenolphthalein as indicator.

Procedure

- Pipette 10 mL of milk.
- Add 5 drops of 1% phenolphthalein.
- Titrate with 0.1 mol/L NaOH until a slight pink color appears.

Solids non-fat (SNF) defines the natural acidity of fresh milk. The higher the solids, that are not fat content, in milk, the higher the natural acidity and vice versa. Developed or real acidity is due to lactic acid, formed as the result of bacterial action on lactose in milk. Hence the titratable acidity of stored milk is equal to the sum of natural acidity and developed acidity. The titratable acidity is expressed as percentage of lactic acid. Colostrum has a high "natural acidity" in part because of its very high protein content.

Part 10. Reagents/solutions

Sodium hydroxide 0.1 mol/L solution

- Dissolve equal parts of sodium hydroxide (sticks or pellets) in equal parts of water in a flask.
- Tightly close the flask with a rubber bung and allow insoluble sodium carbonate to settle down for 3 - 4 days.
- Use the clear supernatant liquid for preparing the standard 0.1 mol/L solution.
- Use 8 mL of stock solution per liter of distilled water.
- Standardize solution accurately against potassium acid phthalate or oxalic acid.

Phenolphthalein indicator 1%

- Weigh 1 g of phenolphthalein in volumetric flask 100 mL.
- Dissolve in ethanol and make up to volume.

Calculations

$$\%Acidity\ of\ milk = [(V_a \times 0.009) / V_m] \times 100\%$$

V_a—volume of 0.1 mol/L NaOH

V_m—volume of milk sample

0.009—conversion factor from V_a of 0.1 mol/L NaOH to gram of lactic acid

Equipment

- Glass burette 50 mL capacity with 0.1 mL graduation
- Round bottomed white porcelain dish, 100 mL
- Pipette 10 mL
- Pipette 1 mL
- Stirrer
- Stirring rods

Part 11. Volatile Fatty Acids

Principle

Fat samples are esterified and evaporated in the GC system; fatty acids are separated in a stationary phase of the installed column depending on the chemical characteristics.

Sample preparation

1) Extraction (Folch et al., 1957)

- 1 mL milk
- 2 mL $Mg(OH)_2$
- 1 mL $CHCl_3$
- Vortex 1 min
- 1 mL $CHCl_3$
- 1 mL H_2O
- Extract the lower part
- Evaporate with N_2

2) Methylation (Precht, 2000)

- (100±20) mg of fat
- 5 mL methyl valerate standard (n-Heptane, containing 0.4 mg/mL methyl valerate)

1 mL of this solution was mixed with:

- 20 μL sodium methylate solution in a sample vial
- Vortex for 3 min.
- Centrifuge for 1 min at 2000 r/min.
- 100 mg sodium hydrogen sulphate monohydrate
- Cover vial with new cap.
- Vortex for 2 min mix.

- Centrifuge for 1 min at 2,000 r/min.
- Use supernatant for GC analysis.

Procedures

1) Switch on procedures

- Switch on device.
- Switch on attached PC.
- Run GC solution program.
- Click on the GC-Device.
- Load the cool-heat method and click on 'Download'.
- Click on 'System on' and let the method work for 5 min.
- Load the heat method and click on 'Download'.
- Run method for 20 min while observing GC parameters on monitor when flame ionization detector (FID) temperature is above 100℃, switch on the detector and ignite the flame clicking the left button on system monitor screen.
- Load the analytical method.
- Click on 'Download'.

2) Injection procedure

Before injecting a sample, there are some steps to follow:

- Click on 'Sample log on'. Fill in the requested data. This is important for post-acquisition data work.
- Fill the GC analysis lab-book with the requested data.
- Do not inject the sample before 'Ready' appears on the PC monitor in blue; LED's on the GC should appear in green.
- Inject the sample and press the 'Start' button on the GC simultaneously.

3) Thermal program

- 50℃ for 3 min
- 5 - 140℃ for 2 min
- 2 - 170℃ for 5 min
- 10 - 220℃ for 2 min
- 5 - 225℃ for 21 min.

4) Injection thermal method

- 1 μL solvent
- 1 μL air
- 1 μL sample
- Inject sample manually after a dwell-time of 5 seconds.
- Remove the syringe after 5 seconds.

5) Switch off procedure

- Load the cool-heat method and click on 'Download'.

- Switch off the flame and then switch off the detector by clicking the left buttons on system monitor screen.
- Wait-till the FID and injector temperature are less than 100℃, then press the "system off" button.
- Close the program.
- Switch-off the GC.
- Close running programs and shut down the PC.

Note: Do not open the oven at any time. This may damage the installed column. If any error message appears, inform the person responsible immediately. Any delay in conveying this information could damage the whole system. Do not inject wet samples. Do not inject more than 1 μL of air.

Calculations

The qualitative and quantitative assessment of fatty acids is calculated by the GC solution software.

Reagents/solutions

- H_2 gas (99.999%)
- N_2 gas (99.999%)
- Sodium methylate 2 mol/L
- Methanol
- Chloroform

Equipment

- Gas chromatograph with FID system
- Vortex mixer
- Syringe 1 μL
- Pasteur pipettes

Part 12. Assembling and Dismantling the Cream Separator

- When the bowl is set, fit the skim milk spout and the cream spout.
- Fit the regulating chamber on top of the bowl.
- Put the float in the regulating chamber.
- Put the supply can in position, making sure that the tap is directly above and at the center of the float.
- Pour warm (body temperature) water into the supply can.
- Turn the crank handle, increasing speed slowly until the operating speed is reached: This will be indicated on the handle or in the manufacturer's manual of operation. The bell on the crank handle will stop ringing when the correct speed is reached.
- Open the tap and allow the warm water to flow into the bowl. This rinses and heats

the bowl and allows a smooth flow of milk and increases separation efficiency.

- Next, put warm milk (37 – 40℃) into the supply can. Repeat steps 6 and 7 above and collect the skim milk and cream separately.
- When all the milk is used up and the flow of cream stops, pour about 3 liters of the separated milk into the supply can to recover residual cream trapped between the discs.
- Continue turning the crank handle and flush the separator with warm water.

Part 13. Yoghurt Making

Using a yogurt maker, oven-method, cooler or slow-cooker:

- Incubate the milk mixture at 110°F (43℃) for 6 – 8 hours without agitating or moving the container. You may be tempted to move and peek in on your yogurt, but the more you move the container you will agitate the curd and lose heat.
- The longer the Yoghurt sits, the more lactic acid is produced and the tarter the taste will then be.
- After fermentation, set the container in the refrigerator for 12 hours to allow the full development of the yoghurt flavor and to allow for further thickening to occur. You will notice a big difference in thickness after refrigeration time. After cooling, if necessary, transfer the fresh homemade yogurt from the pot or Dutch oven to a clean container that takes up less room in the fridge for storage (not necessary if using a glass mason jar that acts as storage container). Enjoy this treat. If you want a thicker yogurt, strain it through cheesecloth.
- Storage: Enjoy for up to 1 week. The flavor will increase during this time since the microorganisms stay active even in the refrigerator.

Part 14. Butter Making

Butter is made from cream that has been separated from whole milk and then cooled; fat droplets clump more easily when hard rather than soft. However, making good butter also depends upon other factors, such as the fat content of the cream and its acidity.

The process can be summarized in 3 steps:

- Churning physically agitates the cream until it ruptures the fragile membranes surrounding the milk fat. Once broken, the fat droplets can join with each other and form clumps of fat.
- As churning continues, larger clusters of fat collect until they begin to form a network with the air bubbles that are generated by the churning; this traps the liquid and produces foam. As the fat clumps increase in size, there are also fewer to enclose the air cells. So, the bubbles pop, run together, and the foam begins to leak. This leakage is called buttermilk.
- The cream separates into butter and buttermilk. The buttermilk is drained off, and the remaining butter is kneaded to form a network of fat crystals that becomes the continuous phase, or dispersion medium, of a water-in-fat emulsion. Working the

butter also creates its desired smoothness. Eventually, the water droplets become so finely dispersed in the fat that butter's texture seems dry. Then it is frozen into cubes, then melted, then frozen again into bigger chunks to sell.

Part 15. Ghee Manufacturing

Ghee is animal fat normally produced from Milk. The followings are the steps taken during ghee making process:

- During pasteurization process milk cream is separated in the cream separator with 50－70% fat content.
- This cream is then churned in butter churn. This makes butter of 84% fat. The whey is drained in whey tank, which is used as by product.
- Butter from the butter churn is taken in the butter trolley.
- This butter either can be stored in the cold room, till sufficient butter is available for further processing.
- Once sufficient butter is available, it is taken in the Ghee Kettle.
- Open the condensate drain valve and remove all the condensate from the steam jacket.
- Steam valve in the ghee kettle is opened slowly and start supplying steam (1.5 bar G.). This is to heat the butter and melt it.
- Close the drain valve and open steam trap and let condensate pass through it.
- Once enough butter is melted, start the agitator of the ghee kettle.
- Continue supplying steam to the kettle.
- The butter oil will start boiling and water from the butter will evaporate.
- Let all the water evaporate.
- Slowly close the steam supply valve and open vent valve on the jacket.
- Let all condensate from the steam jacket get drained.
- Let ghee to cool down.
- Continue agitating ghee with the help of agitator.
- After ghee temperature has come down to around 70℃, drain this in the ghee filter tank.
- With the help of SS strainer, most of burned protein from the butter is filtered out.
- Once all ghee is filtered, start the ghee pump.
- The ghee should be fed to ghee clarifier, where ghee will be clarified.
- The clarified ghee is collected in the balance tank.
- Ghee from the balance tank, with the help of another ghee pump is transferred to jacketed ghee storage tank.
- This storage tank has a water jacket with electric heater. In case of very cold weather, switch on the electric heater. This is to keep ghee in free-flowing condition.
- This is the final product. Pack ghee in consumer packing.
- Clean the complete plant from inside and outside with a warm detergent followed by water and wipe out all traces of butter / ghee.

➢ Clean all pipes/valves.

Part 16. Cottage Cheese Making

Cottage cheese makes a delicious light breakfast or lunch when served with fruit or a salad. This dish is so simple to make at home that there's no reason to pick up a tub at the store. Learn how to make cottage cheese using rennet, vinegar or lemon juice.

Ingredients

One: Use Rennet	Two: Use Vinegar	Three: Use Lemon Juice
1 - quart whole milk	1 gallon (3.8 L) pasteurized skimmed milk	1 - quart whole milk
4 drops liquid rennet	3/4 cup white vinegar	1/2 teaspoon citric acid or lemon juice
1/2 teaspoon salt	1/2 teaspoon salt	1/2 teaspoon salt
6 tablespoons heavy cream or half and half	1/2 cup heavy cream or half and half	6 tablespoons heavy cream or half and half

Using rennet

➢ *Heat the milk*: Pour the milk into a small saucepan and place it over medium heat. Heat the milk slowly, making sure it doesn't boil, until it reaches 85°F (29℃). Use a candy thermometer to monitor the temperature. Turn off the heat when the milk is sufficiently warm.

➢ *Add the rennet*: Place the drops of rennet directly in the milk. Use a spoon to stir the mixture for about 2 minutes.

➢ *Let the mixture stand*: Cover the saucepan with a clean dish towel and let the rennet and milk sit untouched for about 4 hours. The rennet will start reacting with the milk to turn it into cheese.

➢ *Slice the mixture*: Remove the dishcloth and use a knife to make slices in the mixture and break up the curds. Slice several times in one direction, and then make several slices in the opposite direction.

➢ *Cook the mixture*: Add the salt to the saucepan. Turn the burner to medium low. Stir the mixture as it heats to help the curds separate from the whey. Stop as soon as the curds have separated, and the whey looks slightly yellow. Don't overcook the mixture, or the curds will be hard.

➢ *Strain the curds*: Place a piece of cheesecloth or a fine-mesh strainer over a bowl Pour the curds and whey into the cheesecloth to strain the curds from the whey. Keeping the curds in the cheesecloth suspended over a bowl, cover the curds loosely with plastic wrap and place all of it in the refrigerator to let the whey continue to drain for a few hours. Stir it every once in a while, to help it along.

➢ *Serve the cottage cheese*: Place the curds in a clean bowl and add the cream or half and half. Season the curds with more salt to taste.

Using vinegar

➢ *Heat the milk*: Place the milk in a saucepan and put it on the stove. Turn the burner to medium and heat the milk to 120 degrees. Use a candy thermometer to monitor the milk's temperature. Remove it from heat once it is sufficiently warmed.

➢ *Add the vinegar*: Pour the vinegar into the saucepan and stir the mixture slowly for 2 minutes. Cover the pan with a dishcloth and let the mixture rest for 30 minutes.

➢ *Strain the curds from the whey*: Pour the mixture into a colander lined with cheesecloth or a thin dishcloth. Let the whey drain for about five minutes.

➢ *Rinse the curds*: Gather the edges of the cloth and hold the curds under a stream of cold water. Squeeze the curds and move them around until they are all rinsed and cooled.

➢ *Finish the cottage cheese*: Place the curds in a bowl. Add the salt and the cream or half and half. Store in the refrigerator or serve immediately.

Using lemon juice

➢ *Heat the milk*: Place it in a saucepan and heat it until it begins to steam but does not come to a boil. Remove the milk from heat.

➢ *Add the lemon juice*: Pour the lemon juice into the warm milk and stir it slowly for several minutes.

➢ *Let the mixture rest*: Cover the saucepan with a dishcloth and let the curds separate from the whey for about an hour.

➢ *Strain the curds from the whey*: Place a piece of cheesecloth over a bowl and pour the curds and whey into the cheesecloth. Let the curds drain for about 5 minutes.

➢ *Rinse the curds*: Gather the ends of the cheesecloth and hold it under cool water to rinse the curds. Continue until they are completely cooled, then squeeze the cloth to get the curds as dry as possible.

➢ *Finish the cottage cheese*: Place the curds in a bowl and add the salt and cream or half and half.

>>> SELF-CHECK QUESTIONS

Part 1. Choose the best answer from the following alternatives.

1. One does not belong to be an off-flavor of milk?

   A. Weed flavor B. Cow barn

   C. Rancidity D. Pleasant

2. One is not an environmental factor that affects quality of raw milk?

   A. Temperature B. Bacteria

C. Equipment D. Sunlight

3. The off-flavor due to poor ventilated pen of the cow is known as?

A. Oxidized B. Feed

C. Cow-barn D. Rancidity

Part 2. Match accordingly.

A	B
1. Environmental factor	A. Smelling
2. Adulteration	B. Lactometer
3. Organoleptic test	C. 80℃
4. Critical control point	D. Cooling
5. Sanitizing	E. 4 - 6℃
6. Critical control limit	F. Sunlight

Part 3. Fill the blank space.

1. ________ is a reactive quality assurance tool.
2. ________ Means what you still have on credit.
3. ________ is a tool to assess hazards and establish control systems that focus on prevention rather than relying mainly on end-product testing.

Part 4. Essay.

1. What are the methods of milk preservation?
2. Write the formula for finding specific gravity?
3. List the seven important principles of HACCP chronologically?

>>> REFERENCES

Auldist M J, Hubble I B, 1998. Effects of Mastitis on Raw Milk and Dairy Products [J] . Aust. J. Dairy Technol, 53 (1): 28 - 36.

Barkema H, Schukken Y, Lam T, et al. , 1998. Management Practices Associated with Low, Medium, and High Somatic Cell Counts in Bulk Milk [J] . J. Dairy Sci, 81 (7): 1917 - 1927.

Bashir S, Awan M S, Khan S A, et al. , 2013. An Evaluation of Milk Quality in and around Rawalakot Azad Kashmir [J] . Glob. J. Food Sci. Technol, 1 (1): 62 - 68.

Biru G, 1989. Major Bacteria Causing Bovine Mastitis and their Sensitivity to Common Antibiotics [J]. Ethiop J Agric Sci (11): 43 - 49.

Bonnier P, Maas A, Rijks J, 2004. Dairy Cattle Husbandry [M] . Agromisa Foundation, Wageningen, the Netherlands.

CFSAN, 2006. Hazards & Controls Guide for Dairy Foods HACCP [E] . http: //www. fda. gov/downloads/Food/FoodSafety/Product-SpecificInformation/MilkSafety/DairyGradeAVoluntaryHACCP/ UCM292647. pdf.

Chatterjee S N, Bhattacharjee I, Chatterjee S K, et al. , 2006. Microbiological Examination of Milk in

Tarakeswar, India with Special Reference to Coliforms [J]. African Journal of Biotechnology, 5: 1383-1385.

De Wit J N, 2002. Dairy Ingredients in Non-dairy Foods [M] // Roginski H, Fuquay J W, Fox P F. Encyclopedia of Dairy Sciences. Elsevier Ltd: 718-727.

Debele G, Verschuur M, 2014. Assessment of Factors and Factors Affecting Milk Value Chain in Smallholder Dairy Farmers [J]. African Journal of Agricultural Research, 9 (3): 345-352.

Fernandes R, 2009. Microbiology Handbook-Dairy Products [M]. 3rd ed. Royal Society of Chemistry.

Godic-Torkar K, Golc-Teger S, 2008. The Microbiological Quality of Raw Milk After Introducing the Two Days Milk Collecting System [J]. Acta Agric. Slovenica, 92 (1): 61-74.

Goff H D, 2002. Ice Cream and Frozen Desserts-Manufacture [M] // Roginski H, Fuquay J W, Fox P F. Encyclopedia of Dairy Sciences. Elsevier Ltd: 1374-1380.

Goff J P, Horst R L, 1997. Effects of the Addition of Potassium or Sodium, but not Calcium, to Prepartum Rations on Milk Fever in Dairy Cows [J]. J. Dairy Sci, 80: 176-186.

Horton B S, 1995. Commercial Utilization of Minor Milk Components in the Health and Food Industries [J]. J. Dairy Sci, 78 (11): 2584-2589.

ICMSF, 1998. Microorganisms in Foods 6: Microbial Ecology of Food Commoditie [M]. Springer.

Lore T A, Kurwijila L R, Omore A, 2006. Hygienic Milk Production: a Training Guide for Farm-level Workers and Milk Handlers in Eastern Africa [M]. Nairobi, Kenya: International Livestock Research Institute.

Lund B, Baird-Parker A C, Gould G W, 2000. Microbiological Safety and Quality of Food [M]. Springer.

Nielsen N, Larsen T, Bjerring M, et al., 2005. Quarter Health, Milking Interval, and Sampling Time During Milking Affect the Concentration of Milk Constituents [J]. J. Dairy Sci, 88 (9): 3186-3200.

Okpalugo J, Ibrahim K, Izebe K S, et al., 2008. Aspects of Microbial Quality of Some Milk Products in Abuja, Nigeria [J]. Tropical Journal of Pharmaceutical Research, 7 (4): 1169-1177.

Ryser E T, Marth E H, 1999. Listeria: Listeriosis, and Food Safety [M]. 2nd ed. CRC Press.

Srujana G, Reddy A R, Reddy V K, et al., 2011. Microbial Quality of Raw and Pasteurized Milk Samples Collected from Different Places of Warangal District, (A. P.) India [J]. International Journal of Pharma and Bio Sciences, 2 (2): 139-143.

MODULE 9: DAIRY CATTLE HEALTH AND DISEASES PREVENTION ACTIVITIES

>>> INTRODUCTION

Health can be defined as the condition of an animal that enables it to attain acceptable levels of production within the farming system in which it is maintained. The term is relative, however, as what one producer may consider being within a normal range and representing good health may be considered ill health by another. The term 'healthy' certainly does not imply that the animal is free from all disease agents; an animal may be infected by a potentially pathogenic agent but be unaffected by it and remain in good health for an indefinite period (Wilson, 2001). Any condition/factor that interferes with the fullest performance of the animals for growth, development, and production is to be regarded as causing disease. For instance, an animal with a broken leg has a disease because it cannot walk properly, but in practice we tend to exclude malfunctions caused by accidents when referring to diseases (Payne and Wilson, 1999).

Good health care and management impact on the productivity of dairy cattle. Well managed herd health care doesn't only mean treating of an animal when it is sick, but also helping it to avoid becoming ill. If dairy herd desired to be healthy, the dairyman must carry out the appropriate operations of diseases prevention: foundation of disease resistant stock, feeding, sanitation and disinfection, and vaccination programs on common diseases when recommended by a reliable veterinarian (Payne and Wilson, 1999; May 1981).

Accordingly, this module explores the information on common problems and diseases of dairy herd health, and disease prevention and control, including chemical application, waste management systems and disposal activities.

1 IDENTIFYING APPEARANCE AND ROUTE OF DISEASES TRANSMISSION

Researchers have recommended that for efficient and effective herd health care and disease prevention program, it is crucial to identify the normal physical-appearancesand characteristic of

animals; signs of abnormalities, major causes and common routes. Accordingly, the normal animal behavior and appearances; general signs, major causative agents, and transmission route of common dairy cattle diseases will be discussed below.

1.1 Recognizing General Appearances of Healthy and Unhealthy Animal

To inspect and assess whether animals are normal, the external body parts of the animal are important. Fig. 14 reveals the different parts of an animal. Many diseases are the results of a complex interrelationship of the animal, environment and disease agents. The term 'web of causation' has been used to highlight the interaction of the components leading to a disease process. Disease determinants affect the frequency of disease occurrence. It is often possible to identify specific factors, physical factors or microorganisms which are essential for the occurrence of disease. The general name given to these is disease agents (McDowell, 1994)

1.1.1 General Appearance of Animals

Healthy animal looks physically active with a well-built body, proportionate size of organs, conforming to breed characteristics and with body size comparable to the animal age, including:

- Body condition: This will be good or fair depending on the physiological status. Bones and ribs are well covered with muscles and not too lean nor excessively fat.
- Skin: It is smooth and pliable with normally grown skin appendages such as hooves, horns and dewclaws. Skin is free of any injuries or wounds/lesions, and external parasites.
- Hair coat: This is uniformly distributed; smooth and shining, normal colored and well grow in accordance with species and breed characteristics.
- Eyes: It looks bright and shining without any bulging, discharge or discoloration. The Eyelids close the orbit completely and with adequately grown eyebrows.
- Nostrils and Muzzle: Nostrils open without any unusual discharge with respect to volume, color and consistency. The muzzle is moistened and shining; whereas in animals with fever or other general diseases muzzle will become dry.
- Natural orifices: There is no any abnormality of structure or discharges from any of the natural body orifices if the animal is normal.
- Activities: Animal should be alert and active while diseased animal will be droopy and inactive. Healthy animals will show intermittent wagging of tail and earlobes in order to get rid of insects and for heat regulation.
- Appetite: Interest to feeding is an indication of normal health state. Normal acts of prehension, mastication, regurgitation and rumination are important signs of health.
- Posture: Animal able to stand normally bearing weight equally on all the fore legs. There should not be any difficulty in lying-down or getting-up. Diseased-animals will recumbent depending upon the type and severity of disease.
- Excretions and secretions: Act of passing urine and physical features of these

excretions such as quantity, color, odor, consistency is observed. Similarly, secretions like milk, nasal discharge, vaginal mucous, tears and sweat also seek attention to differentiate normal and abnormal features.

- Temperament: Behavior of the animal is important to understand its mental state. Usually animals in heat and those taken to new places will excite or frighten affecting normal behavior. Besides that, other deviations in normal behavior and temperament are indicative of ill health involving pain, irritation or nervous disorders.
- Abnormal acts: Disorders affecting nervous and muscular systems cause various abnormal physical acts (e. g. intermittent lying and getting up, circling). Continued occurrence of such acts often leads to developments of vices in certain animals.
- Production and reproduction: Maintaining normal milk production interns of quality and quantity is an important state of health since optimum production occurs when all the physical, physiological and environmental factors are congenial. Reproduction being an accessory physiological activity, occurrence of normal reproductive activity forms a comprehensive sign of health especially in the case of females.

1. 1. 2 General Signs of Abnormal Appearances

The general signs and guidance to identify whether the disease is present include:

- Skin lesions, Loss of appetite, emaciation or loss of body conditions/production
- Digestion disturbances and diarrhea
- Respiratory signs (sneezing, coughing, frothiest nasal discharges, difficult breathing, grunting)
- Depression, and fever (rough hair coat, drying of muzzle area), discoloration of the visible mucous-membrane
- Nervous signs (unconsciousness, encircling move), isolation from the herd
- Abnormal-gait, lameness, forelimbs abduction, staggering
- Edematous dewlap, swellings of the pre-femoral and pre-scapular lymph-nodes
- Abnormal discharges through natural body orifices
- Increased/decreased in TRP (temperature, respiration and pulse)
- Systemic/local swelling (swelling of the udder and teats), and finally deaths (Fig. 54 and Fig. 55)

1. 2 Determining Factors for Disease Occurrence

Factors that determine disease occurrence in the population are host, agent and environment. A balance between the agent and the host is maintained by environmental factors. If environmental factors are favoring the host at the expense of the agent, the balance tilts toward the host, thus the host outcome the infectious agent, and vice versa. Thereby, the disease spreads within the population (Fig. 9. 1).

1. 2. 1 Host

Resistance to particular infectious agents appears to be highly developed in some species

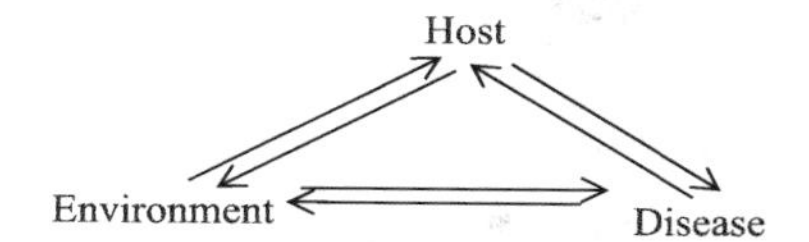

Fig. 9. 1 The host/animals, environments, and disease agents' interaction

of livestock: anthrax can affect a wide variety of mammals, but birds are highly resistant; horses remain uninfected while FMD spreads through groups of cattle, sheep and pigs. It is possible, however, to break down species resistance by using large doses of the disease agent or by challenging the host through an unusual route, e. g. by injection of the disease agent into body cavities.

In any disease outbreak, there are always a number of animals which remain unaffected. They may be not fully exposed to the disease agent or were resistant to it. It must be recognized that considerable genetic variation exists in our livestock so that individuals will vary in their reaction to a disease agent. The genotype of an animal may cause it to be susceptible to the effects of an agent that precipitates a clinical disease.

1. 2. 2 Environment

Environmental factors detrimental to the occurrence of disease can be classified into two like 1) physical factors such as soil, climate (temperature & humidity) and shelter; 2) biological factors like vectors, man, predators, free-living animals and plants. Physical determinants are important predisposing-factors in that those rapid changes in temperature and humidity may precipitate respiratory disease, and rising water table levels may bring bacterial spores to the soil surface.

Physical (a biotic) factors

Physical factors in the environment may act as single agents, e. g. lightning strike, or a fall. The significance of these varies widely with nature of the production system and can be of some importance in extensive systems.

Biological factors

- Plants: A wide variety of pasture and forage species can become toxic to livestock under specific circumstances. The concentration of toxic substances in plants may vary through the year or with the variety. A well-known example in the tropics is the variation in the concentration of the toxin 'mimosine' in different varieties of the valuable shrub legume. Some poisonous plants are ingested accidentally by grazing stock, although animals familiar with such species usually avoid them; only young or newly introduced stock is normally affected. Plant poisoning may occur when feed is short; the poisonous plants are dried in hay or when animals have access to the poisonous roots of the plants.

- Animals: Predation occurs when one species is powerful enough to overwhelm and kill the other. It can be an important source of loss in extensive pastoral systems, notably through the action of large carnivores, but it is also of some importance in more intensive systems with small stock.

1.2.3 Disease Agents

This refers to anything living or an inanimate object (fomite) whose presence is necessary for the occurrence of disease is termed a disease agent. More frequently, agents act together with others as with a bacterial infection invading an animal's lung tissue that has been previously damaged by the ammonia present in a badly ventilated house or by a viral infection. Many diseases that prove difficult to control are multi-factorial, involving a number of agents acting together or in concert. Control of disease is not possible by attacking the precipitating agent but eliminating or reducing the effects of predisposing factors (see the section of disease prevention and control).

Microorganisms are those that need a microscope to be observed in naked eyes can produce diseases to livestock collectively called pathogenic agents, which are discussed as follows:

- Viruses: These are of the smallest minute particles. They multiply inside living cells and are generally readily inactivated by the environment, but they are usually not affected by antibiotics or chemotherapeutic agents.
- Bacteria: Bacteria are microscopic single-celled organisms much larger than viruses, and can multiply outside cells, although some that cause major diseases multiply within cells. They may develop resistant spores which remain a source of disease for many years. Treatment with antibiotics and chemotherapeutic agents may be effective.
- Protozoa: These are single-celled blood parasites of major importance in the tropics. Tick-transmitted haemoparasites of importance are *Theileria parva* causes East Coast fever and *Babesia bigemina* (red-water). Trypanosome causes Trypanosomosis, transmitted by tsetse flies, and precludes most forms of livestock husbandry from many tropical regions, particularly in Africa.
- Fungi: These vary greatly in size, the largest being visible to the naked eye. They are often resistant to environmental influences and are sensitive to a narrow range of therapeutics.
- Parasites: Parasitism occurs when one species, usually smaller, lives at the expense of another. The presence of parasites in small-number may have a negligible effect on the host, but if the balance between the host defenses and parasite invasiveness is disturbed disease may result. Parasites that can affect the animal are categorized as endo-and ectoparasites.

1.3 Common Routes and Methods of Diseases Transmission

1.3.1 Methods of Disease Transmission

There are three main ways by which disease agents can be transmitted from the infected to the susceptible host. These are: 1) direct contact between infected and susceptible host, 2) It may be conveyed between these individuals by means of inanimate objects (indirect contact), and 3) via other animals as carriers, reservoir host and/or vectors. Disease transmission between infected and susceptible host is by contaminated vehicles, personnel (laborer, milker, and attendant), reservoir/carrier animals, birds and wind. Transmission through fomites; like contaminated milking utensils, feeding equipment, contaminated instruments used for animals breeding, contaminated feeding materials and equipment, contaminated instruments used for vaccination and drug administration, etc. (Blood and Radostit, 1999).

1.3.2 Routes for Entry of Pathogenic Agents

Infectious agents have access to enter the body of susceptible host through several routes. Accordingly, the knowledge of this is important to devise the appropriate methods to prevent and control the infections in animals.

Disease entry into the susceptible host involves:

- Oral-route: Ingestion of contaminated feed, water or licking any objects contaminated
- Respiratory-routes: Inhalation of infectious droplets through air; infective aerosols or dust particles
- Cutaneous-route: Transmission of infectious agents through direct skin contact with the surfaces infected/contaminated by infectious agents
- Urogenital route: Venereal disease
- Intra-mammary route: Mastitis
- Dirty-equipment: These can introduce contamination into sealed drug containers, spread diseases between animals, and result in abscess at the points of injection (Bartlett et al., 1990)

2 COMMON DISEASES OF DAIRY CATTLE

There are several diseases that can affect dairy cattle, generally categorized as infectious (caused by pathogenic agents); parasitic; and non-infectious (metabolic) diseases. But only those that can cause the greatest economic losses in the dairy industry were discussed in this chapter.

2.1 Bacterial Diseases

2.1.1 Mastitis

Mastitis is an inflammation of one or more quarter of the udder. It is the most common and costly diseases of dairy animals that can be present in either a clinical or subclinical form. Clinical mastitis is manifested by udder inflammation, abnormal milk, increased milk somatic-cell-counts (SCC), and usually the presence of a pathogen in the udder. The cow may exhibit systemic-effects, such as elevated body temperature and reduced feed-intake, but in most cases, symptoms of the disease are confined to the mammary-gland. Subclinical mastitis is defined as the presence of a pathogen in the udder and elevated SCC without visual-signs of the disease. Depending on the type and severity of clinical mastitis, intra-mammary antibiotic-therapy is administered under the guidance of a valid veterinarian-client (Boeckman and Carlson, 2003). Although antibiotic treatment is not generally used to control subclinical-mastitis, this form of mastitis is more insidious and can result in decreased milk production over the lactation period (Bartlett et al., 1990).

Mastitis is caused by any of a variety of microbes (mostly bacteria) that gain access to the interior of the mammary gland through the teat-opening. Bacteria may also enter through injury to the teat. These microbes live on the cow and its udder, and in the environment—the floor, feces, soil, feedstuffs, water, plants, and milking equipment and utensils.

Mastitis may be attributed to poor management, improper milking procedures, faulty milking equipment, inadequate housing, but also the climate, season, housing type, nutrition and stress; and contaminated treatment equipment and materials. These all interact with the genetic and physiological factors such lactation stage, milk yield, milk-flow rate and pregnancy (Fig. 56 and Fig. 9. 2).

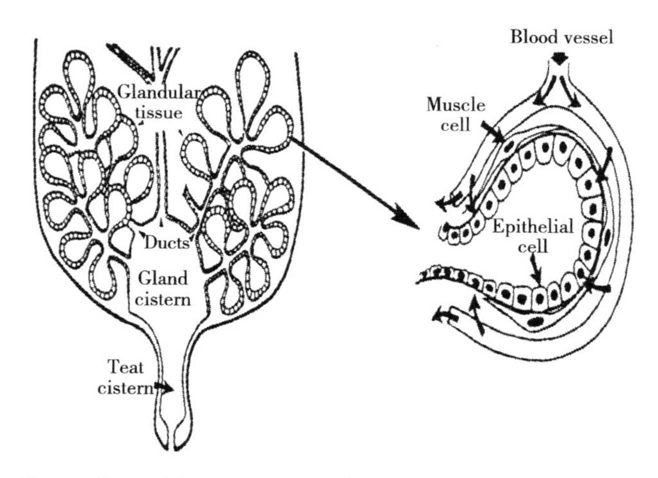

Fig. 9. 2 Entry of mastitis-organisms to the quarter (s) of the mammary system

Mastitis results serious economic problems in the dairy industry that causes losses by:

- Lowering milk production from infected cows
- Increasing the cull rate in the herd
- The cost of treatment
- Loss of infected milk that must be thrown away
- Increased labor cost to treat infected cows
- Possible loss of permit to sell milk if infection becomes serious enough

(1) Dynamics of infection

To understand mastitis infection, you must realize how its level changes within the entire herd. New infections can be brought into the herd in 4 general ways:

- New infections during lactation
- New infections during the dry period
- Infected heifers entering to the herds
- Infected cow purchases

More than 100 types of microbes can cause mastitis; which are grouped into two main types:

- Environmental-bacteria: Commonly present in the cow's environment
- Contagious-bacteria: Spread from infected quarters to other quarters. Contagious pathogens may be introduced into a herd through the introduction of infected stock.

1) Environmental mastitis

The keys to control environmental mastitis are good udder hygiene, correct use of good milking machines or hand milking, pre-milking teat-disinfection, dipping teats after milking and dry cow therapy. The order of milking is heifers first, then uninfected cows, and infected cows last. Despite good preventative procedures, new infections will occur while cows may sometimes recover spontaneously. Infection with environmental bacteria can cause severe mastitis (Bartlett et al., 1990; Diggins and Burndy, 1988).

2) Contagious mastitis

The main mechanism of transmission of contagious mastitis is spread of pathogens from cow to cow at milking. These bacteria live on the teat skin or in the udder and are spread when infected milk contaminates the teat skin of clean quarters or other cows. This can be by milk on milkers' hands or teat cup liners, through splashes or aerosols of milk during stripping, and by cross flow of milk between teat cups. Spread of contagious mastitis can be minimized by good hygiene, keeping teat ends healthy, using milking equipment that is operating well, and disinfecting teat skin after milking.

(2) Clinical symptoms

Mastitis have acute and chronic forms: acute mastitis is manifested by hot, inflamed udder; swollen, hot, hard, and tender quarter; drop in milk production; abnormality in milk: lumpy, stringy, straw-colored, contains blood, yellow clots; cow goes off feed, shows depression, dull eyes, rough hair, chills, and death may result. Whereas, Chronic Mastitis manifested by abnormal milk-clots, flakes, watery, slight-swelling and hardening of udder that comes and goes; sudden decrease in milk production may not show any symptoms, often not treated; sometimes does not respond to treatment; more economic problem than the acute form. Both acute and chronic mastitis may cause permanent udder damage.

(3) Prevention and control of mastitis

1) Cleaning and sanitizing dairy equipment

Cleaning milk harvesting equipment is a separate operation to sanitizing. Both operations are needed to ensure minimal bacterial contamination. Cleaning removes residual milk and dirt from surfaces, while sanitizing removes bacteria from cleaned surfaces (Fig. 57).

The general guide to cleaning is to:

- Remove all loose dirt and debris, and rinse or wet the equipment with cold or warm (not hot) water.
- Hot-wash using a detergent solution to remove surface deposits.
- Rinse with cold water and drain.
- Apply sanitizer to contact surfaces and allow drying.
- Different types of detergents have different roles. Neutral detergents are the most convenient to use as they require no skin protection. Alkali detergents remove protein, fats and carbohydrates, whereas acid detergents are best at removing milk stones and hard-water scale. Good cleaning practices require regular use of both alkali and acid detergents, but they may be less effective when used in cooler water (Fig. 58 and Fig. 59).

2) Follow appropriate milking procedure

Many research findings revealed that appropriate milking procedure significantly important to prevent mastitis to be transmitted among susceptible milking animals. Practical milking involves:

- Calm and relaxed: Cows should be calm and relaxed during milking and this will occur only if the cows are handled gently, routines are consistent, milking equipment is functioning properly and fits the cows milked. Calm cows manure and kick the cups-off less frequently and have better milk-let-down.
- Fore-strip: Good milkers take care to avoid getting milk on their hands as this can spread bacteria from teat to teat. An effective technique is to squeeze the base of each-teat between the thumb and the first two fingers, then pull gently downwards. If no clots, flecks or other abnormalities appear in the first two squirts, move to the next teat. The risk of spreading mastitis is reduced by milkers wearing disposable rubber gloves.
- Clean and dry: Ideally, teat cups should only go onto clean and dry teats. Milking wet teats increases the risk of mastitis and milk contamination with bacteria. Dry the teats wet with a paper-towel first.
- Pre-dip: Pre-milking teat-disinfection is an effective way of decreasing mastitis due to environmental bacteria. Use only a product approved for pre-milking disinfection and applies according to the label directions. The teats should be clean-&-dry before applying sanitizers as organic material will neutralize their effectiveness and moisture will dilute the product. The sanitizer should remain in contact with the teat for 30 seconds, and then removed with paper towels or suitable woven cloths to avoid contaminating the milk. Each cloth must be used for only one cow per milking. Cloths should then be placed in a disinfectant solution, washed and dried before the next milking.

- Milk let-down: The optimum time to apply teat-cups is immediately after milk let-down, seen by the teats becoming plump with milk. Putting the cups on too fast may result in them crawling during the first minute of milking and constricting the base of the teat; this leads to prolonged incomplete milking. Milk letdown usually occurs 60 – 90 seconds after the cows' teats and udder are first touched by the milker or is stimulated by the sights and sounds of milking and the predictability of a calm, consistent milking routine (John, 2011).
- Application: Air leakage through the teat cups during milking should be minimized by first checking that the teat cups hang over the claw correctly. The cluster should be balanced, each teat cup lifted with a 'kink' in the short milk tube until the moment of attachment, and then the units adjusted for proper alignment.
- Detachment: At the end of milking, the vacuum to the cluster should be cut by kinking the milk line so that the teat cups release. Pulling the unit off the cow without cutting the vacuum can damage the teats
- Confirm-and-Teat-dip: Check the cow after milking to avoid under-and over-milking. Disinfect each teat after milking with a spray or dip. Bacteria in milk from infected quarters may contaminate the other teats skin during milking, e. g. after a liner has milked the infected quarter, bacteria may be transferred to the next 5 – 6 cows milked with that cup. After milking, these bacteria multiply on the teat-skin and extend into the teat-canal.

Teat-disinfection after milking reduces by 50% new-infections due to 'contagious' bacteria. Teat dips include an active ingredient to kill bacteria and an emollient to keep the teat skin healthy.

It is concluded that early detection and treatment of clinical mastitis reduces the risk of severe disease-transmission, and chronic infection progression. Gloves, liners and other equipment are to be cleaned with running water and disinfectant solution to remove infected milk. Rinsing with running water for about 30 seconds provides a physical wash followed by a sanitizing dip in disinfecting solution such as 1% Iodophor. Disinfectants take time to kill bacteria; any other unit or cow shouldn't be touched for at least 20 seconds. Drying hands on a paper towel after this will also help to reduce the bacteria that remain (Fig. 60 and Fig. 61).

(4) Treatment of clinical mastitis

Cows should be treated for mastitis when there is heat, swelling or pain in the udder, or there are changes in the milk (wateriness or clots) that persist for more than three squirts of milk. Particular attention should be paid to swollen quarters that do not milk out. Stripping foremilk involves squirting more than three streams of milk—preferably onto a black surface—to look for clots, watery or discolored milk. Quarters with a few small flecks only in the first three squirts may be left untreated and checked again next milking. Milk containing infection may be spread during this procedure, so avoid splashes or sprays of milk, and always use gloves.

It is easy to introduce bacteria into the teat with a treatment nozzle, if the teat end has not been disinfected. Operators can be injured by cows when administering intra-mammary treatments. It is important to take time and have helper; more than one person is often needed to do the job well, especially if cows are not used to having their teats handled.

The steps involved in intra-mammary infusion include:

- Restrain the cow.
- Milk the quarter out completely.
- Ensure that your hands and the teats are clean and dry.
- Put on disposable gloves.
- Completely disinfect the end of the teats to be treated. This is critical. Disinfect by vigorously scrubbing the teat opening with a cotton ball and alcohol (or teat wipes) for a minimum of 10 seconds.
- Check the cotton ball. If there is any dirty color, repeat the scrub using a clean cotton ball until there is no more dirt seen.
- If treating more than one teat, treat the nearest one first, then the more distant teats to reduce the risk of unintentionally contaminating an already disinfected teat.
- Remove the cap of the tube of antibiotic and, without touching its tip, gently insert the nozzle into the teat canal.
- It is not necessary to insert the nozzle to its full depth as this can dilate the teat canal excessively and predispose the cow to mastitis.
- Squeeze the contents of the tube into the teat. Massage it up the teat into the udder.
- Teat-dip treated quarters with freshly made-up teat dip immediately after treatment (Table 9. 1).

Table 9. 1 Keys for control of mastitis

Management task	Specific actions
Milking hygiene	Milk teats that are both clean and dry
Milking machines	Stable milking vacuum, no slipping or squawk of liners
	Shutting off vacuum before removing
Post-milking teat dipping	Immediately after removing teat cups, full teat immersion, not spraying
Drying off	All quarters of all cows after last milking
Treatment of clinical cases	Early detection and treatment; keep records of treatment
Culling	Cull chronic cases
Environment	Clean and dry; un-crowded and well ventilated
Herd replacement	Test new animals before adding to herd; check new animals regularly

2. 1. 2 Brucellosis

Brucellosis also "*Bang's disease*" is a contagious disease of cattle caused by bacteria

"*Brucella abortus*", which affect cattle at all age, and is highly zoonotic. It causes high economic losses due to abortions, calf deaths, decreased milk production, loss of national and international markets and human disease (DPO, 1994; Geering, 1987).

Symptoms

Infection is usually symptomless, except in pregnant females in which the bacteria invade the uterus and cause abortion or premature birth from the seventh month onwards, this is the most commonly observed symptoms of brucellosis. Subcutaneous swellings containing infected fluid are common in African cattle. On the other hand, the birth may be normal, but the cow may fail to clean or expel the afterbirth. Animals that infected often have higher than normal temperature at calving time; reduced milk production. High fever and sweating observed in man infected by this disease (Matthewman, 1993).

Because aborting and calving cows are the main sources of infection, brucellosis is a very serious disease in cattle operations in which adult females are reared intensively, e. g. modern dairy units. In extensive pastoral systems, the chances of contact with an aborting or infected calving cow are reduced and brucellosis is usually of little in significance (Blood et al., 2012).

Great care must be taken by persons coming into contact with animals infected to avoid contracting undulant-fever. The disease may be contracted from the consumption of non-pasteurized milk produced by infected animals. Great danger can exist in handling newborn calves or abnormal fetuses from infected herds.

Prevention and control

Brucellosis finds its way into a herd through any of the following:

- The purchase of infected or exposed animals
- Contact with neighbor's herd over a line fence
- Exposure at livestock shows where an infected animal may be on exhibit
- Livestock trucks that go from farm to farm handling animals and that have not been properly cleaned and disinfected
- Public livestock auctions where proper sanitation is not practiced
- Aborted calves that are dragged to the place by carnivorous animals

Brucellosis may be detected by having a veterinarian's blood test of the herd. If the disease is found to exist, there are two recommended plans for its eradication:

- Test all cows and heifers, removing any reactors from the herd and vaccinating all calves between the ages of 6 – 8 months that are not infected.
- Test at regular intervals at least once each year and eliminate reactors without calf hood vaccination. This plan may be used in very lightly infected herds or as a means to controlling the disease before many animals become infected.

Vaccination programs commonly used in the tropics are: *Brucellosis* (S19) -females should be inoculated during 3 – 8 months of age once in their lifetime.

2.1.3 Tuberculosis

Tuberculosis (TB) is an infectious, chronic, debilitating disease of both animals and man, manifested by formation of nodular granulomas known as *tubercles* in any body part, but lesions are most frequently observed in lymph nodes of: the head, thorax, lung, intestine, liver, spleen, and pleura (Geering and Forman, 1987).

Transmission

Inhalation of infected droplets expelled from the lungs of infected animals and ingestion of contaminated feed or water or milk; and is less commonly intrauterine. TB has two forms: pulmonary form and gastrointestinal/digestive form. Pulmonary form is transmitted by inhalation of droplets; and this type is common in intensive production system when large number of animals are aggregated together in the farm. Whereas gastrointestinal/digestive form transmitted by ingestion of milk, feed or water and it is common in extensive animal production system.

Generalized signs are like progressive emaciation (weight loss), anorexia, and fluctuating fever, and bronchopneumonia, chronic intermittent moist cough with sign or difficulty in breathing enlargement of superficial lymph nodes. Gastrointestinal *TB*-signs are diarrhea, emaciation, enlargement of retropharyngeal lymph node may be soon (Merck Veterinary Manual).

Treatment of tuberculosis

It is not economically feasible because it takes long period and infected animals can transmit the disease during the course of treatment and it may cause drug resistance on human being and animals.

Prevention and control

TB affects dairy cattle in all age groups and is caused by bacterium of the genus Mycobacterium; "*M. bovis*" is known to produce disease in cattle, have many opportunities to infect the animals and develop this illness. Because the tuberculin test is the primary method for diagnosis of TB in cattle, the test has been applied. Tuberculin test with purified protein derivate of bovine origin is one of the most common diagnostic tests (FAO, 1999). Individual cattle must be tested for *Tuberculosis*. The tuberculin test on the caudal fold is a practical method used world-wide. The brucellosis test, by individual serum, is also carried out in endemic areas of the disease (Blood and Radostits, 1990).

2.1.4 Anthrax

Anthrax is recorded as one of the most fatal and highly infectious diseases of all warm-blooded mammals. It is the most peracute disease of both animals and man, which is characterized by septicemia and sudden death with the exudation of tar-like blood from the natural body orifices, failure of blood to clot, absence of rigor mortis, and the presence of swelling of the spleen. Anthrax is one the most zoonotic diseases, and is pathognomonically characterized by fever, dullness or depression, increased heart-rate and respiration, inappetence (partial loss of appetite), anorexia (complete loss of appetite). Anthrax, however, is above all an infection of herbivores, especially cattle, horses and small ruminants, which ingest the pathogens as spores or vegetative forms from the soil.

Anthrax is caused by the bacteria '*Bacillus anthraces*', which is Gram-positive, highly aerobic and spore forming rod-shaped organism, surviving in apparently neutral, alkaline soil at a pH level not lower than 6.0 with enough calcium and relatively high nitrogen content. Drought, lack of fodder, overstocking, and intensive pasture rotation are causes and/or factors which favor an outbreak of anthrax. By grazing down to the last remaining parts of the plant during the dry season, which is covered with soil particles, the animals can be heavily infected with spores of *B. anthraces*. If the animals are crowded in an infected pen, the contaminated dust, which gets inhaled may lead to pulmonary infection (Seifert, 1996).

Clinical signs

Anthrax mostly shows the course of peracute or acute septicemia. After pulmonary infection with contaminated pen dust, the incubation period (IP) is not more than 12 hours. Mostly the animals die without presenting any symptoms of the disease; only secretion of tar-like blood from natural body openings.

After IP, the body temperature rises to 42℃. Animals may remain standing with staring eyes and a hanging head; they become depressed and lie-down; initial excitement is followed by listlessness and recumbence with severe disturbances of the circulation and respiration; occurrences of small, scattered hemorrhages in the visible mucous membrane; the mucosa becomes blue-violent in color; the tongue, the throat, and the abdomen as well as the udder become edematous. There is a loss of appetite rumination is suppressed and acute digestive disturbances may set in. some animals may develop diarrhea, which is usually hemorrhagic. In lactating cows, milk production decreases and small amount of milk still secreted is either blood stained or yellow. Pregnant animals usually abort. Dark-red tar-like blood is secreted from the body openings.

Control and prevention

Vaccinate annually before dry season or early rainy season, treat with antibiotics for sick animals, properly dispose of dead animals (deep buried burn), and control scavenger. Anthrax vaccine should be administered every six months for five years in endemic areas.

Notice: Opening of carcass of the animal died due to anthrax is not allowed in order to avoid contamination with its spore.

2.1.5 Blackleg

It is a disease caused by bacteria "*Clostridial chovie*" that results in sudden death in young rapidly growing cattle. An animal still found alive is depressed, not eating and is lame in one or more legs. There will be swelling of the muscles of the hindquarters, shoulders or neck and crackling due to gas under the skin may be felt. Black-leg is a toxemic bacterial disease characterized by emphysematous swelling of heavy muscles (thigh and shoulder and sometimes cardiac muscle) manifested by crepitation. It affects mostly young cattle (6 months to 2 years old). Animals with well-plane of nutrition are more susceptible to the ethnology. Fortunately, an effective vaccine is readily available and usually included in any routine vaccination program.

Transmission

This disease is transmitted through ingestion of contaminated feed or water, through trauma, parturition and surgery procedures.

Symptoms

- Peracute form: Sudden death without showing premonitory signs
- Acute form: Anorexia, fever, depression, dwelling of heavy muscles which are painful and edematous with crepitation sound through palpation, lameness. Muscle becomes dark through incision.

Prevention and Control

Treatment for sick animal with antibiotics e.g. fortified procaine-penicillin. Good practices for proper disposal of sick animal and contaminated meat, vaccination, and hygiene (Geering, 1987).

2.1.6 Anaplasmosis

Anaplasmosis is a Rickettsial infection of red-blood cells of ruminants with *Anaplasma species*. The disease is mainly seen in cattle and infrequently in buffaloes with *A. marginale* (Blood and Radostits, 1990). In endemic areas, indigenous cattle are commonly infected early in calf-hood and remain infected throughout life. These animals are healthy, although the *Anaplasma* infection may flare-up under stress and cause a mild form of the disease (FAO, 2010).

Transmission

Infection can only follow the inoculation of blood from an infected animal, and the main vectors are ticks; many species can transmit infection, but one-host *Boophilus ticks* are particularly important in cattle. Infection can also be transmitted by biting flies and by blood contaminated instruments and syringes, needles. Some wild ruminants can be infected and thus provide reservoirs of infection as well as infected domestic ruminants.

Symptoms

Cattle from non-endemic area are very susceptible. *Anaplasma marginale* can be detected inside red-blood cells. Over the next few days, the parasitemia increases, resulting in fever. The infection of RBC causes anemia; pale color of visible mucous membrane; but later yellowish in color (jaundice). Other signs are depression, loss of appetite, incoordination movement and labored breathing, abortion, and suffer a drop in milk yield.

Unless treated many animals die within a few days of the fever starting. Survivals may take a long time to recover, some regain full health, and all become carriers of the infection. If examined post-mortem, there is a generalized yellowish discoloration of organs and tissues, enlargements of spleen and distension of the gallbladder with thick brownish green bile, hence, gall sickness these changes can also occur in other diseases, e. g. babesiosis and heart-water.

2. 2 Viral Diseases

2. 2. 1 Foot and Mouth Disease

Foot and mouth disease (FMD) is an acute viral and extremely contagious disease of cloven footed animals manifested by vesicles and erosions in the muzzle, nares, mouth, feet, teats, udder, and papilla of the rumen. There are three viral strains causing FMD, mainly A, O, and C. Three additional strains: SAT-1, SAT-2, and SAT-3 have been isolated from Africa and further strains ASIA-1 from Asia and the Far East (Blood and Radostits, 1990; FAO, 2010).

Transmission

Direct and indirect contact with infected animals and their secretions including saliva, blood, urine, feces, milk and semen, aerosol droplet dispersion, infected animal by-products, swill containing scraps of meat or other animal tissue and fomites and vaccines (Blood and Radostits, 1990).

Symptoms

Before vesicle formation; incubation period 1 – 5 days or longer, morbidity is nearly 100%; mortality is variable depending on the virus strain, and its virulence and susceptibility of the host, 50% in younger animals and 5% in adults. Fever is up to 41. 7℃, dullness, lack of appetite, drastic drop in milk production, and uneasiness and muscle tremor. After vesicle formation there is a mouth smacking and lips quivering, extensive salivation anorexia with slow and careful chewing and swallowing, after some 3 days stomatitis. Similar lesions can be observed on feet, especially inter-digital space and lameness will follow.

Prevention and control

Avoid introducing of livestock products from FMD endemic. If an outbreak occurs in FMD free area, kill all sick and contact animals and disinfection of the area; however, in FMD endemic area, killing of all infected animals is not economical. Applying the principles and practices of bio-security and animal welfare program is very important to prevent and control FMD in the farm. There is no effective treatment for viral infection, but in order to avoid secondary complication provide broad-spectrum antibiotics. However, vaccination is the most important recommended practice to avoid spreading of FMD.

2.2.2 Bovine Viral Diarrhea

Bovine viral diarrhea (BVD) is an infectious viral disease of cattle manifested by an active erosive stomatitis, gastroenteritis, and diarrhea. The disease is transmitted by direct contact with clinically sick or carrier animals, and indirect contact with feedstuffs or fomites and contact with aborted fetuses. Transmission through aerosol droplet dispersion or by insect vectors may also be a possibility. Virus may persist in recovered and chronically ill cattle which are considered a potential source of infection (FAO, 1994).

Symptoms

Incubation is 1 – 3 days; fever, congestion and erosion in the mucus membrane of the oral cavity, depression and anorexia, cough, polypnea and salivation, dehydration, and debilitation foul-smelling, diarrhea, cessation of rumination, reduced milk supply, abortion in pregnant cows, laminitis, and congenital anomalies of the brain (cerebral ataxia) and arthritis in young calves. BVD is prevented and controlled by the same mechanisms applied to FMD, and vesicular stomatitis (Thomas, 2006; Geering, 1987).

2.2.3 Lumpy Skin Disease

It is a febrile disease of cattle characterized by fever and skin-nodules with various sizes. The disease is caused by the capripox virus, which can affect all cattle breeds, transmitted by biting insects/flies like *Stomoxys*, *Culicoides*, *Glossina* and *Musca* (DPO, 1994; Thomas, 2006). Therefore, the incidence of the disease is hip at high fly population. It is less commonly transmitted by contact and fomites. It causes economic loss by reducing milk production and reproductive performance of the dairy animals, permanent lesion on the hide and sometimes it causes death.

Symptoms

Peracute, acute and sub-acute form of the disease may occur generalized nodule formation on the flanks, limbs, muzzle, ventral abdomen, which are painful to touch, depression, anorexia, nodules-necrosis and ulceration, stiff-gait, lacrimation, salivation and nasal-discharge, enlargement of superficial lymph-nodes. Secondary bacterial-complication, myiasis and demodicosis are commonly associated problems (Thomas, 2006).

Treatment and control

Application of antibiotics to combat secondary bacterial complication; ivermectin for external parasite; isolation of sick animals; and appropriate vaccination program is very crucial. Lumpy-skin disease is prevented by applying appropriate program (Thomas, 2006; Geering, 1987).

2.3 Protozoan Diseases

2.3.1 Trypanosomosis

It is a protozoan disease transmitted by tsetse and other biting insect and other mechanical ways. It is manifested by intermittent fever, progressive emaciation and animals if untreated leads to heavy-mortality. It is caused by several-species of blood-parasite "trypanosome", such as *T. vivax* and *T. congolense* are dairy cattle area of concern.

Symptoms

Major signs are anaemia, enlargement of superficial lymph-nodes, weakness and progressive emaciation fever, loss of appetite, stunted growth, decrease fertility, and productivity, abortion (Blood and Radostits, 1990).

Prevention and Control

Vector control-traps and target, insecticide application, and control of tsetse-flies by using Sterile Insect Techniques (SIT); rearing of trypanotolerant breeds, such as West-African Shorthorn cattle, N'Dama cattle, Sanga cattle. Besides, application chemoprophylaxis-prevention; and treatment with anti-trypanosome drugs is most appropriate.

2.3.2 Babesiosis

Babesiosis, also said to be red-water; cattle tick fever, is tick-borne protozoan diseases of domestic livestock caused by Babesia species, which infect red-blood cells. There are various babesiosis in the tropics. Following the bite of an infected ticks, the parasites invade red-blood cells where they multiply, break out and invade more red blood cells in increasing numbers. The first clinical sign is fever for 1 to 3 weeks after the infective tick bite. The invasion and destruction of increasing numbers of red-blood cells causes anaemia, shock and the release of large quantities of red blood cell pigment (hemoglobin) into the circulation.

Affected animals are depressed, lose their appetite, and the eyes and gums are pale from anaemia and yellow (jaundiced) due to bile pigments in the circulation. Unchanged hemoglobin is excreted in the urine, which is colored red as a result. In severe cases of *B. bovis* infection in cattle, brain capillaries become blocked with infected red-blood cells, causing cerebral babesiosis which is manifested by nervous signs, incoordination, paralysis, coma, and death.

Transmission

Babesia is exclusively transmitted by ticks. These ticks acquire Babesia infection from infected animals and transmit the infection to other animals at a subsequent blood meal, usually in the next generation of the tick as the infection can pass through the eggs.

Treatment

The disease can be treated by administration of drugs like acriflavine and quinuronium sulphate. Of the currently available drugs, imidocarb and diminazene aceturate are widely used for the treatment. Some are potentially toxic, and manufacturer's recommendation must be followed carefully. In all animals, for the treatments to be effective, it should be administered as early as possible in the disease.

Prevention and control

The problem controlled by a strict tick control program aimed at the one host *Boophilus* tick. It is often prudent, however, to combine tick control with vaccination because it can be very difficult in practice to ensure that protected cattle are not exposed to infected ticks from indigenous cattle (Thomas, 2006; Geering, 1987).

2.4 Fungal Diseases

Mycotoxicosis

Aflatoxicosis is caused by aflatoxin, which are secondary fungal metabolites, is produced by the fungi *Aspergillus flavus* (AF) and *Aspergillus parasiticus* (AP). According to Esther and Blair (2012) it exists in multiple types; the primary ones being B_1, B_2, G_1, and G_2; among which B_1 is the most prevalent and most toxic. Aspergillus can be detected on commodities such as corn using fluorescence, which differentiates between B_1 and B_2 (blue) or G_1 and G_2 (green). Biomarkers are used to measure the level of exposure of man and animals. Although large amounts of the fungus may be visible to the naked eye, aflatoxins are often toxic at such low levels that contamination may escape visual detection.

Aflatoxins are toxic to humans and animals; their toxicity ranges from acute (death) to chronic effects (liver cancer). Exposure of young animals to aflatoxins can retard growth. Environmental conditions, such as moist soil and warm temperatures promote Aspergillus growth and aflatoxin production. Animals and humans are exposed to aflatoxins through consumption of contaminated products such as meat, dairy products (milk, cheese, and yogurt), or eggs. Major crops affected by aflatoxins include maize and groundnuts (peanuts). Agricultural practices can be modified to reduce and prevent aflatoxin contamination:

➢ Pre-harvest: Management of insects, weeds and nematodes, Planting date and irrigation, crop rotation or fertilization, Use of drought tolerant & locally adapted varieties.

- Harvest: Prevent compromise to the crop by harvesting when mature. For maize, harvest early to prevent completion of the *Aspergillus* life cycle, proper drying and storage in a dry place.

2.5 Parasitic Diseases

2.5.1 Internal Parasites

The types of worms commonly affecting cattle are roundworms, flukes and tapeworms. Roundworms are by far the most important and widespread; while flukes and tapes are being restricted to certain areas only.

Internal parasites of dairy cattle include:

- *Roundworms*: Species like barber's pole (*Haemonchus placei*); hair and black scour worms (*Trichostrongylus* spp.); gastrointestinal nematodes: Hook worm (*Bunostomum phlebotomum*), Nodule worm (*Oesophagostomum radiatum*), and small intestinal worms (*Cooperia* spp.).
- *Liver flukes*: Liver-flukes such as *Fasciola hepatica* and *Fasciola gigantic* both highland and lowland types respectively, of liver flukes cause severe losses throughout much of Ethiopia, where suitable condition for the growth of certain types of host snail are found. The flukes living in the liver bile ducts of livestock lay eggs and pass out in the feces and they heath in water and infect certain species of snails and after some period of development snail release the fluke into water and animals are infected through grazing infected pasture.
- *Tapeworms*: These worms are the predominantly problem in young cattle. Calves are exposed to infective larvae as soon as they begin grazing outside at a few weeks of age. Young animals are more susceptible to internal parasites with cattle over 18 – 24 months frequently immune unless stressed by poor nutrition or raising a calf.

Note: mature cattle from the inland shifted closer to the coast may not have this immunity (Fig. 62).

Clinical signs

The signs of worms include ill-thrift (not doing well), scouring, anaemia and a soft swelling under the jaw (bottle jaw) due to edema. These symptoms do not become obvious unless the infestation is severe, and/or the animals are low in condition due to inadequate food intake.

Prevention and control

A massive worm infection can express itself in the animal as a gradual loss in condition and production, or directly as illness. Although some kinds are relatively harmless until they become adult, thereby causing wounds and obstructions. For worm in disease infections, the golden rule really applies: "it is better to prevent than to treat" .

It is important to bear the points in mind may include:

- Drugs that treat roundworm infections rarely also treat fluke-infections, and vice versa. So, if both groups of worms cause problems in your area, then be careful to treat against both.
- Some worms go through their life cycle developing on the ground and then in cattle. Others, before becoming adults in cattle, need to spend part of it in a different host, which is usually a particular invertebrate (snail, ant, etc.). This host is called the intermediate host.
- Worms generally need humidity (usually warmth) for their development. In the tropics, this mostly corresponds with the rainy seasons (if there is a season distinction). This is why peaks of roundworm infections often start in the rainy season. Therefore, preventive treatment is recommended before and after the rainy season. If the rainy season is longer than three months, a treatment during the rainy season is often also recommended.
- Flukes which cause problems in cattle usually have a snail intermediate host, which are found on marshy land and in stagnant waters. Flukes often become a problem when rivers stop flowing and dry up into small ponds, thereby concentrating the number of infected snails that is usually towards the end of the dry season.

2.5.2 External Parasites

(1) Ticks and tick fever

Various genera and species of ticks, which can affect dairy cattle, are the most serious external parasite of cattle in Ethiopia and the tropics as well. The movement of livestock is strictly controlled to ensure this situation continues. A waybill, travel permit and dipping or spraying may be required (Fig. 9.3).

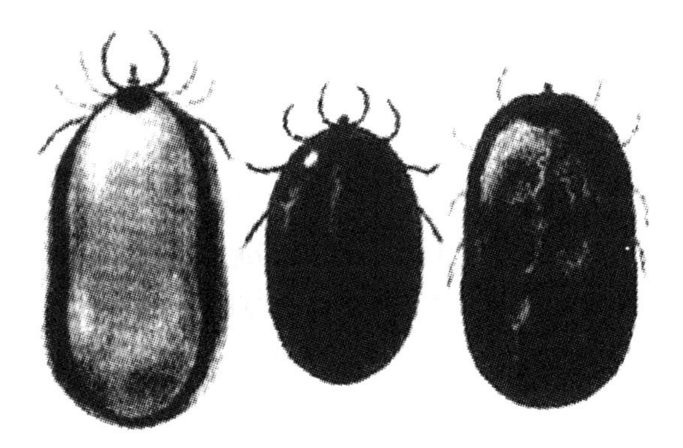

Fig. 9.3 Full-fed and blood engorged female ticks

1) Effects of ticks

Ticks are real nuisances of dairy cattle that sucks blood and infect dairy cattle with nasty diseases are called tick-borne diseases. Although not all kinds of ticks transmit these diseases, they still weaken the animal as a result of causing blood loss. They create wounds which allow bacteria to enter the skin (as in the case of the tick "*Amblyomma variegatum*" and the bacterial disease). As a result, hides lose their value. Ticks may also attack the udder, causing the loss of a teat, thus making a milk cow less productive. The major effects of ticks in livestock production include:

- Blood loss: Removes large quantity of blood from infected animal and leads to anaemia and infects dairy cattle with nasty disease.
- Tick worry: Presence of ticks on animal body make animal to be annoyed. They waste their time of grazing and feeding and become poor body condition growth and production.
- Abscess-formation: Deep bite by certain species of ticks and there may be secondary bacterial complication
- Tick paralysis: It is not common; it is caused by releasing toxin and affects the nerve tissue.
- Disease transmission: It is a very important effect of ticks. Diseases such as cowdriosis (heart-water), Anaplasmosis, Babesiosis, Theileriosis.

2) Prevention and control of ticks

Different tick control measures were discussed below. However, there is no standard answer to the question of the best way to control ticks. In most tropical countries they are difficult to control to an acceptable level without spending more this end than the problem costs. There are preventive and curative treatments that can be effective against most tick-borne diseases, if applied in time (though they remain expensive) (FAO, 1992).

Intensive control of ticks is often only economical when exotic cattle are kept in an attempt to improve the production and productivity: these animals tend to be more susceptible to ticks and disease, but their higher productivity permits the cost of control. Consequently, one should find out which combination of tick and tick-borne disease control measures will be more economical for his farm. It will depend strongly on the kinds of ticks in the region, on the farm situation, and on the quality of the veterinary service available to you (FAO, 1992; Blood, 1993).

To control the ticks, one hand sprays the cows and the stable twice a week. After reading the tick control options below, one may think that after a few months there will be so few ticks left in your backyard, that you will be able to reduce your control measures.

Rearing host resistance to ticks: Some cattle can acquire an ability to reduce the numbers and weight of ticks feeding on them. This is called host resistance to ticks. Some individuals and breeds are better at acquiring resistance than others. Select and slaughter those animals usually have heavy ticks-infestations and keep those which show good host resistance to ticks.

Environmental tick control: The use (strategic or not) of acaricide can contribute to low infestation of ticks on grazing land, but so can environmental tick control measures. These are: zero grazing, pasture rotation, rotation of crops, ploughing or re-seeding of pastures, and cutting or burning of grass. With zero-grazing you use pasture from places where the animals themselves do not go, so infestation of ticks and parasites should decrease. Of course, not all tick populations are equally affected by these different measures, and not all of these measures may be realistic in your farm situation. Environmental tick control is difficult to put into practice if you do not own land or if you have not organized tick control with the other farmers with whom you share the grazing land.

Use of acaricides: Acaricides are chemicals that kill ticks. Things to be considered:

- Toxicity: The acaricide used for tick control are generally also toxic to man and animals if they come into contact with or ingest them in sufficient quantities, hence, these must be used very carefully. This means that if you decide, for example, to sponge the acaricide onto the animal, always wear gloves, avoid sponging with your hand in an upright position, and wash off immediately any acaricide solution that comes onto your skin. The manufacturer's instructions as well as the precautions stated on the label should be followed to the letter. Products should be plainly labeled for animal use. Most acaricide are toxic to fish, hence, these should not be allowed to enter streams or ponds.
- Application: Acaricides can be applied to cattle in several ways, e. g. dip baths and spray races are often used in large herds, and hand-spraying and hand-dressing on small herds. Make sure ears and maxillae (where the legs join the body on the inside) are treated sufficiently. Ear-tags containing acaricide may be used in areas where the brown ear tick (*Rhipicephalus appendiculatus*) is a major problem.
- Timing: To control *Boophilus* ticks (which transmit *Babesiosis* and *Anaplasmosis*), treatment every 21 days reduces the number that can infest the pastures but permits enough tick-feeding to maintain pre-immunity against *Babesiosis* and resistance against ticks (FAO, 2012). To control the other types of ticks of importance to cattle, treatment should usually be once every seven days. Exotic cattle under disease threat of East Coast Fever should even be treated with acaricide twice a week.
- Strategic dipping/spraying: In areas with distinct seasons, there is a time that the number of reproducing adult ticks increases. If you know which season this is, start treating your animals a few weeks in advance and throughout the season using the treatment intervals described above under timing.

(2) Cattle lice

Both sucking and biting lice are found in cattle in the tropics. They may not be easy to see unless the hair coat is parted. They are dark red-brown to gray, from pinhead size to 4. 5 mm, have six legs and move through the hair when not actively feeding. In severe infestations they can be found all over the animal, but lighter infestations are confined to areas like the neck, dewlap, head, udder, tail base. Heaviest infestations occur during winter and in cattle low in condition or with lowered resistance due to worms. Lice usually disappear with the arrival of warmer weather and shedding of the winter coat.

Cattle in poor condition and on poor nutrition are more likely to build up a bigger burden of lice, and the lice are likely to remain on them longer. It seems that healthy cattle in good condition develop some form of natural resistance that controls a lice burden.

Symptoms

Lice irritate the skin when feeding. To seek relief, some cattle will rub against trees and posts, damaging their skin. Although rubbing is a good indicator of lice infestation, cattle will also rub for other reasons, e. g. when shedding their coats. Cattle severely infested with sucking lice may become anemic because of the amount of blood lost.

Prevention and control

Use care when using insecticides on the dairy, use only insecticides approved for dairy animals and facilities follow label directions carefully to avoid illegal residues in the milk.

3 METABOLIC DISEASES AND PROBLEMS RELATED TO CALVING

Metabolic diseases also called diseases of nutritional imbalances or non-infectious diseases. These are common in high milking-potentials dairy cattle. Over/under feeding of dairy cattle may lead to such cases of health problems: milk-fever, ketosis, acidosis, bloat, chock, and vitamins and mineral deficiency diseases (John, 2011; Russel and Constable, 1999).

3.1 Common Metabolic Diseases

According to research work, e. g. Thomas et al. (2006), there are several metabolic diseases-and-disorders that can attack dairy cattle. Some of economically important are discussed in the subheading set-out below:

3.1.1 Milk Fever

Milk fever (also called parturient paresis) is a disease seen commonly in dairy cattle and goats during peak lactation phases or advanced pregnancy. The disease arises due to deficiency of calcium to meet the increased demand of late pregnancy or high milk yield. Usually when the calcium availability through feed becomes deficient, animals' system mobilizes calcium from the bones with the help of certain hormones. When the availability even from this source becomes insufficient, there will develop symptoms of hypocalcemia.

Immediate causes of the condition include actual deficiency of calcium in the feed or presence of excess phosphorus, failure of calcium absorption from the gut and deficiency of parathyroid hormone. Incidence of milk fever increases with age, parity and production level. Cows of the dairy breeds or dairy cross have increased incidence. High blood levels of estrogen inhibit Ca mobilization; this may be a factor on pastures that are high in legumes.

Clinical signs

excitement in early stages, stiff legs, muscle tremors, staggering gait; the cow lies down on sternum, becomes drowsy and unable to rise; cow is down post-calving, and will become depressed with a slow heart rate, decreased rumen activity, low body temperature, and head is turned back onto the flank and cow has a dry muzzle and is constipated. Without treatment, most animals will become more depressed; the cow lies on her side, develops bloat and become comatose, and dies.

Treatment

Slow administration of intravenous commercial calcium solution (usually 300 – 500 mL) is given over 20 – 30 minutes. A second bottle may be given under the skin at the same time. Decrease the rate of milk removal (i. e. , give the calf supplemental feeding so it will not nurse as much from the cow). Cows that are down more than 12 hours require slinging from a hip hoist, 15 – 20 minutes twice daily, to reduce nerve and muscle injury. Animals that do not respond to treatment should be seen by a veterinarian (Cuaneo et al. , 2001).

Prevention

Decrease calcium intake during the last 2 months before calving by reducing legume forages. Cattle allowed to graze on a pasture with high legume content will be at greater risk. Use an intramuscular injection of vitamin A/D pre-calving. It may help to change legume roughage to grass hay 2 to 4 weeks before calving (James, 2001).

3. 1. 2 Ketosis and Fatty-liver

Ketosis is also called *Acetonaemia* in cattle, *pregnancy-toxemia* in sheep. It is mainly caused by a low-level of blood glucose in the body. The late-gestation dry-cow has a rapidly increasing requirement for energy because of fetal growth at a time when her appetite is falling. During the last week of gestation, dry matter intake (DMI) is typically 50 – 70% of earlier intake. Cows typically have a larger decline in DMI than heifers and may eat only 8 – 10 kg/day in the last few days before birth. Over-conditioned cows have a greater drop in DMI at calving and more health problems. Rapid increases in energy concentrations of the ration at this time may contribute to sub-clinical rumen acidosis and ketosis during early lactation (Zollinger and Hansen, 2003; James, 2001).

Ketosis and fatty liver are usually seen shortly after parturition before peak milk production.

Symptom

Cows with clinical ketosis typically display diminished appetite, frequently have hard-dry feces, decreased milk production, reduced body weight, and may display neurologic-signs.

Treatment

It is directed at promoting glucose availability and synthesis, and appetite. Treatment options include intravenous glucose (300 mL of 50% glucose intravenous solution), oral drenching with propylene glycol (oral 240 – 300 mL/day for 3 days) and IM administration of 5 – 20 mg of dexamethasone. Sick cows are more likely to eat hay in preference to mixed rations and providing sick cows with a range of feedstuffs may help to get them back on feed.

Risk factors for ketosis include: increased parity, a prolonged previous calving interval, excessive body condition at parturition, heat & cold stress, and inadequate bunk space, high levels of butyrate in forages, and inadequate housing and free-stalls (John, 2011; James, 2001).

Cows with ketosis have an increased risk of developing metritis, retained placenta, mastitis, and abomasal displacement. Conversely metritis, mastitis, and abomasal-displacement may lead to ketosis through the depression of appetite. Subclinical ketosis is more common than clinical ketosis. Milk fat percentage is increased in cows with subclinical and clinical ketosis, and milk protein percent may be lower in cows with sub-clinical ketosis. This may be the result of a reduced energy supply because milk protein percent is positively associated with net energy balance. First test milk components may be used as a herd-screening test for negative energy balance through the transition period. A large number of cows with a first test fat percent >5% reflects a negative energy balance during late gestation. Problems with the transition period are also reflected by poor production during early lactation (Thomas, 2006; James, 2001; Geering, 1987)

Because ketosis occurs in early lactation, recommendations for prevention focus on the nutritional management of the dry and transition cow. The dry period is divided into two feeding groups: far-off and close-up. The goals of the transition diet (close-up) fed to cows during the 21 – 28 days prior to calving are specifically designed to prevent sub-clinical ketosis by maximizing dry matter intake and providing adequate energy density.

According to Diggins and Bundy (1988), John (2011) and other stakeholders, the general principles relevant to preventing ketosis and feed additives purported to help prevent ketosis are:

- Avoid ketogenic feedstuffs (silages high in butyrate).
- Feed concentrates in the close-up period.
- Avoid over conditioning cows during late lactation and the early dry period.
- Monitor dry matter intake and DCAD in the close-up ration. Excessive use of anionic salts can depress palatability and dry matter intake.
- Niacin fed prior to calving at the rate of 6 – 12 g/day may be helpful in reducing blood levels of BHB.
- Propylene glycol requires repeated daily oral administration (240 – 300 mL).
- Inclusion of monensin (300 – 450 mg/day) in the lactating cow ration during the first 28 days postpartum.

3.1.3 Hypomagnesaemia (Grass Tetany)

Hypomagnesaemia or low blood levels of magnesium may be seen following a sudden change in diet, especially when cattle are put on rapidly-growing, young spring pastures. The amount of magnesium present in these grasses may be normal, suggesting that hypomagnesaemia does not always simply reflect dietary magnesium deficiency. High potassium diets reduce magnesium absorption from the gastrointestinal tract. Weather, pasture growth, and the concentration of magnesium, potassium, and sodium in the diet dictate the risk of disease. Disease is most commonly observed in older cows grazing cereal crops or lush grass-dominant pasture during or following inclement weather. Hypomagnesaemia may also be observed in 2 – 4 – month-old milk-fed calves that have restricted access to grain or forage.

Symptom

Clinical signs observed with hypomagnesaemia include: poor coordination, restlessness, muscular spasms, convulsions and death.

Treatment

Treatment involves the administration of magnesium or a combination of magnesium and calcium solutions intravenously or subcutaneously, but the response may be slower than that with hypocalcemia, and the margin of safety is less. Subcutaneous administration of magnesium may be less stressful than intravenous administration and avoids precipitating seizures.

Prevention and control

It is achieved through grazing management and magnesium supplementation. Pastures fertilized with high levels of potash and nitrogen represent a higher risk, and grasses and cereal crops are at greater risk than legumes. Cold, wet windy conditions increase risk.

3.1.4 Bloat

Cattle have four compartmented stomachs-the rumen, reticulum, omasum and abomasum. The largest is the rumen, a huge fermentation vat full of bacteria. The digestion process produces a lot of gas, normally expelled by belching. Sometimes the belching mechanism does not work-this is due to excessive gas production as a result of grazing lush succulent pastures and fodders.

The rumen is dilated with gas carbon-dioxide (CO_2) and methane (CH_4) characterized by the disturbance of the rumen by a foaming mass of contents, by which the animals cannot eliminate by belching and relieve themselves since only very small part of gas present is free in the upper part of the dorsal sac of the rumen, generally observed in cattle while grazing legume pastures like clover and tree-lucerne. But other types of pastures and feeding methods can cause bloat as well as any physical obstruction of the esophagus, as in choke and some downer cow cases.

It is clear that soluble carbohydrates of green succulent fodders are easily and rapidly fermented, which lead to a rapid accumulation of volatile fatty-acids (VFA), CO_2 and CH_4. This slow absorption of VFA and slow eructation of gas lead to bloat. The rumen distension impedes the blood flow and anorexia develops which is responsible for respiration failure.

Clinical signs

Obvious swelling in upper left flank, sticks up above the level of backline, increasing, distress and discomfort-may kick at flank, get up and down, bellow, rapid breathing-mouth may be open with tongue sticking out, and death occurs rapidly if animal gets down. Sometimes green froth comes out of the nose and mouth, and some animals may have little diarrhea. But some animals will always become sick and need treatment.

Treatment

Do not feed the animals for a few hours; make them move about. In mild cases is with anti-bloat preparations (vegetable oil, solid cooking fat, butter oil/ghee, or even milk, and mineral soap) by mouth. More severe cases require a vet. If there is no time to wait for the vet, trocar-and-canola or even a sharp, clean knife can be punctured into the rumen on the upper left flank where the swelling is greatest. After the gas and froth are released, an anti-bloat preparation is poured through the canola into the rumen to help break down the remaining froth and foam (Note: important to read the label). Vegetable oil (250 – 500 mL) or paraffin oil (100 – 200 mL) can be used in an emergency (John, 2011).

Prevention and control

Various anti-bloat preparations and strategies are available. Consult veterinarian for further practices of prevention. Feed animals with dry grass before you put them on to new wet green pasture; do not give water to animals just before you put them on to wet pasture; and do not put animals on to wet green pasture.

If disease is difficult or expensive to prevent, it may be better to wait for the disease to happen and then ferities, especially if it does not happen often or is not very serious. In addition, give animals clean food, keep animals in clean, dry places, remove feces from animal enclosures often, and avoid keeping animals too crowded together.

3.1.5 Choke

Acute choke is an obstruction in the esophagus (food pipe) and emergency in all species if it also obstructs the airway. In cattle the object is usually stuck high up in the esophagus near the throat, or it can be at the entrance to the windpipe, which is rapidly fatal. The objects causing choke include plastic bags, potatoes and other whole vegetables, or large fruit like mangoes, apples and oranges.

Symptom

➢ Anxiety, distress, head shaking, stretching the neck out, exaggerated mouthing and forced attempts to swallow

- Profuse salivation, maybe regurgitation of food and water
- Unusual lump may be felt along the ventral neck
- Bloat occurs rapidly because of animal's inability to belch
- History—but do not rely on information about the paddock being apparently clean. Cattle have little oral discrimination, meaning they tend to vacuum up the oddest things while grazing.

Treatment

Seek veterinary help, but if life threatening, you will have to do your best. Depending on the state of the animal, either take to yards and restrain in a head-bail or at least secure with a halter to nearest substantial object. Insert between the upper and lower molar teeth the left and right sides of a drink water gag if one is available.

- Grab tongue and pull out while carefully reaching down-back of the mouth as pull out the object.
- An assistant may be able to push the object from the outside up towards the mouth.

3.1.6 Ruminal Acidosis

This occurs when animals are fed an excessive amount of digestible carbohydrate. Common scenarios include animals breaking through fences and consuming grain, accidental overfeeding, and introduction to new commodities. Offending feeds include anything that contains an abundance of highly digestible carbohydrate such as grain, bread, brewers' grain, molasses, potatoes, and bakery by-products. The condition often reflects a failure to acclimatize animals to a ration or disruptions in feeding routine. The amount of feed required to cause illness depends on the nature of the feed, the animal's prior diet, nutritional status of the animal, and the nature of the rumen micro-flora.

Sub-clinical-acidosis (SCRA)

It is a less severe but economically important manifestation of the disease. SCRA may be a problem in dairy cattle when they transition from a relatively low energy ration during the dry period to a high energy ration after calving. Problems associated with SCRA include depressed and variable feed intake, depressed milk production, and an increased incidence of lameness secondary to laminitis. While engorgement with wheat has been shown to decrease rumen pH to 4 or less and to result in critical systemic disease, pH levels <5.5 are sufficient to predispose SCRA (Zollinger and Don Hansen, 2003; James, 2001). Reduced particle size also exacerbates acidosis through increasing the ruminal digestion of starch.

Rumen-micro-flora and papillae take 3 – 5 weeks to adapt to the change from a forage-based ration to a high-energy lactating cow ration. The net energy of a ration can be safely increased in 10% increments. Adaptation to the lactating cow ration may be achieved in part by feeding a transition ration to cows prior to calving. This ration has an energy level in between the dry cow and lactating cow rations and may also be formulated to help prevent milk fever.

If affected animals are examined within a few hours of engorgement, rumen-distension and occasional abdominal discomfort (kicking at the abdomen) may be the only abnormalities observed. In mild cases, cattle are anorexic, but fairly bright and alert and have soft feces. Rumen motility is depressed, but not absent (Godden, 2014).

Affected cattle do not ruminate for 2 – 3 days but will begin to eat by the 3th to 4th day without treatment. In outbreaks of severe engorgement, animals will be recumbent within 24 – 48 hours; some will be staggering and others standing quietly, separated from the herd. Affected cattle have depressed or no rumen motility, fluid-filled distended rumens, diarrhea, sunken eyes, depressed mentation, and may grind their teeth. When acidosis is secondary to grain overload the feces, usually contain incompletely digested grain. The diarrhea is profuse, and the feces have a foul odor (John, 2011).

Treatment

When animals are found engorging themselves, they should be removed from the feed, given access to good quality, palatable hay, be held off water for 12 – 24 hours, and be encouraged to exercise every hour for 12 – 24 hours. Animals that have consumed a toxic amount of grain will become depressed and anorexic in 6 – 8 hours; these animals should be treated individually. Moderately affected cases may be managed by administering 500 g of magnesium hydroxide or magnesium-oxide in 10 L of warm water. Veterinary attention will be required for more severely affected animals.

Prevention of ruminal-acidosis

It is achieved by preventing accidental access to grain and through controlled, gradual introduction with forages is one of the most effective ways of avoiding problems with ruminal-acidosis. If grain is to be fed separately to forages, problems may be avoided by small incremental increases in the amount of grain fed. When feeding grain separate to forage, it is important to consider the social behavior of cattle and to provide sufficient feed bunk space so that the dominant cattle are prevented from eating a disproportionate percentage of the grain fed. Sodium bentonite (2%) and limestone (1%) are buffers that may be added to grain for the first two weeks of feeding. Ionophores alter rumen fermentation and help prevent acidosis by reducing the relative production of lactic acid. Virginiamycin (Escalin) may be added to the ration to reduce the number of acid-producing bacteria in the rumen (Godden, 2014; John, 2011).

3.1.7 Left Displaced Abomasum

Left-sided displacement of the abomasum occurs most commonly in concentrate-fed, large-sized, high-producing adult dairy cattle during the first four weeks of lactation. A left-displaced abomasum may be caused by many factors; anything that causes a cow to go off feed or abruptly change dietary intake just before or soon after calving is a potential cause. Risk factors for abomasal displacement include pre-partum negative energy balance, hypocalcemia, a high body condition score, winter season, and low parity. The gas-filled abomasum becomes displaced under the rumen and upwards along the left abdominal wall. 8% of abomasum displacements occur within the first 3 weeks following parturition.

Clinical signs

Signs include depressed appetite, reduced milk production, ketosis, reduced rumen fill, with 'sprung' ribs, feces range from scant pasty to diarrhea. On palpation of the rumen, it is difficult to feel the doughy dorsal sac in the para-lumbar-fossa as it is displaced medially.

If possible, dry-matter intakes for each ration should be determined. Check for mouldy hay and silage. The neutral detergent fiber of forages should be checked to determine if levels are too low (insufficient roughage) or too high (limiting dry matter intake). The possibility of overcrowding should be investigated, and the as-fed ration should be examined to check for mixing errors and sorting. If anionic salts are fed to eliminate or reduce milk fever, urine pH should be checked. In the week before calving, the urine pH of Holsteins should be 6.2 - 6.8 and, of Jerseys, 5.8 - 6.3. Urine pH <5.8 suggests excessive anionic salts which are likely to depress dry matter intake. The ratio of milk protein to fat may be used as an indicator of negative energy balance during early lactation. Ratios less than 0.71 are associated with a 2 : 1 odds ratio that an abomasal-displacement will occur within the following 3 weeks. Negative energy balance is also reflected by loss in body condition.

Treatment

It requires correction of the abomasal displacement. This may be achieved by rolling the cow (roll from right lateral to dorsal and then left-lateral-recumbency). Approximately 40% of cows will re-displace following rolling. Alternatively, the abomasum can be repositioned surgically and sutured into the correct location. Supportive treatments include oral fluids supplemented with 100 g of lite salt (NaCl / KCl mix), 240 - 320 mL of propylene glycol orally daily for three days and an intravenous injection of 400 mL of 50% dextrose.

3.1.8 Metritis

Metritis results from uterine contamination with bacteria during parturition. In the majority of cattle that experience a normal parturition, this infection is resolved spontaneously. Lochia or normal postpartum uterine fluid, is normally mucoid-and-light-yellow to brown/red. Passage of lochia begins 3 days postpartum, and all of the fluid is normally expelled by 18 days postpartum. With uterine infections, the uterine fluid develops a foul-odor, becomes more abundant and watery.

Symptoms

Symptoms include fever, depressed mentation, anorexia, and reduced milk production.

Rectal examination reveals a large thin-walled uterus that contains a fetid watery uterine discharge. Therapeutic options for uterine infections include hormonal manipulation to promote estrus, anti-inflammatory and antimicrobial therapy. The objective of hormonal therapy in resolving postpartum metritis is to induce estrous-cycles, thereby increasing estrogen levels. Estrogen stimulates uterine tone aiding evacuation of abnormal uterine contents and increases production of mucus that contains host defense compounds (Godden, 2014).

Treatment

Prostaglandin-treatment of cows with peri-partum health-disorders, including RFMs, dystocia, or both, is likely to benefit their reproductive performance. However, blanket treatment of all postpartum cows with prostaglandin is not recommended. Antimicrobial therapy is indicated for the treatment of invasive metritis in which affected cows are systemically ill. Antimicrobials may be administered by intrauterine and or systemic routes. Intrauterine antimicrobial drug use achieves a high concentration of drug in the uterine lumen and on the lining of the uterus, but an inadequate concentration of drug in the deeper tissues. Intrauterine therapy alone is unlikely to achieve good therapeutic outcomes, and systemic treatment of metritis is more efficacious as it provides better drug distribution to all layers of the uterus and ovaries. Veterinary advice should be sought to guide antimicrobial drug use.

4 DISEASE PREVENTION AND CONTROL ACTIVITIES

This duty covers all farming practices for animals are healthy and productive. However, not all of the practices are applicable in all circumstances and may be superseded by national, international or market demands. According to Godden (2014), John (2011), FAO (2010) and stakeholders, the suggested practices and principles are set-out under the headings:

4.1 Establishing Disease Resistant Herd

4.1.1 Choose Breeds Well Suited to the Local Environment and Farming System

Selecting dairy animals that are suited to the local environment will greatly reduce the risks to productivity posed by animal health and welfare problems. Of particular relevance is the animals' ability to adapt to climatic extremes, feed quality, local parasites like ticks and their acquired resistance to endemic disease.

Determine herd size and stocking rate

It is crucial to determine herd size and stocking rate based on management skills, local conditions and the availability of land, infrastructure, feed, and other inputs. Larger herds and higher stocking rates generally require a higher level of organization, infrastructure and skill to manage. Disease burdens can be higher and individual animals requiring intervention can be more difficult to identify and treat.

4. 1. 2 Vaccination Program

Vaccination is the process of administration of vaccine into the body to stimulate immunity. Vaccine is antigen processed to be administered into the body of animals to stimulate immunity without causing disease. Antigen is any foreign organisms, chemicals, or toxin that stimulates the production of antibodies. Vaccination of all animals, as required by local animal health authorities, is a useful tool to limit the impact of disease (FAO, 1992 and 2010).

The protective spectrum of the colostrum can be enhanced by vaccinating the cow against the diseases that may threaten the newborn calf. The antibodies manufactured in response to the vaccine given at the proper time (read directions) appear in the colostrum. Target disease agents are, for example, *Escherichia coli* (*E. coli*), *Clostridium perfringens*, rotavirus, corona virus, infectious bovine rhinotracheitis virus (IBR), bovine virus diarrhea virus (BVD), and others. The effectiveness of some of these vaccines is sometimes questionable, with apparently great results on one farm and poor results on another. Some of the apparent failures of vaccines are due to not following directions or vaccinating cows that are not in good enough condition to mount a good response to the vaccine. Also, the vaccine organisms may differ slightly from the ones carried in the herd, and therefore protection by vaccination may not be optimal in such a case (Randall, 2009; Paul et al., 2001).

4. 2 Preventing Disease Entry

4. 2. 1 Keep a Closed Herd

This means no new animals enter the herd and previously resident animals do not re-enter after they have left the herd. This is difficult to achieve in practice, so strict control of any animal introductions is essential. Increased risk of disease may also occur when animals share grazing or other facilities (Merck Veterinary Manual, 2014).

Screening all dairy herds for diseases before entry to the farm is significant to their area of origin and new location. All animals must have: identification (a birth to death) system to enable trace back to their source; and some form of "Vendor Declaration/Certification" that details the health status of animals and any appropriate tests, treatments, vaccinations or other procedures that have been or are being carried out. Reject sick animals, and it is good to practice deworming all introduced animals for internal parasites on arrival.

➢ Ensure animal transport on and off the farm does not introduce disease: Explore information whether the animal is healthy or diseased. Sick or infirm animals shouldn't be transported alive. The disposal of diseased and dead animals is done in a way as to minimize the risk of disease spread. E. g. transport vehicles don't move dead or diseased animals from one farm to another, without taking appropriate actions to minimize the risk of spreading disease.

- Monitor risks from adjoining-land and have secure boundaries: Be aware of local (endemic) diseases and/or exotic diseases which have the potential to affect the health of the herd or flock, especially from neighboring farms. Contain animals appropriately to ensure there is no risk of disease spread between farms and within farms.
- Where possible, limit the access of people and wildlife to the farm: People (and vehicles) visiting may spread disease between the farms. Keep tanker/milk-pickup access and public tracks, clear-of fecal contamination. Restrict access to and put in-place appropriate processes to minimize spread of disease. Visitors to the farm should wear clean protective clothing and clean, disinfected footwear if entering areas that pose a high risk of transferring disease onto or from the farm. Keep records of all-visitors as appropriate.
- Perform vermin control activities: Ensure appropriate vermin controls in place where they are breeding. Their breeding sites should be eliminated, especially since these sites harbor disease-pathogens, such as manure-heaps, livestock disposal sites, milking shed, feed and water storages, and animal housing areas. Vermin vary geographically, but can include indigenous-animals, rodents, birds and insects.

4.2.2 Implementing Herd Health Program

Herd health program covers all aspects of animal husbandry and other farm activities relevant to the health and safety of animals and their products. This aims to keep animals healthy and productive, may include disease screening, vaccination, controlling measures and waste management system as required, including parasites. Prophylactic treatments may be required as protective measures when no viable alternative strategy exists. Hence, programs to be developed in consultation with skilled veterinarians:

- Using identification system: All dairy animals need to be identified easily by all people who come in contact with them. The methods used to be permanent, allowing individual animals to be uniquely identified from birth to death (see module 4).
- Regular observation and inspection of animals: This is used to detect abnormalities and/or infectious diseases. Some useful tools may include rectal-thermometers, observing behavior and body condition of animals and examining foremilk (John, 2011).
- Attend and keep sick animals isolated: Treat all diseased, injured and poor-health by proven methods after accurate diagnosis. Prompt treatment can limit the spread of infectious agents accordingly treat diseased animals to minimize the prevalence of infections and the source of pathogens and the spread of contagious disease. Provide separate facilities and milk sick animals last.
- Separate milk from sick animals and animals under treatment: Follow appropriate procedures to separate milk from sick animals and animals under treatment. This milk is not suitable for human consumption and if stored on farm should be clearly labelled as

such. Clean milking equipment and utensils thoroughly to avoid cross contamination.

- Manage animal diseases of public health importance: Follow local regulations and OIE recommendations to control zoonosis aim to keep diseases of public health significance at a level of animal populations that is not hazardous to people. Avoid direct transmission to people through animal management and hygiene practices. Ensure the safe disposal of animal waste and carcasses. Prevent the contamination of milk with feces and urine or other animal wastes. Do not use milk from sick animals for human consumption.
- Perform herd-health records keeping: It is important that staff, veterinarians and others involved with handling of dairy animals on the farm know what treatments have been given to which animals. Put in place an appropriate system to readily identify treated animals, and record appropriate details in accordance with local regulations and to manage withholding periods for milk and meat (Seifert, 1996; May, 1981).

4.3 Appropriate Use of Chemicals and Veterinary Medicines

4.3.1 Using Chemicals and Medicines as Directed

Chemicals and veterinary medicines for use in dairy production by the relevant authority may categorize as dairy: detergents, disinfectants, sanitizers antibiotics and herbicides, pesticides and fungicides (FAO, 2012). It is important to manage the use of these to prevent their adverse effect on OHS animals and users, and/or the safety of food products. Be aware of chemicals that can leave residues in milk. Dairy producers should:

- Use chemicals only for approved purposes, e.g. lactating animals are not treated with medicines not recommended for animals producing milk supplied for processing or otherwise used for human consumption.
- Read the label of directions as it will contain all the information about the legal and safe use of the chemicals.
- Follow the advice given on the label and any chemical data sheet or risk assessment; and observe the specified withholding periods.

4.3.2 Only Use Veterinary Medicines as Prescribed by Veterinarians

Veterinary medicines pose risks to humans, animals and food safety and are subject to special controls on their supply and use. Use only approved veterinary medicines, at the recommended dose according to the label directions, or as prescribed by a veterinarian. Relevant withholding periods must be observed (John, 2011).

All veterinary medicines and chemicals for treatment have a withholding period stated on the label. If no withholding time is stated or no labelling instructions exist, the product should not be used. Use of veterinary medicines contrary to the label recommendations is termed 'off-label use' and poses additional risks (Godden, 2014).

Store chemicals and veterinary medicines securely to ensure they are not used inappropriately or do not intentionally contaminate milk and feed. Check and observe product expiry dates. Dispose chemicals and their containers in a way that will not cause contamination to animals or the farm environment.

4.3.3 Antiseptics, Disinfectants, and Wound Dressings

Disinfectants are strong chemicals for cleaning contaminated things such as feed bowls, knives, and places where infected animals have been. It is dangerous to put strong disinfectants onto wounds or animal skin. Antiseptics are weaker chemicals for putting onto wounds; they kill microbes but should not be so strong that they kill an animal's flesh as well (Forse, 1999).

Do not mix different kinds of antiseptics or disinfectants together and when you mix them with water use clean water. Whenever, you use water to put on wounds, use clean water-boil it and let to cool.

> Warning: many disinfectants are poisonous. Do not let people or animals drink them. Be careful when you throw them away, they shouldn't get into water that people or animals drink.

According to several stakeholders, including Seifert (1996) and Forse (1999) there are different kinds of antiseptics disinfectants and wound dressings. Some of important chemicals and their application are discussed below:

1) Alcohol

Alcohol may be ethyl alcohol (ethanol or methyl alcohol). It is a good disinfectant. Mixed with at least the same amount of water it is an antiseptic but avoid putting undiluted alcohol on wounds. It is useful for cleaning animals' skin or your hands before doing an operation (it is not the best disinfectant for sterilizing knives and instruments because it doesn't kill microbe spores).

2) Ash

Some people use clean ash from a wood fire to put on wounds to stop bleeding, reduce infection and prevent fly damage. Clean ash from a fire is sterile and doesn't cause an infection. Many other antiseptics control infection better.

3) Antibiotic sprays and powders

Many antibiotics come in spray cans for putting on wounds. The antibiotic is often mixed with some color something to dry wounds up. These sprays work, but they are expensive.

Mix gentian-violet and tetracycline (10%) powder to make a good cheap substitute and is good putting on wounds.

4) Bleach (hypochlorite)

Hypochlorite is a common kind of bleach. Most kinds of bleach are good disinfectants that they kill many bacteria and viruses. Bleach is cheap and easy to find. You must mix with water to use it on wounds.

Procedures to prepare:

- Mix 20 mL/liter of water to make a careful disinfectant wash. Useful for washing the udder or very infected wounds.
- Mix at least 200 mL/liter of water to disinfect buildings and equipment.

5) Boracic acid

- Mix 20 g/Liter of water for washing wounds.
- Mix 10 g/Liter of water for washing infected eyes.

6) Caustic Soda (Sodium hydroxide)

This is a very strong disinfectant that kills infections, many microbe spores and viruses that other disinfectants do not kill, such as Foot and mouth viruses. It is useful for cleaning up contaminated things and places where infected animals have been. Be careful, this can burn your skin and damage metal.

Mix 10 – 20 g/Liter of water for disinfecting buildings. Useful for foot-rot and other infections of the feet.

7) Formalin

Sold as a 40 % solution of formaldehyde. Follow the direction of the manufacturer.

- Useful for treating of infection of the feet
- For a footbath use a 1 – 2% solution of formaldehyde

8) Hydrogen peroxide (H_2O_2)

It is usually available as 3% or 6 % used for wounds.

- Mix 300 mL, 3% solution/L of water.
- Mix 150 mL, 6% solution/L of water.
- Useful for deep wounds. H_2O_2 makes a froth that pushes dirt out of deep-wounds and abscesses. Sores in the mouth.

9) Insecticide dressings

Some people use dressing made from plants to kill flies and fly eggs. Some of these work, but they do not usually kill as well as modern insecticides.

10) Iodine

It is often available as tincture (weak solution in alcohol). This is a dark brown fluid. To make an iodine-tincture:

- Mix 20 g of iodine crystal, 25 g of potassium iodide and 25 mL of water/Liter of alcohol. Useful for wounds, putting on the navel of newborn animals to stop/prevent infection. Wash the navel thoroughly with the liquid.
- Mix tincture of iodine petroleum jelly (Vaseline) to make an oily cream to put on sores (Forse, 1999). It is useful for udder washing and teat dipping to control mastitis.

11) Savlon

It is a mixture of chlorhexidine and cetrimide. It comes as a wound dressing or as concentrated antiseptics to mix with water.

- Mix about 5 mL/L of water for putting on wounds.
- Mix about 30 mL/L of clean water for cleaning instruments or follow the directions to dilute the concentrate. Do not keep the instruments just Savlon for long because they will rust.

12) Potassium permanganate

It comes as a dark-nearly black powder and it turns dark-blue/red when you mix with water. It is a useful antiseptic or disinfectant.

- Mix 1 g/Liter of water. The liquid should only have a pale color. Useful for putting on wounds or washing the mouth.
- Mix 10 g/Liter of water. It is useful for disinfecting things instrument and clean them.

13) Washing soda (Sodium carbonate)

It is useful for disinfecting buildings and contaminated things. It kills viruses causing foot and mouth disease.

14) Neem trees (*Azadirachta indica*)

Many people soak the leaves and other parts of neem tree in water (some people boil the leaves first and let the liquid cool).

General directions to follow for preparation and application of these chemicals:

- Follow Manufacturer's specification carefully.
- Dilute chemicals in water at correct concentration level.
- Do not mix disinfectants which make these ineffective.
- Disinfectants become ineffective if they contain a build-up of dirt and organic matter and have to be changed regularly.

4.4 Waste Management and Environment Control Activities

4.4.1 Waste Management Plan

To minimize the risk of causing pollution, it is important to develop a smart waste management plan to identify when, where and at what rate to spread manures, slurry, and other organic-wastes (FAO & IDF, 2011). Waste management plan have considerations:

- Compliance with local regulations or contractual obligations.
- Avoid possible pollution of watercourses, ponds, lakes, reservoirs, wells, boreholes, underground water from applying wastes to shallow soils and/or fissured rock.
- Avoid potential pollution of habitat such as woodlands, protected/recognized flora/fauna zones.
- Maintain adequate buffer zones near vulnerable or sensitive areas such as water sources, habitat areas and the like.
- Timing and level of application on sloppy ground, heavy/impermeable soils and areas subject to flooding and optimum application levels in areas that already have high soil fertility.

- Current weather and soil conditions at the time of application like frost, frozen ground, heavy rainfall and waterlogged soils, and national and regional environmental controls.

4.4.2 Implementing Waste Management System

Applying practices to reduce, reuse or recycle dairy farm wastes as appropriate to the local environment is components of good dairy farming practices to reduce water and energy consumption by properly maintaining equipment and infrastructure in many farms. Opportunities to recycle plastics, drums and other consumables should also be investigated (FAO, 2010).

Management of wastes storage and disposal

It is important to minimize environmental impacts in the dairy farm through:

- Properly constructing and handling site waste storage areas such as manure heaps, slurry stores and farm dumps.
- Giving due consideration to the local amenity with regard to sight, smell, and the risk to the environment from pollution and vermin.
- Regularly inspecting permanent slurry stores and manure heaps for signs of leaks and impending structural failure to minimize the risk of runoff polluting the environment.
- Properly disposing other wastes such as waste milk, dead livestock, plastic silage wrap, farm chemicals and fertilizers to prevent pollution of the environment and any potential disease occurrence.
- Eliminate potential breeding sites of flies and other disease carrying vermin.

Avoid adverse impact on the environment

Dairy farmers should adopt systems that avoid the potential for contamination of the local environment. To this end:

- Locate storage facilities for oil, silage liquor, soiled water and other polluting substances in a safe place and take precautions to ensure accidents do not result in the pollution of local water supplies.
- Avoid disposing of agricultural or veterinary chemicals where there is potential for them to enter the local environment.

Managing dairy run-off on farm

Use agricultural and veterinary chemicals and fertilizers appropriately to avoid contamination of the local environment:

- Protect the environment by only using approved agricultural and veterinary chemicals and medicines according to the directions on the label.
- Store farm chemicals, preferably away from milk storage areas.
- Safely dispose chemicals and containers that are expired and defective.
- Consider biological and other non-chemical approaches to controlling farm pests,

such as eliminating pest breeding sites.

- Apply integrated pest-management practices where appropriate; and fertilizers in a manner that minimizes the risks of off-site nutrient impacts.
- Avoid using fertilizers that contain toxins, heavy metals or other contaminants. Ensure the safe disposal or reuse of empty fertilizer bags.

4.4.3 Disposing Waste and Carcasses

Before handling a carcass, consider the diseases that can be passed to humans (anthrax, brucellosis, rabies, ringworm and mange are the most common ones). If the animal died unexpectedly, a post-mortem will reveal the cause of death and guide the means of disposal. Post-mortems should be performed by a qualified veterinarian. If anthrax is suspected the carcass should be burned and no post-mortem should be carried out (FAO, 2012).

- Disposal by burning dead body (carcass): Dig two trenches having 2 m length, 40 cm width and 40 cm depth in a cross. The trenches will provide oxygen to the fire. Place two iron bars so they lie across one of the trenches. Place strong wooden posts across the bars. Place the carcass and a heap of fuel (wood and straw soaked in waste oil) on the wooden posts. Light the fire and burn the carcass, and
- Disposal by burying: Dig a hole (2 m×1.5 m×2 m) long, width and depth. Put the carcass in the hole and cover with soil and logs or large stones to stop wild animals or dogs digging it up again.

4.5 Animal Welfare Program

Animal welfare means the application of sensible and sensitive animal husbandry practices in the farm. It is primarily concerned with the wellbeing of the animal to perceive high animal welfare standards as an indicator that food is safe, healthy and of high quality (FAO and IDF, 2011). Proper handling and care of the animal according to animal welfare program described below.

4.5.1 Ensuring Animals Free from Thirst, Hunger and Malnutrition

Dairy animals must have access to sufficient feed and water depending on their requirement. Hence, it is necessary to perform the activities:

- Adjusting stocking-rates: Due consideration is given to the animals' number, physiological-needs and feeds nutrient quality when determining stocking rates, and animals have access to sufficient water, feed and fodder supply each day.
- Protect animals from toxic plants and harmful substances, and provide clean water
- Regularly clean watering troughs and inspect these to be fully functional
- All reasonable steps are taken to minimize the risks of the water supply, freezing or overheating, as appropriate
- keep stock adequate water-supplies from effluent and chemical treatments Runoff pasture and forage-crops

➢ Make sure that grazing plant material is free from toxic plants

4.5.2 Ensuring Animals Free from Discomfort

Design and construct buildings and handling facilities as to protect hazards: Consideration should be given to the free flow of animals when designing and building animal housing and/or milking sheds (Godden, 2014). It is recommended to avoid dead ends, steep and slippery pathways and ensure dairy buildings are safely wired and properly earthed.

Provide adequate ventilation for animals housed: For sufficient fresh air, removing humidity, heat dissipation, and preventing build-up of gases: CO_2, NH_3 or slurry gases

Control adverse conditions and the consequences thereof: Protect animals from cold or heat stress, unseasonal change; consider shade or alternative means of cooling (misters and sprays).

Provide housing, windbreaks, and additional feed in cold conditions.

Plans to protect animals against emergencies: Back-up power-supplies and natural disasters (fire, drought, snow and flood). Provide high ground in case of flood, adequate firebreaks and have evacuation provisions.

Ensure suitable flooring and safe footing in the housing and animal traffic-areas: Excessively rough concrete surfaces with sharp protrusions and stones can cause excessive wear or penetrations to the sole of the hoof, results lameness which leads to secondary hoof infections (foot-rote).

Avoid unnecessary-pain: Especially for procedures cause suffering e. g. dehorning, castration.

4.5.3 Keeping Good Hygiene of the Farm Area

Good hygiene will prevent transmission of diseases, but waste management is a significant logistical problem where cattle are managed intensively. According to John (2011) udder health is largely influenced by the level of hygiene practiced before, during and after milking. Milking wet, dirty udders increases the risk of mastitis, wiping cows' teats with a dirty cloth transmits pathogens between cows, and failing to sanitize teats after milking provides opportunity for disease to spread. Young calves are particularly susceptible to pathogens that cause diarrhea, but the risk of disease is reduced by feeding adequate colostrum, good milk handling practices, and reducing exposure to manure from other animals or strictly following up of hygienic procedures.

4.6 Management of Wound and Fractures

4.6.1 Wound Management

A wound is a break in the skin/body of the animal, usually caused by a sharp object. Wounds are caused accidentally or by parasites and other animals (e. g. fights and bites). When left untreated, the exposed tissues may become infected. Therefore, treat the animal according to the following procedure:

➢ Keep the animals quiet and stop it from moving around.

- ➢ Stop any bleeding, and clip hair or wool away from the edges of the wound.
- ➢ Remove all foreign objects. Wash the wound thoroughly with plenty of clean water (water should be boiled, cooled and salt or a mild antiseptic-added).
- ➢ Dry the wound with a clean cloth.
- ➢ Put a wound dressing or antibiotic powder on the wound.
- ➢ Dressing a wound when there are a lot of flies, that repels or kills fly eggs and larvae.
- ➢ Encourage wounds to drain pus by pressure and incision if necessary.
- ➢ If the wound doesn't heal, becomes black and smells bad, & cut away the dead flesh
- ➢ Wash the wound with antiseptic and treat with antibiotic powder.

4.6.2 Management of Fractures and Joint-dislocation

Fractures (usually in the legs) result from falling into holes, falling over heavy farm implements or jumping over fences. For large, heavy animals face fractures where the bone breaks high up in the leg, it is better to slaughter the animal for meat. If treating is found important follow the procedure below:

- ➢ If the bone has come through the skin, clean the wound and give local anesthesia by injection.
- ➢ Arrange the leg so that the broken ends of the bone touch in their normal positions as far as possible.
- ➢ Tie a piece of wood (a splint) to the leg to keep the bones in position.
- ➢ Confine the animal to reduce movement during the healing period.

4.7 Equipment, Tools, and Materials

There are several types of materials, tools, and equipment required for the prevention, treatment and control of dairy cattle diseases in general. Some of these are described as follows:

Personal protective equipment: Boots (protect the foot against external calamities), sun-protection (sun-hat, sunscreen), overalls, gaunt/gloves, rubber apron, protective eye and ear wear (give protection of the eyes and ears); respirator or face-mask (give protection of the face).

Instruments and equipment: Knapsack sprayers-spraying chemicals and drugs, syringe and needle (administer vaccine, antibiotics), balling-gun (swallow bolus), drenching gun; stomach-tub, trocar-and-canola; hoof-trimmer; hoof-pick/hoof knife, dehorners, Burdizzo, emasculator, thermometer, surgical needles and blade, vaginal-speculum, iceboxes, forceps and scissors, surgical blade, microscopes, sensitive balances, graduated beakers cylinders, pipettes, test tubes, test-tubes rack, universal bottles and Petri dishes, autoclaves, hot-air/dry oven, centrifuges, refrigerators (Forse, 1999).

Facilities and equipment for waste disposal activities: Waste disposal cart, waste disposal and incineration pit, effluent sewage (slurry), crushes restraining and casting ropes, nose-ring, etc.

Materials, drugs and chemicals for health care:

- Materials: Clean potable water supply; soapy detergents; drapes and cotton-bandage, gaunt; towels and drapes; first aid-kits; and case-book, etc.
- Drugs and chemicals: vaccines (live and killed vaccines), anthelmintic boluses, reagents and chemicals, antibiotics acaricide and pesticides, fluid-therapy and vitamins, vegetable-oils, and fluid-therapy, antiseptics and disinfectants.

>>> SELF-CHECK QUESTIONS

Part 1. Multiple Choices.

1. The smallest sized particles that can produce disease in animals is termed as
   A. Bacteria　　B. Viruses
   C. Protozoa　　D. All
2. Which one of the following is a single-celled blood parasite of major importance in the dairy industry of the tropics?
   A. Bacteria　　B. Viruses
   C. Protozoa　　D. Fungi
3. Inanimate objects that transmit disease agents from infected to susceptible animals as
   A. Living organisms　　B. Animals
   C. Carrier agents　　D. Fomites
4. Among the following ________ is not used for transmission of diseases from infected to susceptible host animals.
   A. Oral route　　B. Respiratory routes
   C. Subcutaneous route　　D. Cardiovascular
5. Which one of the following is different from another?
   A. Anthrax　　B. Brucellosis
   C. Ketosis　　D. FMD
6. Among the tools below ________ is not useful to detect and diagnose infectious disease in cows.
   A. Rectal thermometers　　B. Observing body condition
   C. Examination of foremilk　　D. Feeding of cows
7. Herd health management program focused on all of the following except
   A. Disease screening　　B. Vaccination
   C. Deworming　　D. Waste management system
8. The best techniques to diagnose *Tuberculosis* in a cow is
   A. ELIZA test
   B. Tuberculin test

C. Sterile insect technique
D. California-mastitis-test

9. Method of disease prevention through immunization of an animal is known as
A. Vaccination B. Examination
C. Inoculation D. Characterization

10. Which one of the following diseases of cattle is caused by bacteria?
A. Milk-fever B. Dystocia
C. Black-leg D. Chock

11. Herd health management program includes all activities below except
A. Routine husbandry activities
B. Disease prevention measures
C. Milk harvesting & handling activates
D. None of the above

12. Which of the following may be a factor for *dystocia* to occur in cows?
A. Higher calf birth weight
B. Sex of calf
C. Size and breed of the dam
D. All are correct

13. Among below one is not important to prevent disease entry to the farm.
A. Ensure transport of animal doesn't introduce disease
B. Where possible limits access of people and wildlife into the farm
C. Vermin control program
D. None

14. Waste materials in dairy farm include all below except
A. Animal manures
B. Slurry and other organic wastes
C. Dead carcass
D. Milk products

15. An inflammation of the inner-wall of the uterus in the cows is referred as
A. Pyometra B. Endometritis
C. Metritis D. Arthritis

16. Control of Trypanosomosis may not include
A. Tsetse-flies, traps-&-target
B. Insecticide application
C. Appropriate vaccination program
D. Rearing trypanotolerant cattle

17. Brucellosis is one the most economically important diseases of dairy industry that finds its way into a herd through any of the following but
A. Purchase infected or exposed animals

B. Contact with neighbor's herd over a line fence

C. Public livestock auctions where sanitation practiced

D. Aborted calves dragged on place

Part 2. Matching.

A	B
1. Consumption of legume pastures	A. Deworming
2. IV injection of calcium-solution	B. Chock
3. Swallowing of potato and mango	C. Liver-flukes
4. Treatment of Internal Parasitic	D. Milk-fever
5. Clean equipment and utensils	E. Slurry
6. Supportive treatment for sick cows	F. Tsetse-flies control
7. *Moniezia* and *Ascaris*	G. N' Dama and Sanga
8. *Fasciola hepatica* and *F. gigantica*	H. Acaricide
9. Sterile-Insect-Technique	I. Fluids, steroids, glucose
10. Trypanotolerant breeds	J. Vermin species
11. Drugs killing ticks and mites	K. Bloat
12. Hypoglycemia	L. Avoid cross contamination
13. Animal manure heaps	M. Internal parasites
14. Rodents, birds and insects	N. Tapeworms
15. Dry-cow therapy	O. Ketosis
	P. Mastitis treatment

Part 3. Fill in the blank spaces.

1. ________ is any condition that disrupts the normal function of an animal's body.

2. ________ drugs or chemicals which used to kill ectoparasites of animals

3. Microorganisms which can produce diseases on the body of dairy cattle is called ________

4. The most commonly used diagnostic-test for TB in dairy cattle is termed as ________

5. The generic name of bacterium which cause TB in dairy cattle is termed ________

Part 4. Describe the following briefly and precisely.

1. Describe the external body parts of a dairy cow

2. Outline the major economic important diseases of dairy cattle? How do you identify whether an animal is sick?

3. Describe the differences and similarities of micro-organisms and parasites.

4. Define the term wound, wound management, and fracture.

5. If a dairy cow in your working area may be injured by something else, then how do

you go through to treat the cow? Discuss all the necessary steps of wound management.

6. Mention materials and equipment required for managements and treatments of wound or fractures for dairy cattle.

7. What do you do, if you may face with damage or injury of large animals? Why?

8. Discuss the procedures to treat fractures of dairy animals.

9. Why and how do you burn a carcass in the working place?

10. What does disposal by burying mean?

11. Develop herd health program for small scale dairy production.

>>> REFERENCES

Andrews A H, Blowey R W, Boyd H, et al., 1992. Bovine Medicine Diseases and Husbandry of Cattle [M]. Blackwell Scientific Publications.

Arnes T R, Sivula N J, Marsh W E, 1995. Health Programs for Dairy Heifers [E]. http://hdl. handle. net/11299/118780.

Bartlett P C, Miller G Y, Anderson C R, et al., 1990. Milk production and somatic-cell-count in Michigan dairy herds [J]. J. Dairy Sci, 73 (10): 2794.

Blood D C, Radostits O M, 1989. Veterinary Medicine [M]. 7th ed. Balliere Tindall, London.

Charles K G, Margaret N L, Camillus A, 2012. Dairy Farmers Training Manual [E]. https://zh. scribd. com/document/176262485/Dairy-Farmers-Training-Manual.

Diggings R V, Bundy C E, 1988. Dairy Production [M]. 2nd ed. Prentice-Hall, Inc.

FAO and IDF, 2011. Guide to Good Dairy Farming Practice [M]. Animal Production and Health Guidelines. No. 8. Rome.

Forse B, 1999. Where there is no Vet [M]. Oxfam Publishing.

Hoffman P, 1995. Optimum Growth Rate for Holstein Replacement Heifers [E]. http://fyi. uwex. edu/heifermgmt/files/2015/02/optimumgrowthrates. pdf

House J, 2011. A Guide to Dairy Herd Management [M]. Meat & Livestock Australia Limited.

Morris T, Keilty M, 2006. Alternative Health Practices for Livestock [M]. Wiley-Blackwell.

Richard W, 1993. Dairying. Tropical Agriculturalist [M]. Macmillan Press, London.

Sanderson W B, 2004. Challenges and Opportunities for Tropical Dairying [C]. Proceedings of the Animal Science Congress.

Seifert H S H, 1996. Tropical Animal Health [M]. 2nd ed. Kluwer Academic Publishers.

Staal S J, Pratt A N, Jabbar M, 2006. Dairy Development for the Resource Poor; A Comparison of Dairy Policies and Development in South Asia and East Africa [E]. http://www. fao. org/ag/againfo/programmes/en/pplpi/docarc/wp44 _ 2. pdf.

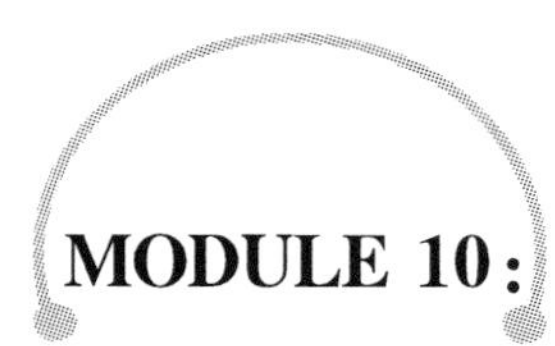

MODULE 10: PERFORM DAIRY FARM RECORD KEEPING

>>> INTRODUCTION

Record refers to all information about the dairy farm activity kept in order to identify whether dairy animals are productive, the farm is running successfully, and to take reliable managerial measures (Syed, 2011).

Keeping track of what is happening on the dairy farm requires well organized recording system. Good farm management requires having a good, useful set of farm records. Good records do not ensure the farm will be successful; however, success is unlikely without them. Farm record is, like the report cards that the students receive at school, farmers can tell how well they are managing their operation compared to other producers in their classes. They can also see the strengths and weaknesses in their operation. Having accurate facts and figures is most useful when borrowing money, seeking government support and completing tax returns. From reliable farm records, equity (or proportion of total assets actually owned) can be updated to assist with future farm investment programs.

Records must be useful, Records must be kept in such a form that they can be easily converted into information, Record keeping systems must be simple, Duplication must be avoided as much as possible, and Records must lead to actions being taken. They must be processed or summarized in order to provide information. Information is required to make sound management decisions, yet poor record systems are common and costly.

The information that records provide serve many purposes. A few are as follows: day-to-day management decisions, financial accounting and taxes, measure progress, troubleshoot problems, genetic evaluation, enterprise evaluation, aid in recovery of stolen property, planning future actions, and research (Stewart, 2005).

Therefore, this module covers the necessary information to record data on dairy production, financial transactions, resources, animals' feeds and feeding, animal health, reproduction and breeding, dairy product marketing, and overall dairy environmental control programs as set-out in the under headings.

1 ADVANTAGES OF DAIRY RECORDING

Records provide the farm manager with data, information and knowledge. There are seven major advantages of keeping records as used as (Arzeno, 2004):

Service tool

The types of services provided are income tax calculations, estate planning, business arrangement reconciliation and obtaining and managing credit. Further, the records system can provide a basis for developing equitable business arrangements for operating agreements, partnerships, and corporations. Records also help in obtaining and effectively using credit by showing factors relating to the profitability, liquidity and solvency of the farm business.

Diagnostic tool

That provides data for financial analysis and other diagnostic instruments, such as identifying the strengths and weaknesses of the business. Records can help to determine the absolute and relative profitability of the business by identifying the strengths and weaknesses of the business. Thus, the manager can see strong points and capitalize on them while recognizing weak points and taking corrective steps. The records system can be used as an indicator of progress from both the business management and financial management standpoints.

Business indicator

Records can show the manager changes in size, productivity and efficiency, and organizational factors unique to his business and farms similar to his. He can measure actual performance in comparison with budgeted performance and/or standards of performance for his type of business.

Financial indicator of progress

A series of records are necessary to monitor progress. Records help the farm manager and/or his lender to measure the changes in the financial condition of the business and to compare actual and planned performance. This needs to be done on a regular basis, so problems can be worked on as soon as they develop.

Forward planning device for short and long-term planning

Past records can be used as a basis for projecting cash flows. The manager can then compare actual performance with the plan. The records system can provide cost information and coefficients of production unique to his situation for budgeting in both the short and long run. He can project short and long-term credit needs and repayment capacities.

Improve economic returns /Keep good financial records

- Income tax reporting: A good set of records is required for the preparation of complete and accurate tax documents. Poor records often lead to preparing income tax returns that result in either underpayment or overpayment of taxes. This might get the tax reporter in trouble if there is an unexpected IRS audit of records.
- Obtaining credit: If you decide to borrow money for your farm business operation, the loan officer or bank will ask to see your financial records, including a balance sheet, an income statement and a cash flow statement. The creditor will require these statements in order to determine your repayment capacity (Arzeno, 2004).

Management tool

Accurate financial records, along with production data, will help the farm business operator analyze the information and make the necessary adjustments to operate more efficiently, thus increasing profitability. Such analysis will help you plan for the future, and it will pinpoint the weaknesses of your farm business and allow you to act accordingly. Further, the manager can schedule purchases of inputs, compare various inputs as to costs and returns, select the kinds and sizes of enterprises, and determine capital generation capacities of different alternatives. In these volatile economic times, forward planning is becoming increasingly crucial; good computer tools are available to help with the task (Brannstrom, 2011).

2 IDENTIFYING MATERIALS, TOOLS AND EQUIPMENT

It is important to keep records of all activities from farm up to table. The goal of these records is to keep track of the quality and quantity of milk that is produced and delivered by the individual members of the group. This information is needed to determine the amount of money each farmer will receive. As a matter of accuracy these records should be kept in a book and not on loose papers. The book should be kept in a safe place.

Based on a level of the farm professionals utilize different kinds of utensils. Some of these are stationeries (book, pen, pencil, paper and rubber), computer, table, cabinet, recording room, weight balances, chair, animals ID card, thermometer, hygrometer, vehicles and so on.

3 IDENTIFY THE TYPES OF DAIRY RECORDS

Livestock records to keep with dairying being such an intensive form of livestock production, keeping track of individual animals is very important. Such information will be essential in preparing realistic budgets for future farm developments, rather than depend on generic estimates of farm performance (Humphreys et al., 2006).

3.1 Types of Dairy Production Farm Data

Some of the key recording data to be kept in the dairy production enterprise include:

- Calving dates: To follow through different stages of each cow's lactation and to assess weight for age of young stock. Also, to update annual livestock inventory as stock change classification, e.g. from calves to yearlings. They are also useful to identify cows that are due to be mated
- Daily milk yields: For closer animal observations if they suddenly and unexpectedly change
- Daily herd milk yield: To check up on milk payments and to fine-tune feeding programs
- Regular milk composition data: If provided by the cooperative or processor, to closely monitor the effects of diet
- Mastitis treatment: For individual cows and other treatments requiring milk not being sold. The drug withholding period must be followed to ensure milk quality is not compromised
- Routine monitoring of feed (forages as well as concentrates) offered and actually consumed: Which can indicate if cows are in heat or sub clinically sick
- Live weight and body condition of adult cows: To monitor milking performance during the entire lactation and better plan feeding programs
- Live weight and body condition of young stock: To monitor feeding management required to achieve growth targets
- Dates when each cow is in heat: To manage artificial insemination programs as well as to predict expected dates of calving
- Dates and results of pregnancy diagnoses: If undertaken, to predict expected calving dates
- Animal sickness, veterinary visits and drug treatment: To follow through animals' responses to treatment. With replacement heifers, it also provides a guide as to whether the heifer's lifetime productivity might be compromised
- Routine vaccination and drenching: To ensure they are timely and to plan future programs
- Stock purchases and sales of culls: To update livestock inventory
- Stock deaths and probable causes: To update livestock inventory and also monitor general herd health
- Age when culled from the milking herd, reason for culling and number of lactations while in milking herd
- Milk and concentrate intakes of young calves: To plan weaning and calculate total

rearing costs

- Yields of forage crops: To better utilize fertilizers and plan forage purchases
- Other dairy enterprise sales: Such as stock fattened for sale, cow manure and any excess forages, for accounting purposes

3.2 Types of Dairy Records and Accounts Required in Dairy Business

- Income and expense ledgers: An income and expense statement is required for tax purposes and can also perform other functions as business management tool. Income and expenses should be recorded and totaled on a current basis for the greatest accuracy and to provide a running account, so problems can be identified when corrective action may still be taken.
- Depreciation: A depreciation record is required for tax filing purposes. It provides a list of all depreciable capital assets used in the farm business and can be used in constructing the inventory record and financial statement. Depreciation records are almost always kept conforming to IRS regulations and recently have had little relevance to financial decision making. It may be necessary to maintain a second set of depreciation records that better reflects true economic costs for analysis and budgeting purposes.
- Profit and Loss Statement: A profit and loss statement takes the income, expense, and inventory records and ties them together to provide the manager with information about his return to labor, management and equity capital for the year and over a series of years. Accurate measurements of feed and livestock inventories are vital to making correct accrual adjustments for calculating net farm income.
- Cash flow: A cash flow record provides a history of how cash moves in the business and can also serve as a device to project cash flow in future periods. The record of actual cash flow can be a monitoring device to measure actual performance against planned performance.
- Production: Production records for the individual crops, livestock, and other enterprises are essential for the manager to evaluate his performance as a production manager. Accurate measurements of input quantities consumed, and outputs produced in each enterprise allow for construction of meaningful enterprise budgets.
- Enterprise account: Enterprise accounts are vital records to help evaluate the performance of each of the several enterprises making up the total farm business. At a minimum they should be broken down between the crop and livestock enterprises. Ideally, each enterprise should be in a separate account. Enterprise accounts can help farm managers see which of their enterprises make the most profit, determine which production methods within an enterprise perform economically under the individual farm conditions, and help determine what level of output is economically appropriate for each enterprise.

- Feed: Feed is one of the largest inputs for livestock enterprises. Almost all homegrown feeds have a farm gate value. Managers should be very concerned about feed utilization and efficiency in the various livestock enterprises. A good record of purchased and farm-grown feed is essential for the manager to make economically sound decisions about the livestock program as well as to properly credit crop enterprises for their share of input into the livestock enterprise. For example, using appropriate transfer prices, corn could be "sold" to the dairy cow enterprise which "buys" the corn from the corn enterprise at its "farm gate price". This concept of transfer pricing is often misunderstood in farm accounting. Farmers tend to view crops harvested and in storage as free goods which are then transferred to the dairy enterprise which is "where the farm profits are made." The obvious fallacy of this logic is that the actual costs of producing the crops are not recognized and tend to make the dairy enterprise look much better than it actually is. Sophisticated managers, familiar with their costs of producing crops, have been using linear programming models to formulate "least cost" nutritionally balanced rations for many years. Although the tools for doing least cost balancing have been available for years, the concept has not been widely used in Wisconsin, again because of the difficulty in determining appropriate "costs" to assign homegrown feeds.
- Labor: With labor costs going up and greater scarcity of competent farm help, more attention must be paid to effective labor use in the business. Constant study of labor use can make fewer hours accomplish the same task. Time study records can help provide this information. Also, labor records are needed for Social Security purposes, Workmen's Compensation, W-2 forms, etc.
- Equipment: Larger, more expensive, complicated equipment, with its high fixed costs plus large operating costs, calls for special analysis on many farms. It may pay managers to keep an individual set of financial records for these types of machines. This information can serve as a guide to when to buy, sell, and trade equipment, as well as the basis for making decisions about appropriate rates to charge for doing custom work.
- Experimental: With exploding technological developments, managers are continually faced with decisions concerning whether new technology can be profitably used in their business. Records of trial projects in crop and livestock enterprises can provide a guide to the physical as well as financial performance of that technology for a specific farm situation.
- Production-field Records: Increased concerns about ground water contamination and chemical carry-over make an accurate record of chemical and cultural practices on each field absolutely essential. As farm sizes continue to grow and the number of potential chemicals available for disease and insect control increases, even the sharpest memory will fail. Field records will be a valuable tool for the manager to analyze the

effectiveness of fertility and weed and disease control programs.

- Production-livestock records: Most successful dairymen have recognized the importance and value of good production and breeding records for the dairy herd. More than half of Wisconsin's herds belong to the Wisconsin Dairy Herd-improvement Association. Those who choose to be on "official" records are visited once a month by afield man from the cooperative. The field man gathers breeding information and production records from the farmer and generates a production roster listing each cow in the herd, her current production level, estimated total lactation production level, and estimated breeding dates.
- Forward budgeting: A pro forma farm plan is a basic planning and control document necessary to judge the progress of the operation throughout the year. This planning document should be done on a quarterly basis so that variances in target receipts and expenses can be monitored and corrected in a timely manner.
- Long range budgeting: When the business is just starting, or when it is contemplating major alternative investments, long run budgeting is necessary to determine if the proposed alternatives can be economically successful. Several computerized simulation models have been written to help farmers and lenders do long range financial planning. According to the FINPACK library, University of Minnesota Centre for Farm Financial Management is an excellent tool for both analyzing historic farm financial performance and developing long range financial projections. Once a strategic decision has been made to make a significant capital investment, detailed transitional plans are needed to forecast the cash flow requirements of the plan between the current period and the long-range goal. Again, this process is greatly assisted by computer software (Becker et al., 2005).

$$Net\ farm\ income\ ratio = Net\ farm\ income\ /\ Gross\ farm\ income$$

3.3 Formats Appropriate to Keep Different Types of Records

Dairy farming is a complex business which demands accurate records and careful financial management. Both financial and production records are required in order to provide the information on which the farm manager can make critical decisions (Brannstrom, 2011).

The formats for the major types of records are: breeding records, production records, feeding records, health records, and financial records.

3.3.1 Breeding Records

- Measure the productive efficiency of the herd
- Breed
- Birth date
- Heat dates
- Enable culling and selection
- Earliest breeding date

- Fertility (age at first service, age at first calving, date of calving, number of services per conception)
- Pedigree/parentage (dam name, grand dam, sire name, grand sire)
- Pregnancy examination
- Expected calving date
- Drying off date (Table 10. 1)

Table 10. 1　Breeding record format

Owner's name: ______ Region: ______ Wereda: ______ Kebele: ______							
Dam ID.	Dam breed	Dam birth	Sir ID.	Sir breed	Mating date	Calving date	Remarks

3. 3. 2　Production (Performance) Records

Calving record

Maintains the information associated with the production performance of cattle and the animal (Table 10. 2).

Table 10. 2　Performance recording format

Owner's name: ______ Region: ______									Wereda: ______ Kebele: ______						
Pre-weaning									After weaning						Remarks
Calf ID.	Birth date	Sex	Birth type	Dam ID.	Dam breed	Sire ID.	Sire breed	Birth weight (kg)	Weaning date	Weaning weight (kg)	Type of rearing	Weaning group	Market weight	Body condition	

Individual cow record

Individual cow record is shown in Table 10. 3.

Table 10. 3 Individual cow records

Identification number:
Birthday:
Sire of the cow:
Dam of the cow:
Maternal GS of the dam:
Maternal GD of the dam:
Record signature:

Breeding/Preg. test/Caving												
Calving sequence	Calving date	Post-parturition oestrus/date	Breeding record							Result of Preg. test	Calving date	Remarks
			1	2	3	4	5	6	7			

Milk production record

Milk production record is shown in Table 10. 4.

Table 10. 4 Milk production records

Farmer:		Month:					
		Year:					
Cow's Name	Date calving	Milk yield						Monthly total (14 - daily total 1+ daily total 2)
		First week of Calving			Second of the month			
		AM	PM	Daily total 1	AM	PM	Daily total 2	

3. 3. 3 Feeding Records

The amount of feed given as well as the type of feed should be recorded. Feeding records should be used the most for day-to-day management, evaluating pasture management practices and for planning of activities in the future: available fodder on farm, quantity fed, concentrate supplemented, minerals, leftover (per head and per feed, if possible) and spoilage (per batch) consumption (per day and per kg) (Table 10. 5).

Table 10. 5 Layout of feeding record

Owner's name: ________				Region: ________		
Wereda name: ________				Kebele: ________		
No.	Date	Type of feed	Composition	Cost/kg	Total cost	Remarks

3.3.4 Health Records

Health record provides the overall health information about the animals in the herd. With the use of records, veterinarians can gain additional information about the probable causes of ill health in an individual animal: vaccination, dipping/spraying, treatment, deworming, and postmortem.

Herd Record

Herd record is shown in Table 10.6.

Table 10.6 Animal health records

Owner's name ______________ Wereda name: ______________
Region: ______________ Kebele: ______________

No.	ID/ Name	Date of observation	Major signs observed	Suspected diseases	Treatment given			Response	Remarks
					Medicament	Duration	Dosage		

Individual-Cow-Card of health record

Individual cow card is shown in Table 10.7.

Table 10.7 Individual cow card

Name of animal		Name of sire	
Date of birth		Name of dam	
Health			
Date	Vaccination	Date	Treatment
Remarks			

3.3.5 Financial Records

Economic records are of paramount interest in providing the farmer with information concerning the profitability of his farm: expenses of feed, health, transportation from selling farm products; milk, live animal; income; profit; and loan/credit (Table 10.8).

Table 10.8 Financial records

Owner's name: ______			Region: ______			
Wereda name: ______			Kebele: ______			
No.	Income		Expenses		Differences (Income-expenses)	Remarks

Producers should maintain the following financial statements: a balance sheet, a statement of owner equity, an income statement, and a cash flow statement.

Balance sheet is a statement of the financial condition of a business. It shows the status of the farm business assets, liabilities and owners' equity at a specific time. It is a snapshot, not a motion picture, and must be analyzed with reference to comparative prior balance sheets. In summary, it shows what is owned (assets) and what is owed (liabilities), and the difference between them, which is called NET WORTH. The term "balance" comes from the requirement that the ledger is in balance through the basic accounting equation of (Agustin, 2004):

$$Assets = liabilities + owner\ equity$$

$$Working\ capital = Current\ assets - Current\ liabilities$$

4 COLLECT, MANAGE, ANALYSES AND INTERPRET THE RECORDS

Farmers will remember significant events on their farm to the date, but often find it difficult to remember exactly when events relating to their individual animals occurred, especially if they have many. However, precise recording is necessary if you want to improve your economic returns (Bonnier et al., 2004). Example: Six weeks after giving birth, Betty (a cow) has an impressive daily milk production of 18 liters and a total milk production per lactation of 3240 liters. However, she doesn't seem to become pregnant quickly: her calving interval is close to 3 years.

Blue (another cow) does not stand out with her peak production of 12 liters a day and total milk production of 2880 liters. But she gives birth to a calf every 2 - year. Intuitively, you may think of Betty as your finest milking cow. Your intuition was wrong, and administration could help you to see why. If you keep a record of both cows' milk production and calving dates, you can calculate the average daily milk production of both animals. You may be surprised to see that Blue is bringing you more benefit (see the next table).

Reporting and Recommended Measures

Information Technology systems in the dairy industry are becoming a necessary requirement accompanied by the advancements of science and technology. Therefore, many computerized dairy recording systems were developed and have been applied in large dairy farms. But the utilization of a computerized recording system by small scale farmers is very rare and not given the priority. Among the main reasons for the lack of usage are due to the cost of buying the system which is expensive, complexity of the program, and the level of computer literacy, where this constraint appeared to be important during the 90 s (Taylor et al., 1991), but it is now less significant given the current trend of computer education being introduced as early as in pre-schools, initiatives from the government with the introduction of Computer Literacy Programs (Razaq et al., 2009) and reduction in the cost of obtaining computers and related peripherals. So, in order to increase the usage of a computerized recording system among the small scale dairy farmers in Malaysia, this paper proposes a simple, low cost and easy to use computerized recording system framework. This framework is hereon referred as Individual Cow Recording and Analysis System (ICRAS©) (Fig. 10.1).

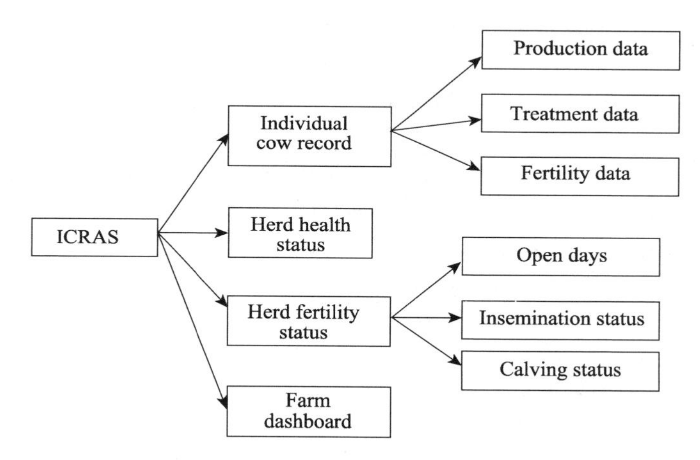

Fig. 10.1 ICRAS© was designed to follow a logical sequence of steps

This framework was initially designed for a farm of 2 – 30 cows since the target users are small scale farmers. However, it has also the capability to capacitate records of bigger farms with more than 100 cows. ICRAS© was designed based on four core elements that a dairy record should have. These elements were divided into four main modules, which are the current farm statistics, herd fertility management, herd health management and individual cow records (Jeyabalan, 2010).

>>> SELF-CHECK QUESTIONS

1. What are the main purposes of record keeping?

2. List at least four materials, tools and equipment, which should involve for dairy recording?

3. Mention the types of dairy records?
4. What does individual cow recording and analysis system (ICRAS) mean?
5. Show the milk production report format?

>>> REFERENCES

Agustin A, 2004. Record Keeping in Farm Management [E] . http: //agbiopubs. sdstate. edu/articles/ExEx5054. pdf.

Brannstrom A J, 2011. Using Farm Records Effectively for Business and Financial Management [E]. http: //cdp. wisc. edu/pdf/Using Dairy Farm Records08. pdf.

Humphreys J, Treacy M, McNamara K, 2006. Nutrient management on intensive dairy farms in the southwest of Ireland [E] . http: //t-stor. teagasc. ie/handle/11019/978.

Jeyabalan V, 2010. Individual Cow Recording and Analysis System for Small Scale Dairy Farmers in Malaysia [J] . International Journal of Computer Applications, 8 (11): 33 - 38.

Razaq A A, Norhasni Z A, Jamaludin B, et al. , 2009. Computer Usage and Achievement among Adults in Rural Area Malaysia [J], Journal of Social Sciences, 5 (1): 1 - 8.

Shah N A, 2009. Advantages and Disadvantages of the Waterfall Model You Ought to Know [E] . http: //www. buzzle. com/articles/waterfall-model-advantages-and-disadvantages. html.

Azhar A S, 2011. Hay Making [E] . http: //www. uvas. edu. pk/doc/advisory _ services/Dairy-farming/english/5 - Hay _ making. pdf.

GLOSSARY

Acaricide: Drugs that kills ticks and mites.

Acarina: Ticks and mites.

Acreage: The area of a piece of land measured in acres.

Acute disease: Is one where the disease in the individual flares up quickly and is soon resolved either through death or recovery, e. g. blackleg or anthrax. The course of an *acute*: take 1/2 days-to-2 weeks before termination; *sub-acute*: 2 weeks to 1 month; and *chronic*: more than one month.

Additive: A substance which is added to animal feeding stuffs to provide antibiotics, mineral supplements, vitamins or hormones.

Aetiology (causative agent): Refers to the factors responsible for disease occurrence.

Agro industry: An industry dealing with the supply, processing and distribution of farm products.

Antibiotics: Chemical products of micro-organisms that kill or inhibit the growth of other M/Os used to treat infectious diseases; broad-spectrum antibiotics effective against a wide range of bacteria.

Artery: An efferent blood vessel from the heart, conveying blood away from the heart regardless of oxygenation status.

Aseptic: Free from bacterial contamination, sterile; used to describe a type of food processing and packaging characterized by non-refrigerated storage and long shelf-life products.

Balance sheet: This is a statement of the financial condition of a business. It shows the status of the farm business assets, liabilities and owners' equity at a specific time.

Balanced diet: A diet which provides all the nutrients needed in the correct proportions.

Baler: A machine that picks up loose hay or straw and compresses it into bales of even size and weight, and then ties them with twine.

Barn: A building used to store hay, grain, or animals on a farm.

Batch: A temporary grouping of cows representing location on test day, i. e. pen or milking group.

Bleach: To make something whiter or lighter in color, or to become whiter or lighter in color.

Breed: A group of animals of a specific species which have been developed by people over a period so that they have desirable characteristics.

Bulk-milk: Raw milk from a dairy farm, as stored in a bulk tank; in contrast with packaged milk. Other bulk products, such as condensed skim and cream, may also be transported in bulk form.

Bulk-tank: A large tank used to store milk until it is shipped to the processing plant.

Bull: The uncastrated adult male of domesticated cattle.

Bullock: The castrated adult male of domesticated cattle.

Butter: A softy fatty food stuff made by churning the cream of milk.

Butter soft, creamy spread: A soft, pale yellow, fatty food made by churning cream and used for cooking and spreading on food.

Calcium: A metallic chemical element naturally present in limestone and chalk. It is essential for biological processes. Calcium is essential for various bodily processes such as blood clotting and is a major component of bones and teeth.

Calf: A younger of a cow, less than one year old. A male calf is known as a bull calf; a female calf is a heifer calf. The meat of calves fed on a milk diet is known as veal.

Carotene: An orange or red pigment in carrots, egg yolk and some natural oils, which is converted by animal's liver into vitamin A.

Casein: Major protein of milk Coagulated by the action of rennet or acids. The major protein contained in milk and the primary protein in cheese. Also, a protein curd or dried product made from milk casein curd.

Cash flow statement: A sound balance sheet and/or a high net farm income does not necessarily mean that the operation can meet financial obligations.

Castration: The removal of the essential sex organs, testes and ovaries, from male or female animals. This allows bullocks and heifers to be housed together, as castrated animals are more docile.

Cattle: Domestic farm animals raised for their milk, meat and hide. Class: Bovidae.

Cereal: A type of grass which is cultivated for its grains. Cereals are used for animal feed.

Cheese: A high-quality food rich in nutrients. Cheese is made from the milk solids or curds and prepared in different ways. Cottage, Swiss, American and Cheddar are types of cheeses. It can range from hard to semisoft and from mildly acidic to sharp.

Chronic-disease: Diseases that run a long course of at least a week, but some can last many months or years; such as contagious bovine pleura-pneumonia has a much longer time scale, lasting months or even years.

Churn: To agitate rapidly and repetitively.

Churner: The vessel in which cream is churned to make butter.

Class 1 milk: The (theoretically) highest priced category in a classified pricing plan. Includes milk used in fluid whole milk and generally includes related fluid products. Milk products in this category are milk, skim milk, low-fat milks, milk drinks, half-and-half, and filled milk. A very similar classification is made under the federal milk marketing order system, where it is termed "class Ⅰ".

Class 1 price: The minimum price that handlers must pay for milk going into Class 1 uses (see Class 1 milk). It is announced every month in California (class 1 is announced every month in federally regulated milk markets as well). The level of the Class 1 price depends on national commodity markets; it uses the higher of the national cheese price or the butter/powder prices.

Class 2 milk: An intermediate product category in the classified pricing plan. This pricing category includes cottage cheese (all types), fluid cream products, yogurt, buttermilk and eggnog. Under federal milk marketing orders, it is termed class 2 and includes most of the above products as well as ice cream and other frozen dairy desserts.

Class 2 price: Minimum price handlers must pay for milk used in Class 2 category. In general, the Class 2 price is tied directly to the Class 4a price by a differential. The Class 4a price is based on prices of bulk butter and bulk nonfat dry milk.

Class 3 milk: An intermediate product category in the classified pricing plan. This pricing category includes ice cream, ice cream mixes and other frozen dairy desserts. Under federal milk marketing orders, it is termed class 2 and includes all of the Class 2 products as well as the dairy desserts. In federal orders, Class 3 refers to a manufacturing product category that includes all cheeses except cottage cheese.

Class 4a Milk: A manufacturing product category in the classified pricing plan that includes butter and powdered milk products. In federal orders, this product category is referred to as "Class 4".

Class 4b Milk: A manufacturing product category in the classified pricing plan that includes all cheeses except cottage cheese. In federal orders, this product category is referred to as "Class 3".

Claw: It is the vessel which used to collect milk from all of the teat cups.

Clinical-signs: Detectable change in tissue structure or function as a result of disease.

Colostrum: A yellowish fluid that is rich in antibodies and minerals produced by a mother after giving birth and prior to the production of true milk. It provides newborns with immunity to infections.

Combine: A large farm machine that harvests grain crops like oats, wheat and barley.

Concentrates: Animal feeding stuffs with a high nutrient relative to their bulk.

Condition scoring: A method of assessing the state of body condition of animals; scores range from 0－5 for cattle and 1－9 for sows.

Contagious disease: Disease that spread by contact. A disease w/c can be transmitted from one to another by close contact.

Controlled grazing: A system of grazing in which the number of livestock is linked to the pasture available, with moveable fences being erected to restrict the area being grazed.

Cottonseed cake, cottonseed meal: A residue of cottonseed after the extraction of oil, used as a feeding stuff.

Critical control point: A step in a process at which a control can be applied, which is essential to prevent, eliminate, or reduce a food safety hazard to an acceptable level.

Critical-limit: A criterion which separates acceptability from unacceptability.

Crude-fiber: A term used in analyzing foodstuffs, as a measure of digestibility. Fiber is necessary for good digestion, and lack of it can lead to diseases in the intestines.

Crude-protein: An approximate measure of the protein content of foods.

Cull: An animal the owner has determined to be inferior and removed from the herd.

Cull value production level: Test day value of a product that is 60% of the current month's herd average test-day value product.

Culling rate: The percentage of cows that are removed from a herd.

Dairy: A place/building where milk is kept, and milk products are made. A company which produces cream, butter, cheese and other milk products.

Dairy cows: A cows and heifers kept for milk production and for rearing calves to replace older cows in a dairy herd.

Dairy farm: A farm which is principally engaged in milk production.

Dairy herd: A herd of dairy cows.

Dairy products: Foods prepared from milk, e. g. butter, cream, cheese or yoghurt.

Dairying: An agricultural system which involves the production of milk and other dairy products from cows kept on special farms.

Dairyman: A person who works with dairy cattle or a person employed in a commercial dairy.

Dehorn: To remove the horns of an animal, done by disbudding when the animal is young.

Diagnosis: Refers the determination of nature of a disease or causative assent, e. g. history taking, symptoms and laboratory tests.

Diet: The amount and type of food eaten.

Diet formulation: The combining of different types of feedstuffs or nutrients so as to form a healthy and balanced diet for an animal.

Dietary fiber: Same as roughage. Dietary fiber is found in cereals, nuts, fruit and some green vegetables. It is believed to be necessary to help digestion and to avoid developing constipation, obesity and appendicitis.

Digestibility: The proportion of food which is digested and is therefore of value to the animal which eats it.

Digestion: The process by which food is broken down and converted into elements which can be absorbed by the body.

Digestive enzymes: Enzymes which speed up the process of digestion.

Digestive system: The set of organs in the body associated with the digestion of food.

Disaccharide: A sugar composed of two monosaccharide units.

Disease: Any condition by which the individual is suturing from discomfort or a condition an individual shows structural, functional or chemical deviation from normal.

Drugs: These are medicines and chemicals applicable for the treatments of animals.

Dry matter intake (DMI): The amount of feed that an animal consumes or requires, discounting its water content.

Endemic disease: It is one which is present at much the same level in the population at all times, e. g. Mange in camels or FMD in some livestock populations in the tropics.

Energy: The force or strength to carry out activities.

Ensilage ensiling: The process of making silage for cattle by cutting grass and other green plants and storing it in silos.

Ensile: To make silage from something.

Enzyme: A protein substance produced by living cells, which promotes a biochemical reaction in living organisms.

Epidemic: Is the occurrence of a disease above its usual frequency in the population and a pandemic is a very widespread epidemic affecting a continent or the whole world, such as occurred when render-pest swept across the African continent towards the end of the nineteenth century.

Equipment: Farm machinery, implements, and livestock conveyances; does not include vehicles for personal or business transport.

Farm cost: The total cost of running a farm, including wages, feed, fuel, electricity, veterinary fees machinery, taxes and more.

Farm records: Records are kept on each dairy cow telling how much milk they make, birth and calving dates, vaccinations and more. There are also records showing the farm costs and income.

Farmstead cheese: A somewhat loosely used term to indicate cheese made by dairy farmer, at or near place of milk production.

Fat free: A product containing less than 0. 5 g of fat /reference amount and /labelled serving. The product must contain no added ingredient that is fat or understood to contain fat.

Fat%: The percent of fat in the milk, calculated by fat (in lb) / milk (in lb) × 100%.

Fat/butterfat: Stored energy in the milk that rises to the top unless the milk is homogenized. Whole milk is about 4% fats. Low fat milk is 1% fat. When the fat, or cream, is removed from the milk it can be used to make butter, cheese, ice cream or other dairy products.

Fatal: Organisms and/or chemical agents that have the affinity to kill the host organisms.

Fat-corrected milk: Estimated quantity of milk calculated on a 4% butterfat energy basis. It is a means of evaluating milk production records of different dairy animals and breeds on a common energy basis.

Feed: Any naturally occurring material suitable for feeding animals.

Feed additive: A supplement added to the feed of farm livestock, particularly pigs and poultry, to promote growth, e. g. an antibiotic or hormone.

Feeding: The action of giving animals food to eat.

Fermentation: Chemical conversion into simpler substances, the breakdown of carbohydrates by microorganisms.

Fertilizer: A substance containing nutrients that will help plants grow. Manure is a good natural fertilizer recycling the nutrients that the animal does not use.

Fiber: A long, narrow plant cell with thickened walls that forms part of a plant's supporting tissue, especially in the outer region of a stem.

Fibrous: Made of a mass of fibers.

Filled-milk: Milk from which natural milk fat has been removed and replaced with other fats or oils from plant sources.

First aid: The assistance given to any person suffering a sudden illness or injury, with care provided to preserve life, prevent the condition from worsening, and/or promote recovery.

Flavored milk: A subclass of fluid (packaged milks) to which flavoring has been added, such as chocolate, strawberry and vanilla.

Fodder: Plant material or a crop which is grown to give to animals as food, e. g. grass or clover.

Forage: A crop planted for animals to eat in the field. Forage crops are highly digestible and palatable and are either very quick growing or very high-yielding.

Freshness date: The date stamped on each milk container that tells the last date the milk is guaranteed to be fresh.

Fungi: Organisms widespread in nature that reproduce by forming spores; a few species are infectious and pathogenic to animals.

Gestation period: The period from conception to birth, when a female mammal has lived young in her womb.

Grade-A milk: Also called fluid grade milk or market milk, produced and processed under the strictest sanitary regulations prescribed, inspected, and approved by public health authorities.

Grade-B milk: Also called manufacturing grade milk, produced and processed with sanitary regulations prescribed, inspected, and approved by public health authorities for milk to be used for manufactured products only.

Grain: The seed of certain grassy plants, high in nutrients that can be ground into flour or meal and used to feed dairy cows, people or other animals.

Grass: A flowering monocotyledon of which there are a great many genera, including wheat, barley, rice, oats. Grasses are an important food for herbivores and humans.

Grazing: Eating grass in a field usually enclosed by a fence.

Grazing management: Looking at the way in which land is grazed and seeing how it can be done most efficiently.

Grazing systems: Different methods of pasture management.

Grazing-land: An area of land covered with low-growing plants suitable for animals to feed on.

Groundnut cake: The residue left after oil extraction from groundnuts, a valuable protein concentrates for livestock.

HACCP: The Hazard Analysis and Critical Control Point system adopted by the Codex Alimentarius Commission.

HACCP plan: A documented system, prepared in accordance with all the principles of HACCP, to ensure control of significant food safety hazards in a food handling process. A HACCP Plan will always include one or more critical control points.

Hard products: Generally used in referring to the more storable manufactured dairy products, such as butter, non-fat dry milk, cheeses other than cottage cheese, and evaporated or condensed milk.

Hay: Grass mowed and dried before it has flowered, used for feeding animals. Hay is cut before the grass flowers and at this stage in its growth it is a nutritious fodder. If it is mowed after it has flowered, it is called "straw", and is of less use as a food and so is used for bedding.

Haylage: A feed for cows that is made by storing fresh cut grass in a plastic bag or silo.

Hazard: Something that can cause harm if not controlled.

Hazard and control points: Terms borrowed from Hazard Analysis Critical Control Point (HACCP) programs to denote points of risk, and the manner of addressing them.

Health: A condition in which individual is in a complete harmony with its environment or surrounding.

Helminths: These are parasites that are living in the body of animals.

Herd: A number of animals such as cattle kept together on a farm or looked after by a farmer.

Hereditary: Referring to a genetically controlled characteristic that is passed from parent to offspring.

Heredity: The transfer of genetically controlled characteristics from parent to offspring.

Homogenization: To emulsify the fat particles in milk or cream in order to give it an even consistency and prevent cream from separating from the rest of the milk.

Homogenized-milk: Milk that has been treated to ensure breakup of fat globules. Homogenization prevents the cream portion of milk from separating from the skim portion. A test of adequate homogenization is such that after 48 hours of quiescent storage at room temperature, no visible cream separation occurs in the milk.

Hormones: A chemical agents that is synthesized and secreted by circumscribed and specialized glands and circulated by blood to other parts of the body to stimulate specific tissue.

HTST: High-temperature, short-time pasteurization. It is the most commonly used process for pasteurizing milk. See pasteurization.

Hygiene: Conditions and practices that help to maintain health and prevent the spread of diseases.

Ice cream: A frozen dessert made from milk and cream usually mixed with fruit or flavoring.

Immunity: This refers to the capacity of the body of animals to defend against disease causative agents.

Income statement: This is a summary of revenues and expenses for a given accounting period.

Incubation period: It is a period at which infectious agent multiplies to produce disease to the host and manifestation of clinical signs offer establishing in the host. And it is depended on pathogenesis of the agent, resistance of the host, dose of the agent and route of entry.

Infectious: Refers to living agents including bacteria, virus, Rickettsia, protozoa, helminths, arthropods.

Infectious disease: These are diseases involving the entry and development of a living organism into the body of the host.

Infestation: Denotes the invasion of an animal by disease coursing agent/organisms that may/not end with disease. It is where the organism is present on the surface of the host (e. g. Ticks and lice), but the term is also used to describe invasion of the gastrointestinal tract.

Inoculation: The act of inoculating or an injection against a particular disease.

Insecticide: Insecticides are the substances or preparations used for killing insects.

Inspection: A careful check to see if something is in the correct condition or if there are problems.

Lactation: The production of milk as food for young. The period during which young are nourished with milk from the mother's mammary glands.

Lactic acid: A sugar which forms in cells and tissue, and is also present in sour milk, cheese and yoghurt.

Lactose: Is the major disaccharide sugar in milk, which composed of 2 sugars, glucose and galactose. It is an entire product of the mammary gland.

Legume: A member of the plant family that produces seeds in pods, e. g. peas and beans. Family: Leguminosae. Or a dry seed from a single carpel, which splits into two halves, e. g. a pea. There are many species of legume, including trees, and some are particularly valuable because they have root nodules that contain nitrogen-fixing bacteria.

Liquidity: Measures the ability to meet financial obligations as they come due without disrupting the normal operations of the farm business.

Livestock: Cattle and other farm animals reared to produce meat, milk or other products.

Maintenance: The upkeep of machinery in proper working condition at all times.

Mammal: An animal that gives live birth, has fur, and feeds their babies with milk.

Mammary glands: Glands in females that produce milk. In cows, sheep and goats, the glands are located in the udder.

Manure spreader: A machine that is usually towed by a tractor and spreads manure on the fields as a natural fertilizer.

Mastitis: Inflammatory condition of the udder.

Mechanics: People that are trained to keep the farm machines running.

Metabolism: The chemical processes of breaking down or building up organic compounds in organisms.

Metabolizable Energy (ME): The proportion of energy from feed which is used by an animal through its metabolism. ME is the measure of energy following digestion, after the alimentary gases and urinary losses have been subtracted.

Metabolize: To break down or build up organic compounds by metabolism.

Methane gas: A flammable gas that can be produced from cow manure in a system called a digester. Farmers can use methane to produce energy to run their farm.

Milk: A pale liquid produced by the mammary glands of mammals, which is the primary source of nutrition for infant mammals before they are able to digest other types of food.

Milk ejection: Milk ejection is the process through which milk is released from the alveoli and flows into the ducts, the gland cistern and the teat cistern, where it can be removed by the calf or the human milker.

Milk secretion: Milk secretion is the process of manufacture of milk from the raw materials in the blood by the alveolar cells and the storage of that milk in the cavity of the alveolus.

Milking machine: A machine that milks the cow by gently pulsating and using mild suction to extract milk.

Milking parlor: Central location where animals, typically in a free-stall barn, are milked and which contains the milking machine. As opposed to tie-stall or stanchion setups where the milker moves to each cow individually, in a parlor the cows travel to the milkers, usually in groups of 2 or more.

Milk-truck: A large, refrigerated tank truck used to transport milk to the processing plant.

Mineral: An inorganic solid substance with a characteristic chemical composition that occurs naturally. The most important minerals required by the body are: calcium (found in cheese, milk and green vegetables) which helps the growth of bones and encourages blood clotting; iron (found in bread and liver) which helps produce red blood cells; phosphorus (found in bread and fish) which helps in the growth of bones and the metabolism of fats; and iodine (found in fish) which is essential to the functioning of the thyroid gland.

Molasses: A dark brown syrup, a by-product of sugar production left after sugar has been separated. It is used as a binding agent in compound animal feeds and is also added to silage.

Morbidity rate: It refers to the total no of animals-in the pop× 100 e. g. No animals in farm "A" =200, No of animals affected by anthrax = 50, No animals died due to anthrax=25.

Mortality rate: Total number of animals died off in a time period within total no of pop×100.

Noninfectious: Refers to non-living agents such as heat, cold, chemicals.

Nut: Any hard-edible seed contained in a fibrous or woody shell, e. g. groundnuts.

Nutrient: A substance that an organism needs to allow it to grow, thrive and reproduce, e. g. carbon, hydrogen, oxygen, nitrogen, phosphorus, potassium, calcium, magnesium or sulphur.

Nutritive: Referring to a substance that provides the necessary components for growth and health.

Nutritive value: The degree to which a food is valuable in promoting health.

Oilseed cake: A feeding stuff concentrate, high in protein, made from the residue of seeds which have been crushed to produce oil.

Overfeed: To give animals too much feed.

Overgrazing: The practice of grazing a pasture so much that it loses nutrients and is no longer able to provide food for livestock.

Oxytocin: A hormone which acts on the smooth muscle cell in the uterus and smooth muscle surrounding the alveoli and ducts of mammary glands.

Paddock: A small enclosed field, usually near farm buildings.

Paddock grazing: A rotational grazing system which uses paddocks of equal area for grazing, followed by a rest period.

Palatability: The extent to which something is good to eat.

Palatable: Good to eat.

Parasites: Living organisms that live in and/or the host organisms to get survival; and may harm the host organisms.

Parturition: The act of giving birth to offspring, when the fetus leaves the uterus, called by different names according to the animal.

Pasteurization: The process of heating the milk, juice or other foods over a certain period of time it to kill harmful bacteria that can cause disease without changing the taste. This was discovered by Louis Pasteur in 1864.

Pastoralism: Branch of agriculture in which "Pastoralist" mobile their herds in search of fresh pasture and water (in contrast to pastoral farming, in which non-nomadic farmers grow crops and improve pastures for their livestock).

Pasture/ Pastureland: Land covered with grass or other small plants used by farmers as a feeding place for animals.

Pasture-management: The control of pasture by grazing, cutting, reseeding and similar techniques.

Pathogenesis: It denotes a progressive development of disease process from the time it is initiated to its final conclusion in recovery or death.

Pathogens: Any micro-organisms that produce disease in the host body.

Pellet: A form of feeding stuff, usually mash, which has been moistened and pressed to form small grains.

Pelleted seed: A seed coated with clay to produce pellets of uniform size and density.

Pelleting: It is done to make the sowing of very fine seed easier.

Peracute: It refers to the course of the onset of pathogenic agents produces to produce disease in the host body. It is very rapid diseases, take almost a few hours.

Pipeline system: A system of pipes used to transport milk from the cow to the bulk tank. Each cow has a stall in the barn. A large stainless-steel milk pipe and a smaller vacuum pipe go around the barn close to the ceiling and have spigots between the stalls. The farmer moves a small milking machine with plastic hoses to each cow.

Predisposing factors: Refers to factors which make an individual more susceptible to a disease by reducing animal's immunity. Poor ventilation, damp weather, inadequate or excess feed to the host animals.

Processing plant: A place where milk is shipped to be bottled or made into other dairy products.

Prognosis: Refers to the probable outcome of a disease in a living individual (fate of diseased animal; death or recovery).

Protein: A nitrogen compound formed by the condensation of amino acids that is present in and is an essential part of living cells and is necessary for the growth and repair of the body's tissue.

Pulsator: An air valve that creates 'pulsation' or the opening and closing of the liner.

Quality control of product: A system for achieving or maintaining the desired level of quality in a manufactured product by inspecting samples and assessing what changes may be needed in the manufacturing process.

Ration: An amount of food given to an animal or person.

Ration-formulation: The process of, putting together different types of feedstuff in order to provide the amount of nutrients required by a particular animal or type of animal.

Record (1): Summary of lactation, production, and health information for an individual animal or an entire dairy herd. A record can be for an individual lactation or a series of lactations representing the life of an animal.

Record (2): Total amount of milk, butterfat, and protein for lactation or the lifetime of individual animals or an entire dairy herd.

Rotational grazing: The movement of livestock around a number of fields or paddocks in an ordered sequence and also called on-off grazing.

Roughage: Fibrous matter in food, which cannot be digested, also called dietary fiber. Animal feeding stuffs with high fiber content, e. g. hay or straw.

Ruminant: An animal that has a stomach with several chambers, e. g. a cow. Ruminants have stomachs with four sections, the rumen, the reticulum, the omasum and the abomasum.

Rumination: The process by which food taken into the stomach of a ruminant is returned to the mouth, chewed again and then swallowed.

Sanitation: An overarching set of practices that reduce the presence of organic material and debris as well as the presence, survivability, and infectivity of disease agents.

Silage: Food for cattle formed of grass and other green plants, cut and stored in silos. Silage is made by fermenting a crop with high moisture content under anaerobic conditions. It may be made from a variety of crops, the most common being grass and maize, although grass and clover mixtures, grain cereals, kale, root tops, sugar beet pulp and potatoes can also be used. Trials indicate that very high-quality grass silage can be fed to adult pigs.

Silo: A large container for storing grain or silage.

Solvency: Measures the liabilities of the business relative to the amount of owners' equity invested in the business.

Somatic cell count (SCC): A count of the white blood cells in 1 ml of milk. A normal count is generally considered to be 100, 000 or less.

Somatic cell count score (SCS): This is a linear score that is assigned based upon the raw count and has a direct relationship to milk loss.

Stocking density: The number of animals kept in a specific area of land.

Stocking rate: A measure of the carrying capacity of an area in terms of the number of livestock in it at a given time, e. g. The number of animals per hectare.

Straw: The dry stems and leaves of crops such as wheat and oilseed rape left after the grains have been removed or grass which is mowed after flowering.

Supplement: Something added in order to make something more complete.

Teat: The projection of mammary glands from which, on female mammals, milk is secreted. It is a nipple on an udder. In cattle there are four quarters of the udder, each drained by a teat.

Teat cups: A soft rubber liner that is mounted in a metal or plastic shell.

Tilth: A good light, crumbling soil prepared to be suitable for growing. The body of the animal with teats from which the milk is sucked.

Total mixed ration (TMR): Grain, hay and silage blended together to make a nutritious feed for dairy cows.

Towels: It is a cloth used for wiping, especially one used for drying wet.

Tractor: A powerful vehicle that can pull a plough or other farm tools.

Transmissible or communicable diseases: These are diseases which can pass from an infected to a susceptible animal.

Udder: The mammary gland of an animal, which secretes milk. It takes the form of a bag under the body of the animal with teats from which the milk is sucked. It is a single mass hanging beneath the animal, consisting of pairs of mammary glands.

Underfeeding: The action of giving an animal less feed than it needs.

Vaccination: It is the processes of immunization of the animals to be protected against disease agents. It is also referred to giving appropriate vaccines to animals against disease agents.

Vaccines: Fortified foreign agents prepared from micro-organisms to immunize the body of animals against disease producing micro-organisms. These can be live or attenuated/killed.

Vectors: Arthropods capable of carrying an infectious agent from one host to the other.

Venereal diseases: These are diseases, such as vibriosis as a classic example, where intimate contact is required. A number of viral infections spread by the airborne route are also contagious as they require close contact due to their fragility. Skin diseases such as mange and ringworm can be spread by casual contact between animals, but the robust nature of the organisms responsible for these diseases allows them to survive in the environment, thus creating the possibility of spread by indirect contact.

Veterinarian: An animals' doctor.

Vitamin: A substance not produced in the body, but found in most foods, and needed for good health. These are a group of chemical compounds found in a variety of foodstuffs, which are necessary for the healthy regulation of physical processes in an animal's body.

Weight: Indicator of the amount or volume of milk produced or the weight of an animal.

Yogurt: Thick custard like dairy product produced by the bacterial fermentation of milk.

Zoonoses: These are diseases that can be communicable to human being from animals through direct or indirect contact.

图书在版编目（CIP）数据

奶牛生产＝DAIRY CATTLE PRODUCTION：英文/埃塞俄比亚农业职业教育系列教材编委会组编．—北京：中国农业出版社，2018.6
ISBN 978-7-109-23351-5

Ⅰ.①奶…　Ⅱ.①埃…　Ⅲ.①乳牛—饲养管理—职业教育—教材—英文　Ⅳ.①S823.9

中国版本图书馆CIP数据核字（2018）第092710号

中国农业出版社出版
（北京市朝阳区麦子店街18号楼）
（邮政编码 100125）
责任编辑　肖　邦　黄向阳

北京通州皇家印刷厂印刷　　新华书店北京发行所发行
2018年6月第1版　　2018年6月北京第1次印刷

开本：787mm×1092mm　1/16　　印张：26　　插页：6
字数：673千字
定价：135.00元

Fig.1 Holstein cow and bull
(China National Commission of Animal Genetic Resources, 2011)

Fig. 2 Jersey cow and bull
(China National Commission of Animal Genetic Resources, 2011)

Fig. 3 Milking Shorthorn cow and bull
(China National Commission of Animal Genetic Resources, 2011)

Fig. 4 Simmental cow and bull
(China National Commission of Animal Genetic Resources, 2011)

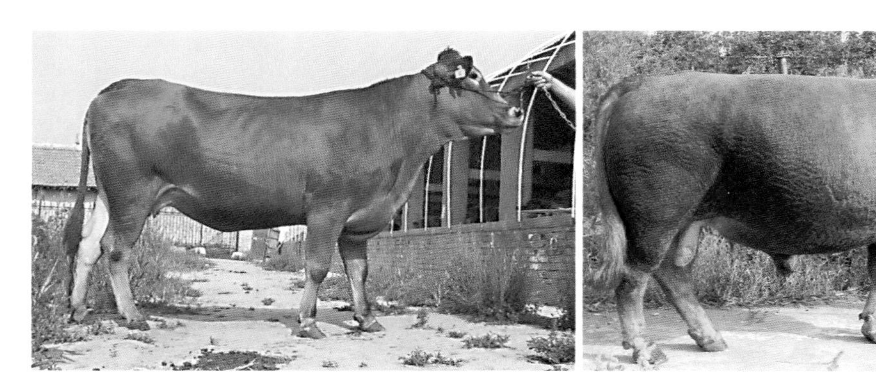

Fig. 5 Milking Devon cow and bull
(China National Commission of Animal Genetic Resources, 2011)

Fig. 6 Boran

Fig. 7 Fogera

Fig. 8 Horro

Fig. 9 Arsi

Fig. 10 Barka

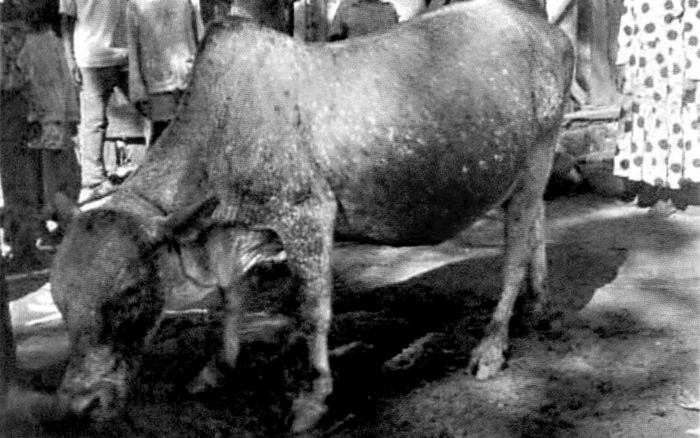

Fig. 11 Sheko

Fig. 12 Ogden

Fig. 13 Senga

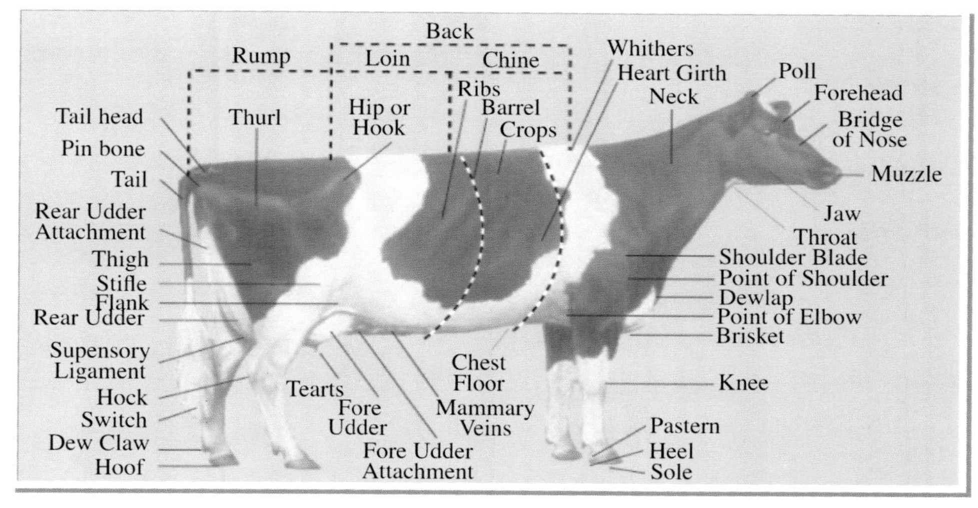

Fig. 14 Body parts of a dairy cow (Paulson et al., 2012)

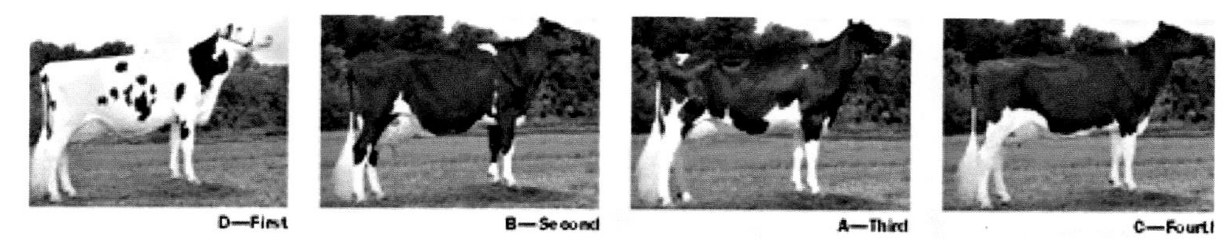

Fig. 15 Four classes of Holstein cows for judgment (Hoard's Dairyman, 2012)

Fig. 16 Tail to tail arrangement system (Wang Zhisheng and Liu Changsong, 2013)

Fig. 17 Face to face (head to head) arrangement system (Wang Zhisheng and Liu Changsong, 2013)

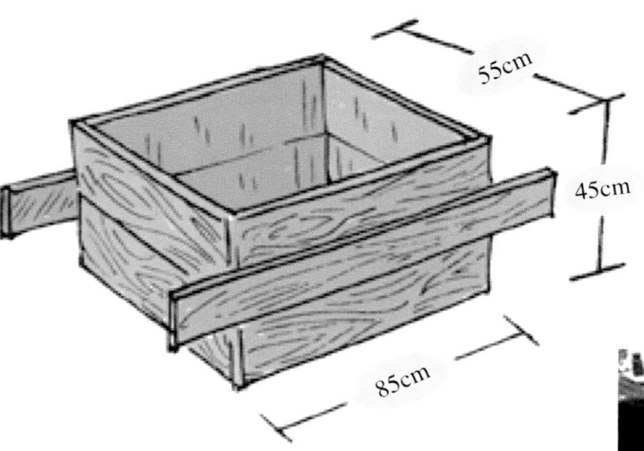

Fig. 18 Bailing box

Fig. 19 Chopping fodder for ensiling

Fig. 20 Swinging cow brush installed, and brush being used by cows (Wang Zhisheng and Liu Changsong, 2013)

Fig. 21 Standing to be mounted by a bull or another cow is the only conclusive sign that a cow is in standing estrus and ready to be bred (The lower cow is in standing heat) (Wang Zhisheng and Liu Changsong, 2013)

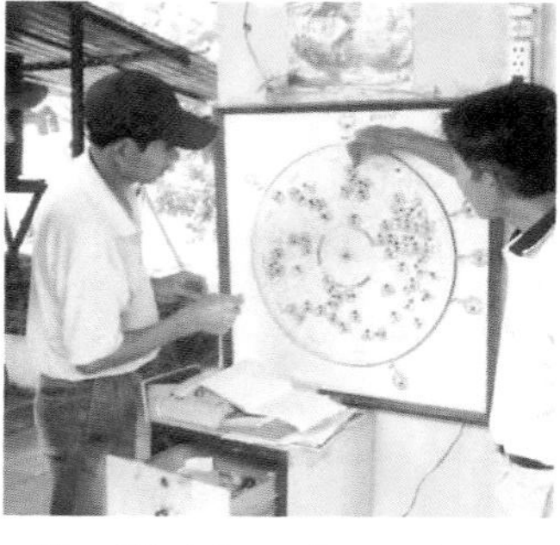

Fig .22 A breeding wheel (Moran, 2012)

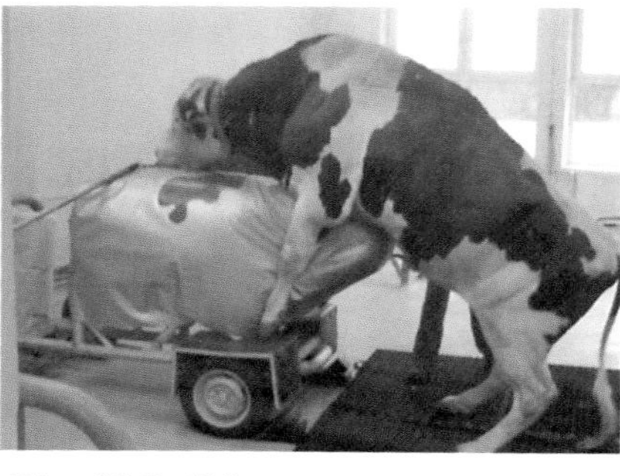

Fig. 23 Bull jump on a dummy cow (Barszcz and Wiesetek, 2012)

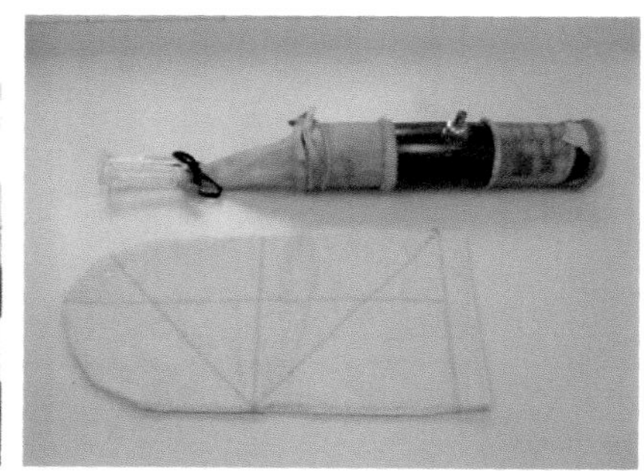

Fig. 24 Structure of an artificial vagina (Barszcz and Wiesetek, 2012)

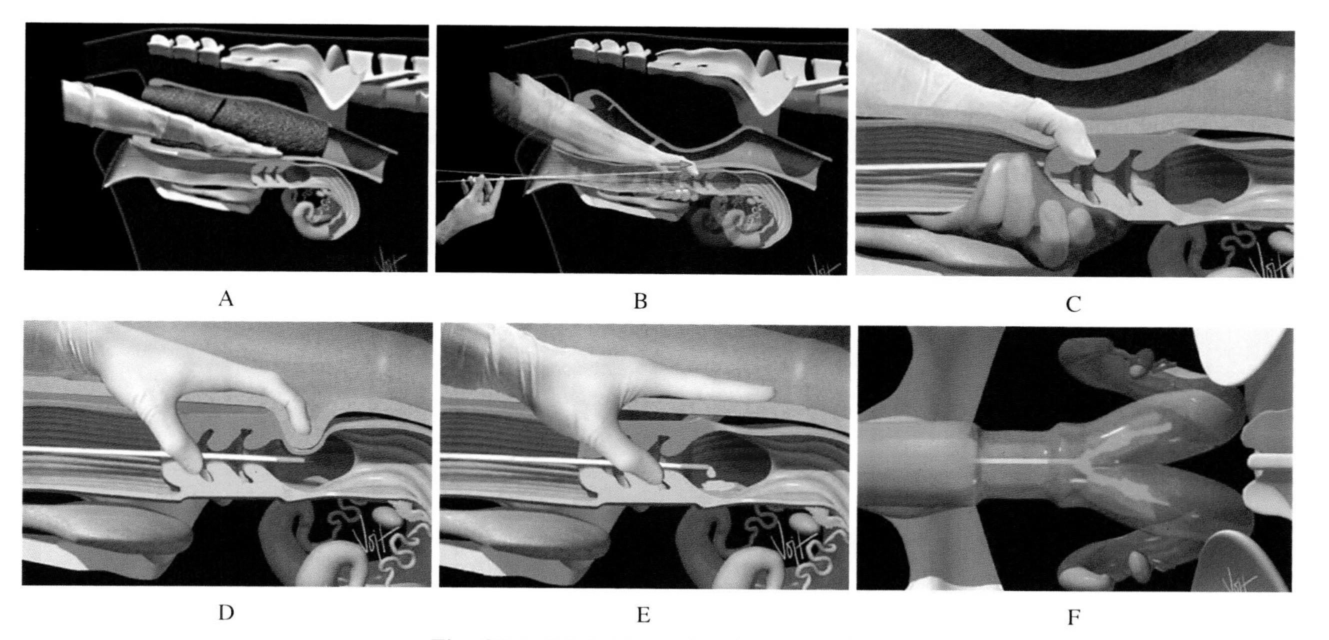

Fig. 25 Artificial insemination procedure

A. Use of hands and inserting into the rectum B. Grasp the cervix C. Grasp the external opening to the cervix D. Work the gun through the cervix E. Releasing the semen F. Contractions transport spermatozoa to horns and oviducts (Delanette and Nebel, 2015)

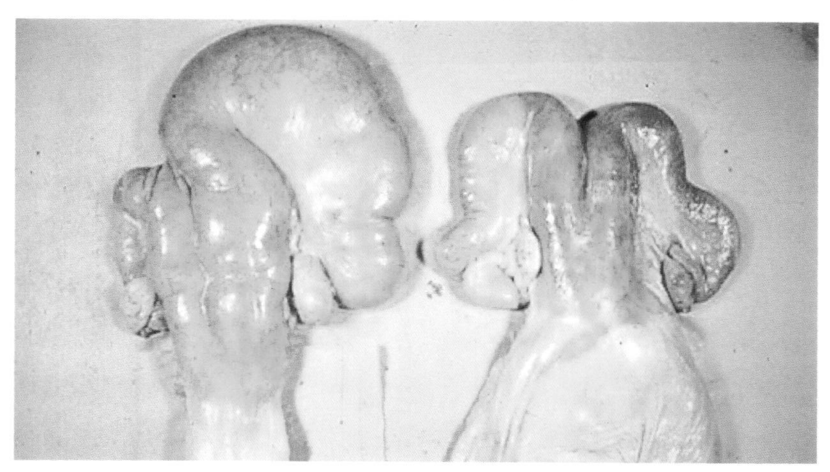

Fig. 26 Dorsal view of a pregnant on left and non-pregnant reproductive tract on the right

Fig .27 The cows and healthy newborn-calf in a clean-dry maternity pen

(Courtesy of Tyler, Iowa State University)

Fig. 28 Pulling the calf
(Wang Zhisheng and Liu Changsong, 2013)

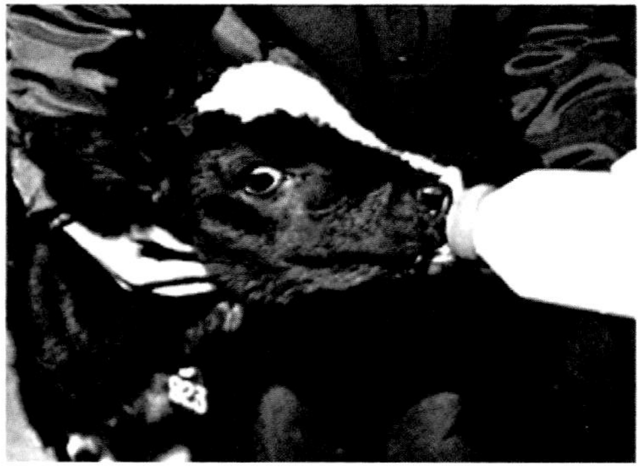

Fig. 29 Feeding of the calf with colostrum
(Alage Dairy Farm, Ethiopia)

Fig. 30 Bucket feeding of the calf
(Wang Zhisheng and Liu Changsong, 2013)

Fig. 31 The calf introduced into solid feed
(USAID, 2012)

Fig. 32 Hutch of the calf from 2-3-month age
(Courtesy of Patrick Hoffman, University of Wisconsin)

Fig. 33 Removable calf's pen
(Courtesy of Patrick Hoffman, University of Wisconsin)

Fig. 34 Ear-tag and chemicals applicator

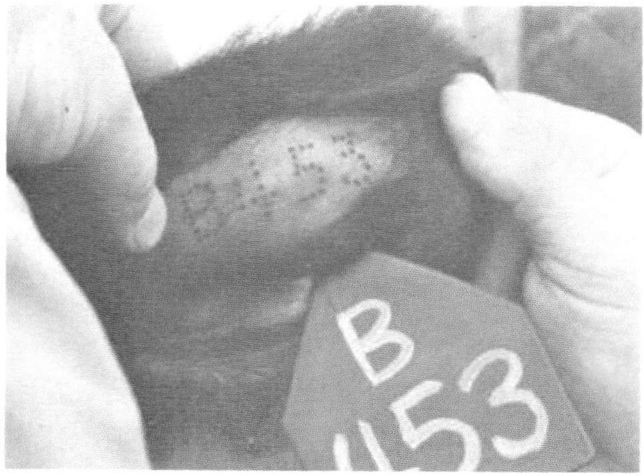

Fig. 35 Branding

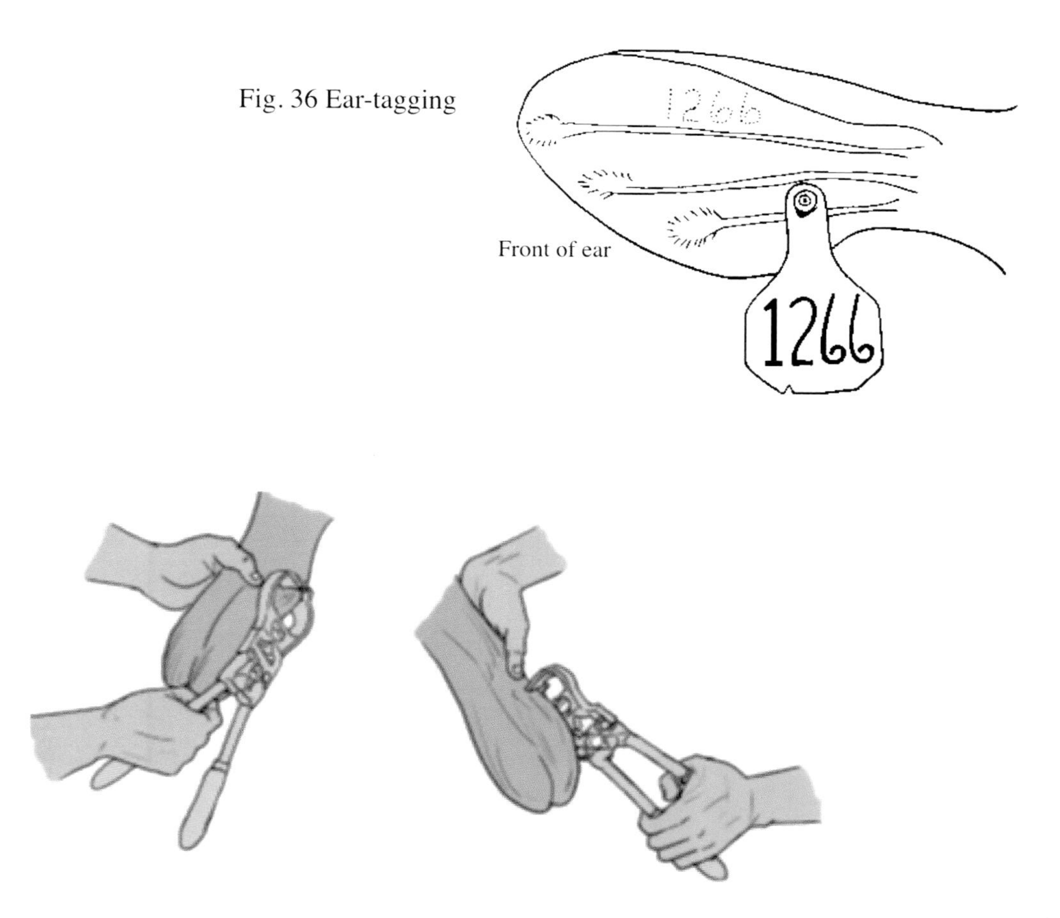

Fig. 36 Ear-tagging

Fig. 37 Burdizzo castration
(USAID, 2012)

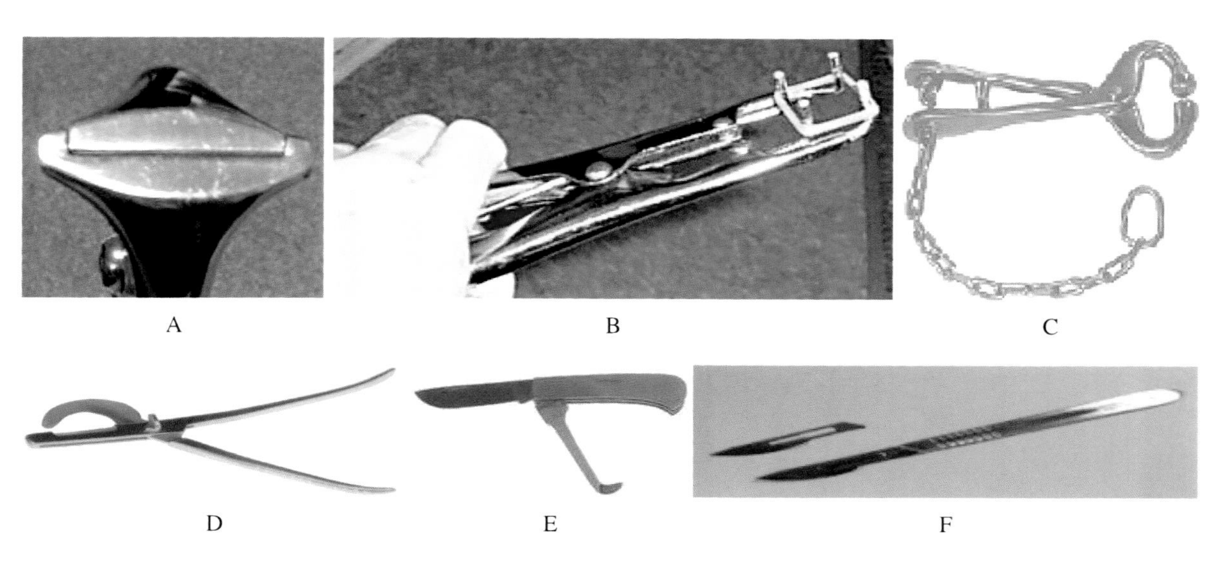

Fig. 38 Various instruments and equipment used for castration
A. Burdizzo B. Elastrator C. Restrain ring D. Emasculator E. Knife F. Surgical blade
(Forse, 1999)

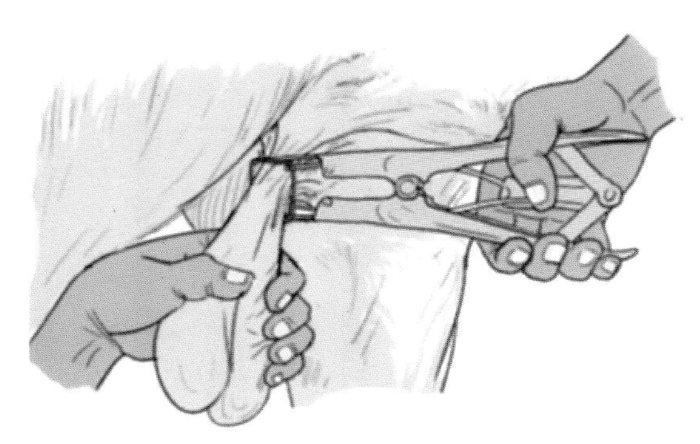

Fig .39 Elastrator ring method of castration
(Patrick Hoffman, 1995)

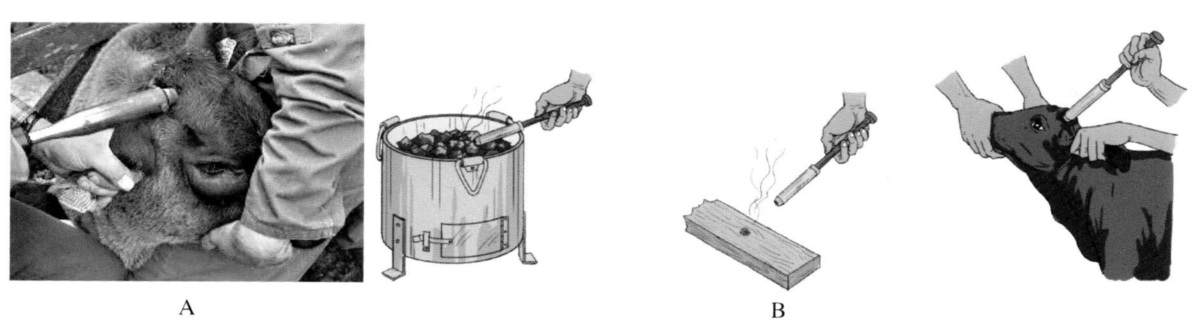

A B

Fig. 40 Dehorning
A. Electrical hot-iron method B. Manual hot-iron method
(Courtesy of Mark Kirkpatrick, Pfizer)

Fig. 41 Gas-dehorners cauterize the blood supply to the horn bud and effectively dehorn without leaving an open wound
(Courtesy of Mark Kirkpatrick, Pfizer)

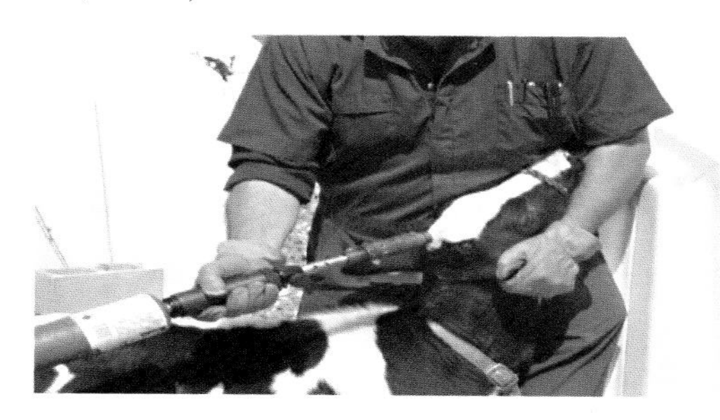

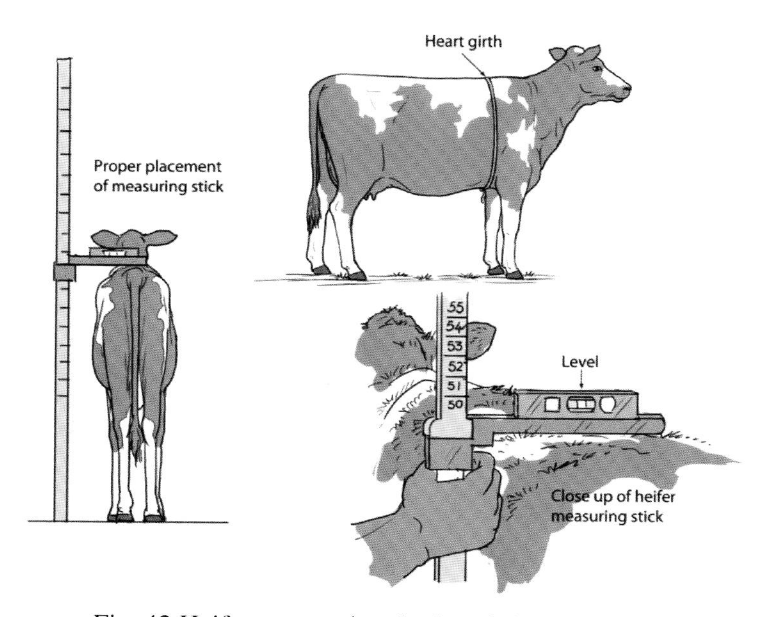

Fig. 42 Heifers measuring: bodyweight and height

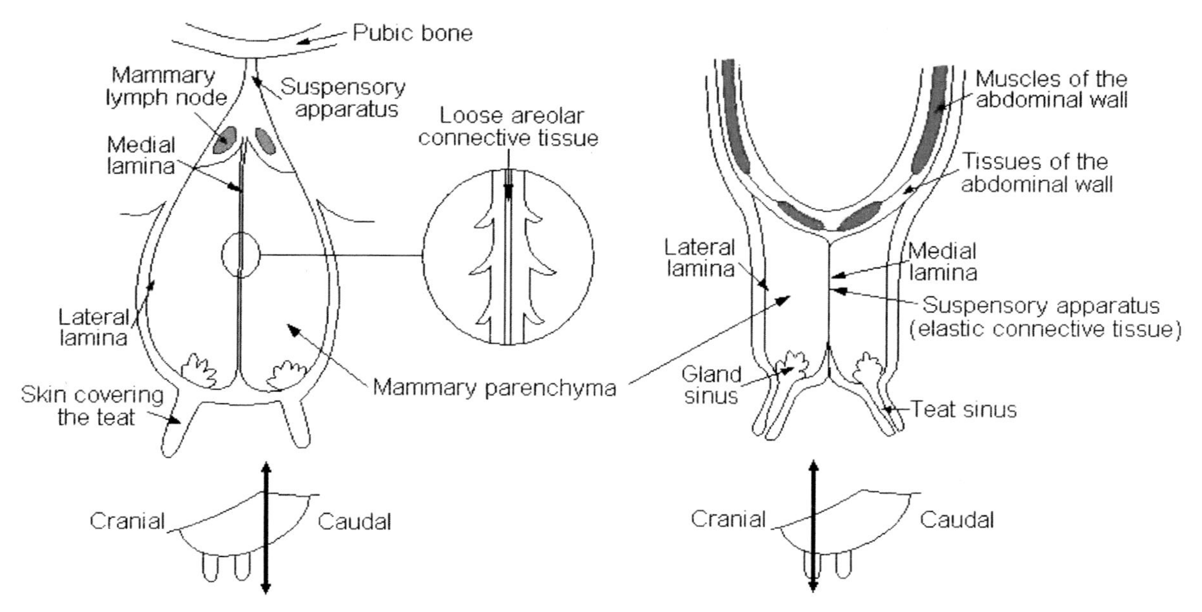

Fig. 43 Suspensory apparatus of caudal & cranial udder of cow (FAO, 1999)

Fig. 44 The milking claw is designed to harvest milk with the least amount of damage to teat end tissues

Fig. 45 Teat cup liners must be replaced at regular intervals to maintain optimal function of the milking system

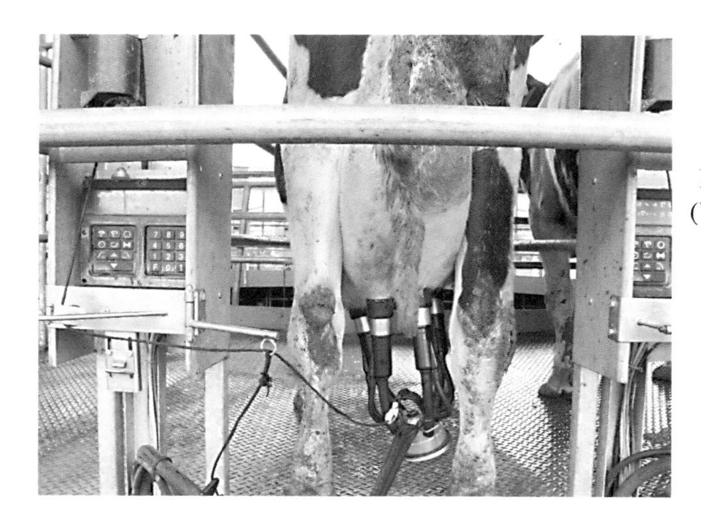

Fig. 46 Milking cows by milking machine
(Wang Zhisheng and Liu Changsong, 2013)

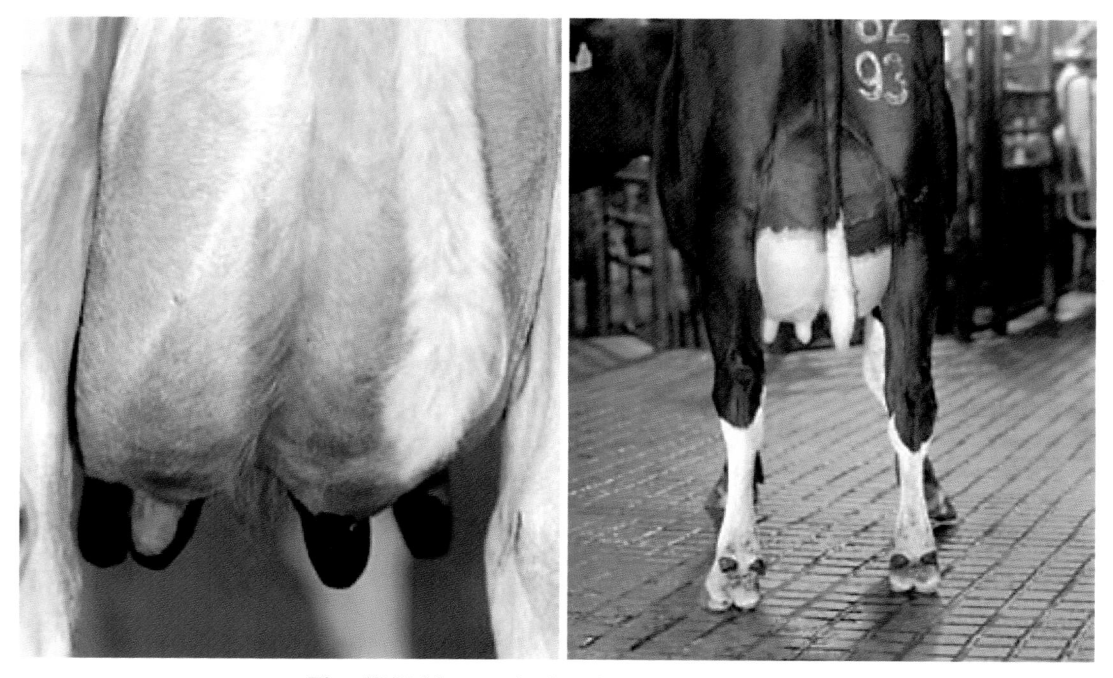

Fig. 47 Udder washed and ready for milking

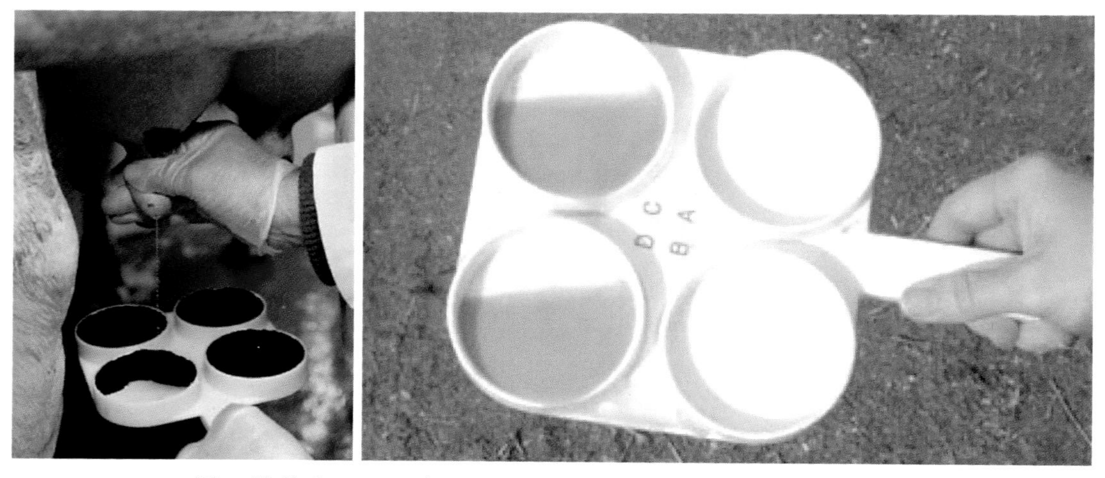

Fig. 48 Strip out each teat by using strip cups for mastitis check
(Courtesy of Bernadette O'Brien)

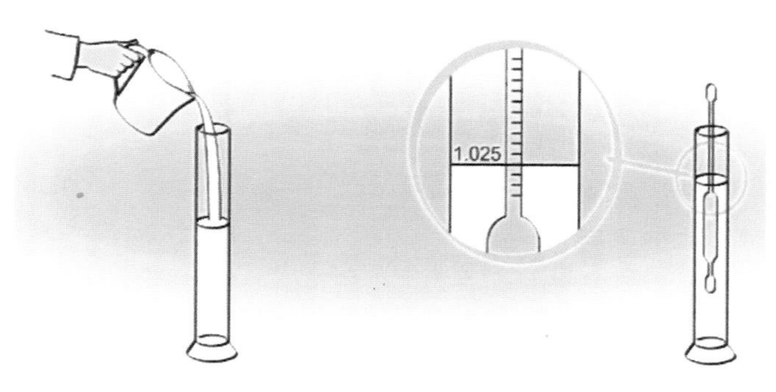

Fig. 49 Using the lactometer to detect addition foreign matters

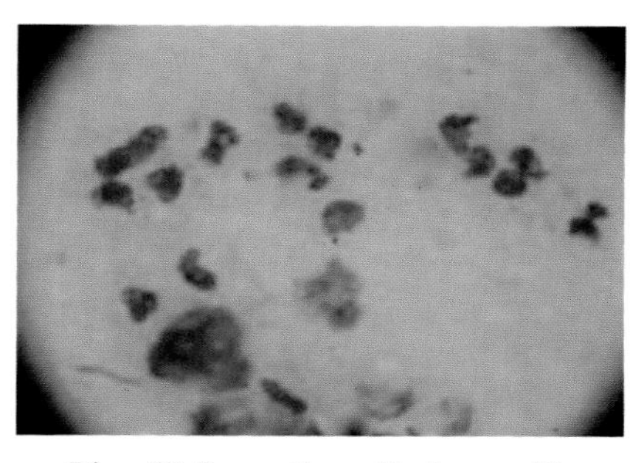

Fig. 50 Somatic cells in a milk sample under the microscope

Fig. 51 Electric butter churners

Fig. 52 Manual butter churners

Fig. 53 Hand paddle butter churner

Fig. 54 The characteristics appearances of diseased animals
(Courtesy of Infovets)

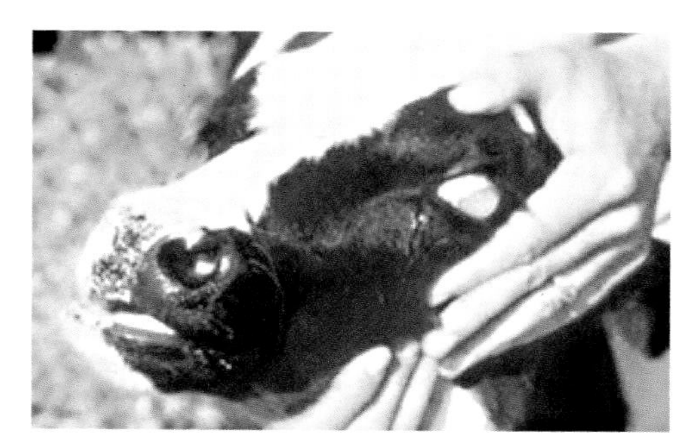

Fig. 55 Abnormities in the eyes and respiration (DPO, 1999)

Fig. 56 Entry of mastitis-organisms to the quarter(s) of the mammary system (Wang Zhisheng and Liu Changsong, 2013)

Fig. 57 Clean equipment before applying the cups to the next cow (John, 2001)

Fig. 58 Using the same cloth and dirty water to wipe teats on multiple cows before milking can spread bacteria through the herd

Fig. 59 Good milking hygiene practice, with separate cleaning towels for each cow

Fig. 60 Teat-spray upwards from beneath to teats, not the side by using sprays about 20 mL of prepared teat-disinfectant/(cow·milking)

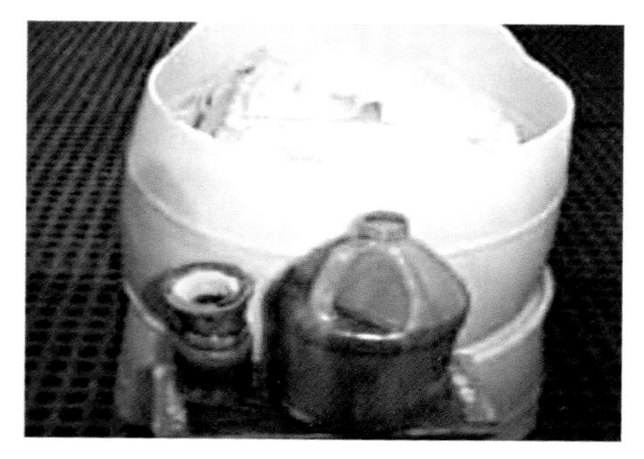

Fig. 61 Teat-dip applied with a dip-cup refilled from a bottle. The white bucket here contains clean cloth-towels to wipe teats before milking

Fig. 62 Internal parasites: tapeworm